Chicago Public Library
C
P
L
REFERENCE
Form 178 rev. 1-94

Structure and Dynamics of Electronic Excited States

Springer

Berlin
Heidelberg
New York
Barcelona
Hong Kong
London
Milan
Paris
Singapore
Tokyo

J. Laane · H. Takahashi · A. Bandrauk (Eds.)

Structure and Dynamics of Electronic Excited States

With 165 Figures and 36 Tables

Professor Jaan Laane
Texas A&M University, Department of Chemistry
College Station, 77843-3255 Texas, USA

Professor Hiroaki Takahashi
Waseda University, School of Science and Engineering
Department of Chemistry, 169 Tokyo, Japan

Professor Andre Bandrauk
University de Sherbrooke, Faculty des Sciences
Sherbrooke (Québec), Canada J1K2R1

ISBN 3-540-63908-x Springer-Verlag Berlin Heidelberg New York

Library of Congress Cataloging-in-Publication Data

Structure and dynamics of electronic excited states / J. Laane, H. Takahashi, A. Bandrauk (eds.).
p. cm. Includes bibliographical references.
ISBN 354063908-x (Berlin : alk. paper)
1. Excited state chemistry. I. Laane, Jaan. II. Takahashi, H. (Hiroaki), 1935-
III. Bandrauk, Andre D.
QD461.5 .S77 1999
541.2'2--ddc21 98-40876

Typesetting: Camera-ready by authors
Cover layout: design & production, Heidelberg
Production: ProduServ GmbH Verlagsservice, Berlin
SPIN: 10569048 57/3020-5 4 3 2 1 0 - Printed on acid -free paper

Preface

In December 1995 a special symposium on the "Structure, Dynamics, and Control of Electronic Excited States" was held in Honolulu as part of the Pacifichem95 meeting. The symposium was co-chaired by Professor Jaan Laane (USA) and Professor Andre Bandrauk (Canada). They were assisted by Professor Hiroaki Takahashi and Professor Keitaro Yoshihara (Japan) who served on the organizing committee. True to its name, the symposium was exciting and dynamic for the one hundred participants. Following the sympsium, with the encouragement of the participants, the organizers decided to edit a text highlighting the work of many of the invited speakers. This volume is the result of this endeavor.

The twelve chapters presented here describe work carried out in the USA, Japan, and Canada, from where the Pacifichem participants primarily came. All of the chapters deal with excited states. Several cover the determination of molecular structures in the excited state and many describe the dynamical processes occuring there. These are the basis of the photochemistry taking place in molecules. The relationships between structure and dynamics are investigated and the development of laser control for reactions is considered. Taken as a whole, these chapters present an in-depth view of the on-going research in this rapidly evolving area of chemistry.

College Station, TX
April 1998

Table of Contents

Structure

Vibrational Potential Energy Surfaces of Non-Rigid Molecules in Excited Electronic States

Jaan Laane
Department of Chemistry, Texas A&M University,
College Station, TX 77843 USA
E-mail: laane@chemvx.tamu.edu

Abstract. Low-frequency vibrational spectroscopy of non-rigid molecules has been used for many years to determine the potential energy surfaces which govern the conformational changes in electronic ground states. Examples of such studies are presented. Jet-cooled fluorescence spectroscopy has more recently been used to determine potential energy surfaces for electronic excited states. Results are presented for the $S_1(n,\pi^*)$ electronic excited states of seven cyclic ketones. In each case the carbonyl out-of-plane wagging vibration and the out-of-plane ring motions were examined. 2-Cyclopenten-1-one is conjugated and retains a single minimum potential function for the carbonyl wag in its excited state. However, cyclobutanone, cyclopentanone, and the other ketones studied all have double-minimum potential energy functions with barriers to inversion ranging from 659 to 1940 cm^{-1} and wagging angles ranging from 20° to 41°. The data have also been analyzed to determine the effect of the $n \rightarrow \pi^*$ transition on the conformations and ring bending/twisting potential energy surfaces.

The jet-cooled fluorescence spectra of *trans*-stilbene and its methyl and methoxy derivatives have also been analyzed. Two-dimensional potential energy surfaces for the phenyl torsions and one-dimensional surfaces for the C=C torsions were determined. These provide insight into understanding the photoisomerization of these molecules.

1 Introduction

Conformational processes in non-rigid molecules often follow vibrational pathways which are well represented by potential energy surfaces. If these surfaces can be defined in terms of one or two vibrational coordinates which are separable from the other vibrations through the high-low frequency approximation, they can often be determined from far-infrared and/or low-frequency Raman data. Common examples of one-dimensional potential functions which have been determined from spectroscopic data include the internal rotation of ethane and the inversion of ammonia.

For the past three decades we have been investigating the potential energy functions of cyclic (four-, five-, and six- membered rings) and bicyclic molecules. Our work, along with that of others, has periodically been reviewed [1–9]. The value in determining the potential energy functions for these various types of molecules is that they yield the molecular conformations, energy differences between different structures, and information on the forces responsible for the molecular configurations. More recently we have extended these studies to electronic excited states, where we have investigated carbonyl wagging motions of cyclic ketones, internal rotations of stilbenes, and out-of-plane motions of ring molecules utilizing fluorescence excitation spectroscopy (FES) [10–12]. Just as with the electronic ground states, it has been possible to calculate potential energy functions for these motions in the $S_1(n,\pi^*)$ and $S_1(\pi,\pi^*)$ electronic states. These make it possible to determine the configurations of the carbonyl groups and the skeletal atoms in the S_1 states.

In this chapter we will first review how low-frequency vibrational spectroscopy has been used to determine potential energy functions for electronic ground states and then show that the same principles are applicable to excited states. Instrumentation for fluorescence studies will be briefly discussed, and then the experimental results for seven cylic ketones and two stilbene derivatives will be presented.

2 Theory and Ground State Potential Energy Functions

2.1 Vibrational Hamiltonian

The Hamiltonian for the vibrations of a molecule in center-of-mass coordinates has the form

$$\mathcal{H}(q_1,q_2,...) = (-\hbar^2/2)\sum_k\sum_\ell \partial/\partial q_k g_{k\ell}(q_1,q_2,...)\partial/\partial q_\ell + V(q_1,q_2,...) + V' \tag{1}$$

where $V(q_1, q_2,...)$ is the potential energy defined in terms of the vibrational coordinates q_i. The $g_{k\ell}$ is the kinetic energy (reciprocal reduced mass) function, and V' is the "pseudopotential" which has been shown to be negligible [13]. For an isolated ring-puckering motion, for example, the Hamiltonian can be assumed to be one-dimensional. For the two out-of-plane motions of cyclopentane and its derivatives, a two-dimensional representation is necessary with q_1 and q_2 representing the ring-bending and ring-twisting vibrations, respectively. The reduction of (1) to a one- or two-dimensional problem is based on the approximation that none of the other vibrations in the molecule interacts with the vibrations of interest. This has been shown to be valid in most cases, but the low-frequency modes are often not entirely uncoupled from all the other motion [14–18].

2.2 Calculation of Kinetic Energy Functions

The vibrational coordinate of interest is represented by defining the motion of each atom in the molecule in terms of that coordinate (e. g., in terms of the carbonyl wagging coordinate). We have used vector methods to define the various out-of-plane motions and developed computer programs for the calculation of reciprocal reduced masses g_{ij} as a function of vibrational coordinates. These kinetic energy functions for one-dimension are typically fitted to polynomials of the form

$$g_{44} = g_{44}^{(0)} + g_{44}^{(2)} x^2 + g_{44}^{(4)} x^4 + g_{44}^{(6)} x^6, \tag{2}$$

where the $g_{44}^{(k)}$ are the coefficients, and where $g_{44}^{(0)}$ is the reciprocal reduced mass for the planar structure of a ring molecule. In some cases odd-powered terms must also be utilized. It should be noted the i, j = 1–3 subscripts for g_{ij} are reserved for the molecular rotations. The methodology for these calculations for various types of ring molecules has been published [19–23].

2.3 Calculation of Energy Levels and Determination of Potential Energy Surfaces

R. P. Bell [24] predicted that four-membered rings should have ring-puckering vibrations which have quartic potential energy functions. Laane [25] has derived the origin of this function and has shown that the angle strain in cyclobutane should give a potential energy function of the form

$$V(x) = ax^4 + bx^2, \tag{3}$$

where the potential energy parameters are defined in terms of the ring-angle-bending force constant k_ϕ, the initial angle strain S_0, and the C–C bond distance R, by

$$a = 128k_\phi/R^4 \tag{4}$$

and

$$b = 32k_\phi S_0/R^2\ . \tag{5}$$

In addition, torsional forces contribute to the function. For example, molecules such as cyclobutane and cyclopentene, which would have eclipsing methylene groups for their planar structures, will have substantial *negative* contributions to the quadratic term so that the constant b becomes negative.

The Schrödinger equation with a potential function of the form given in (3) can be solved only by numerical approximation methods. Fortunately, these matrix diagonalization techniques can give very accurate results when appropriate basis sets are used. In order to simplify the analysis of data, Laane [26] published a set of tables for determining the eigenvalues of the mixed quartic/quadratic potential function of (3). Utilization of these tables is simplified by use of a transformation to the reduced (dimensionless) coordinate Z:

$$Z = (2\mu/\hbar^2)^{1/6} a^{1/6} x, \tag{6}$$

where μ is the reduced mass. This results in the reduced Schrödinger equation

$$-d^2\psi/dZ^2 + (Z^4 + BZ^2)\psi = \lambda\psi, \tag{7}$$

where the energy levels E are given by

$$E = A\lambda \tag{8}$$

and where

$$A = (\hbar^2/2\mu)^{2/3} a^{1/3} \tag{9}$$

and

$$B = (2\mu/\hbar^2)^{1/3} a^{-2/3} b\ . \tag{10}$$

The variation of the eigenvalues λ versus B (both in figure and tabular form) can be found elsewhere [26]. When $B = 0$, pure quartic oscillator levels result and the energy minimum is at $x = Z = 0$. For $B > 0$ the potential function has a single minimum and is a mixed quartic/quadratic function. For negative B the potential function has a double minimum with a barrier equal to $AB^2/4 = b^2/(4a)$. In this region pairs of levels begin to merge and become degenerate. Thus, molecules of this type will have irregular patterns of spectroscopic bands for the lower quantum transitions.

For two-dimensional problems where both vibrations are symmetric, the appropriate potential energy function is

$$V(x_1x_2) = a_1x_1^4 + b_1x_1^2 + a_2x_2^4 + b_2x_2^2 + cx_1^2x_2^2 \tag{11}$$

where x_1 and x_2 are the two coordinates. The calculation of energy levels for this potential energy function requires the use of the product of two basis sets and the diagonalization of matrices with typical dimensions of about 150. The methods have been outlined [27] and utilized in many of our publications.

2.4 Examples for the Ground State

One-Dimensional Functions. As an example of a recent study completed in our laboratory [28], Figs. 1 and 2 show the far-infrared spectra and one-dimensional potential energy function for the ring-puckering vibration determined for 1,3-dioxole. The far-infrared spectrum results primarily from single quantum ($\Delta v = 1$) transitions, while the Raman spectra (not shown) result from $\Delta v = 2$ transitions. The two types of spectra complement each other very nicely and allow the quantum transitions to be assigned unambiguously. Adjustment of the potential energy parameters a and b in (3) results in the potential function

$$V(\text{cm}^{-1}) = 1.59\times10^6x^4 - 4.18\times10^4x^2 \tag{12}$$

which is shown in Fig. 2. This corresponds to a value of $B = -5.47$ in (10). The scaling factor $A = 36.7$ cm^{-1}. The frequencies calculated for this function have an average deviation from the observed values of only 0.7%. The barrier to planarity (or inversion) is 275 cm^{-1}, and the equilibrium dihedral angle is 24°. This result is highly unusual since this is the only unsaturated five-membered ring molecule without any CH_2–CH_2 torsional interactions that is non planar. In the case of 1,3-dioxole the nonplanarity arises from the anomeric effect [29,30] present in molecules with O–C–O linkages. Here the –O–C–O–C= torsional angles have minimum energies at 90° and tend to twist the CH_2 group out of the plane of the rest of the molecule. Hence, this effect forces the ring to be puckered.

A second example of a molecule which can be studied one-dimensionally is 4H-pyran [31]. Figure 3 shows its far-infrared spectrum while Fig. 4 shows its ring-puckering potential energy function. Here the function has a single minimum, indicating that the molecule, unlike 1,3-dioxole, is planar. The observed transition frequencies can be seen to comprise a regular sequence of closely spaced bands. For this molecule both a and b in (3) are positive.

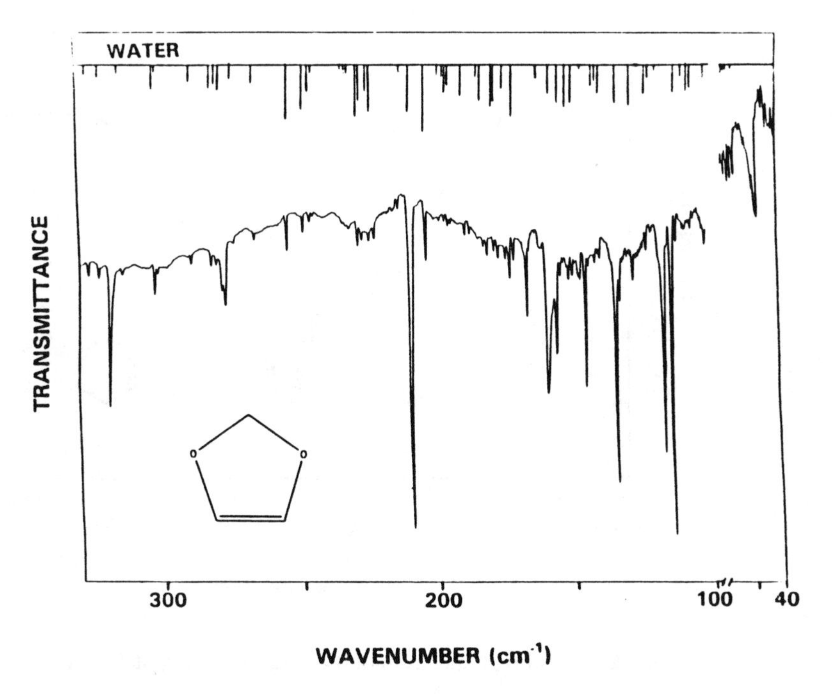

Fig. 1. Far-infrared spectrum of 1,3-dioxole

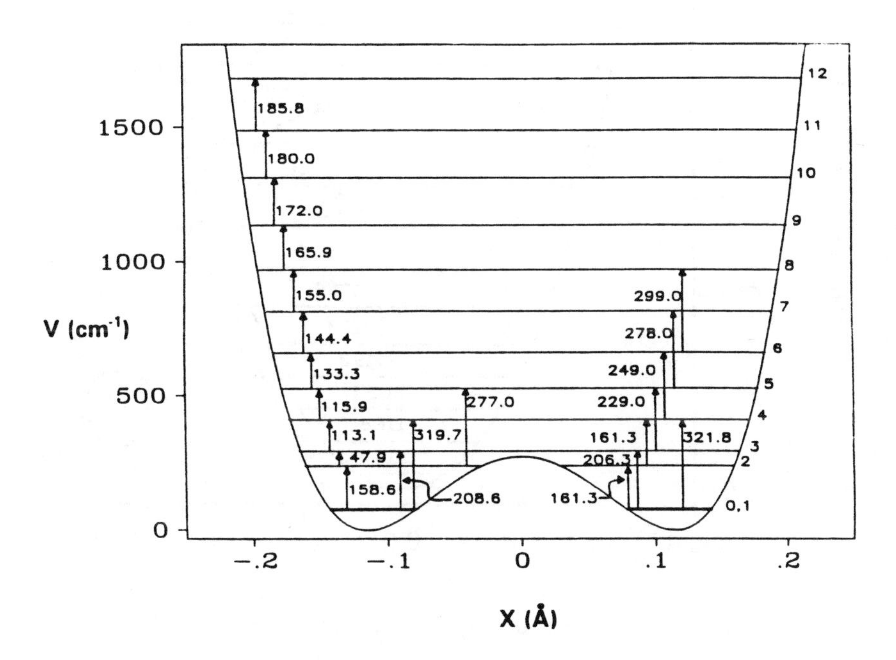

Fig. 2. Ring-puckering potential energy function for 1,3-dioxole

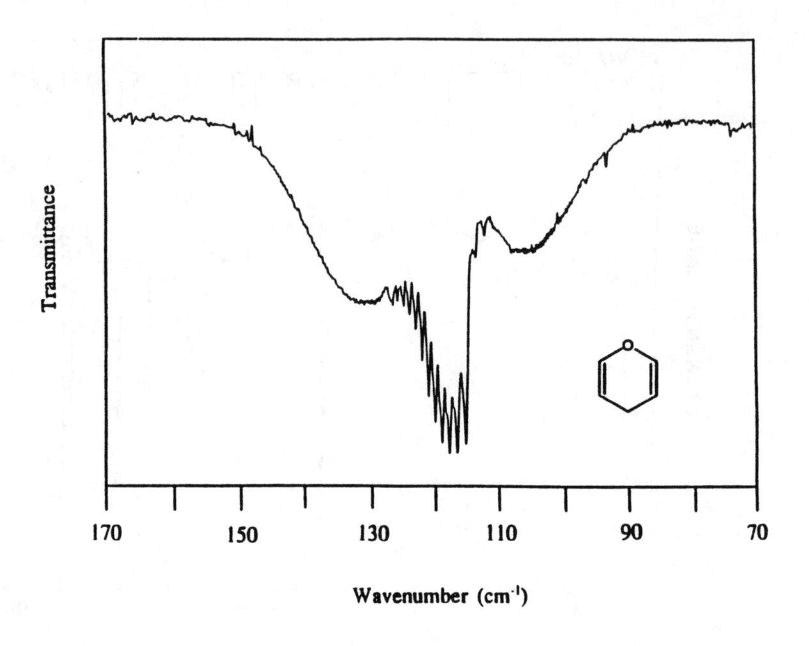

Fig. 3. Far-infrared spectrum of the ring-puckering of 4H-pyran

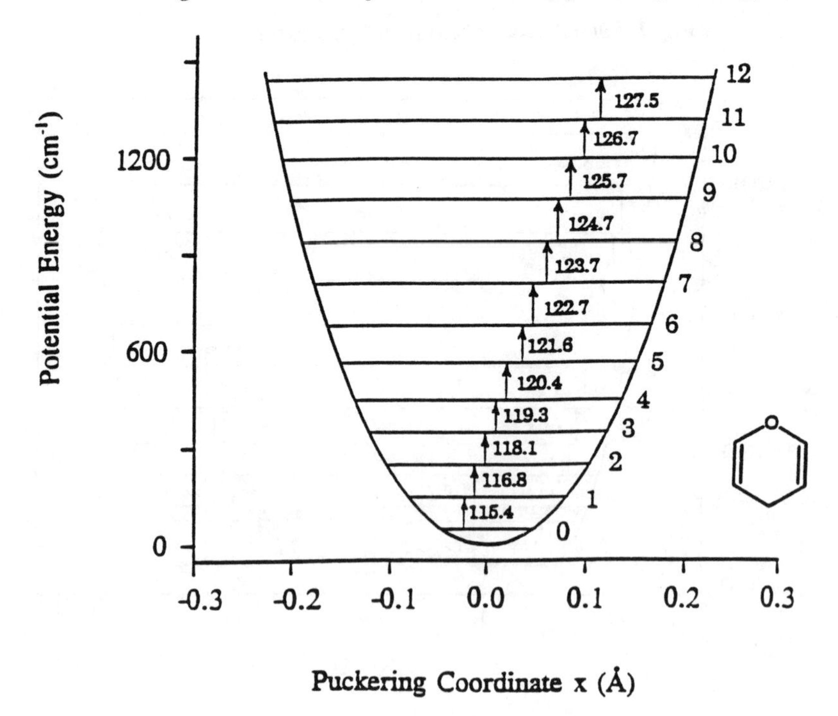

Fig. 4. Ring-puckering potential energy function for 4H-pyran

Two-Dimensional Potential Energy Surfaces (PESs). When two low-frequency vibrations, such as the ring-bending and ring-twisting modes of cyclopentane, interact, a two-dimensional potential energy surface (PES) of the type in (11) can often be used to interpret the spectra. Figure 5 shows the potential energy function determined for cyclopentane [32]. This is a special case with "free pseudorotation" [33] where ten bent and ten twisted conformations all have almost exactly the same steric energy. This PES gives rise to pseudorotational energy states approximated by

$$E = n^2 B \quad \text{for} \quad B = \hbar^2/2mq_0^2 \quad \text{and } n = 0, 1, 2,... \tag{13}$$

where m is the effective mass of a methylene group in cyclopentane and q_0 is the equilibrium value of the radial coordinate, which measures motion out of the ring plane. The pseudorotational constant B was determined to be 2.60 cm^{-1} so that these levels occur at very low frequencies. The PES also gives rise to "radial bands" which occur near 280 cm^{-1}. For a less symmetric molecule such as 1,3-oxathiolane, a barrier to pseudorotation results. The two-dimensional potential energy surface is then best examined in terms of ring-bending and twisting coordinates. Figure 6 shows the ring-bending spectrum of 1,3-oxathiolane [34], and Fig. 7 shows the PES for this molecule. This molecule has two minima corresponding to two equivalent twisted conformations, and it has two saddle points at 570 cm^{-1} corresponding to bent forms. The planar structure at the origin of the coordinate system has an energy maximum of 2289 cm^{-1}.

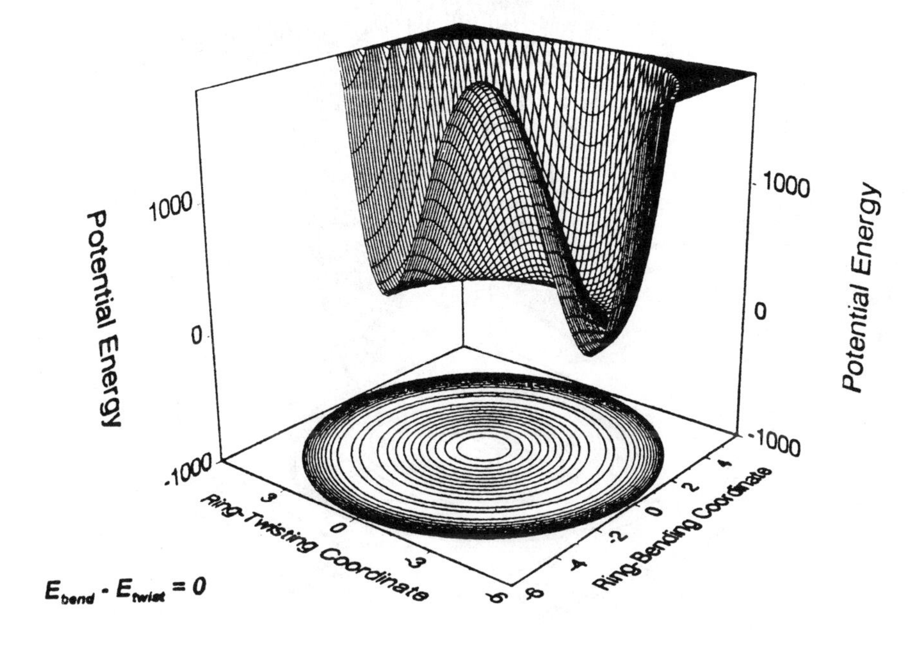

Fig. 5. Potential energy surface for the out-of-plane ring modes of cyclopentane

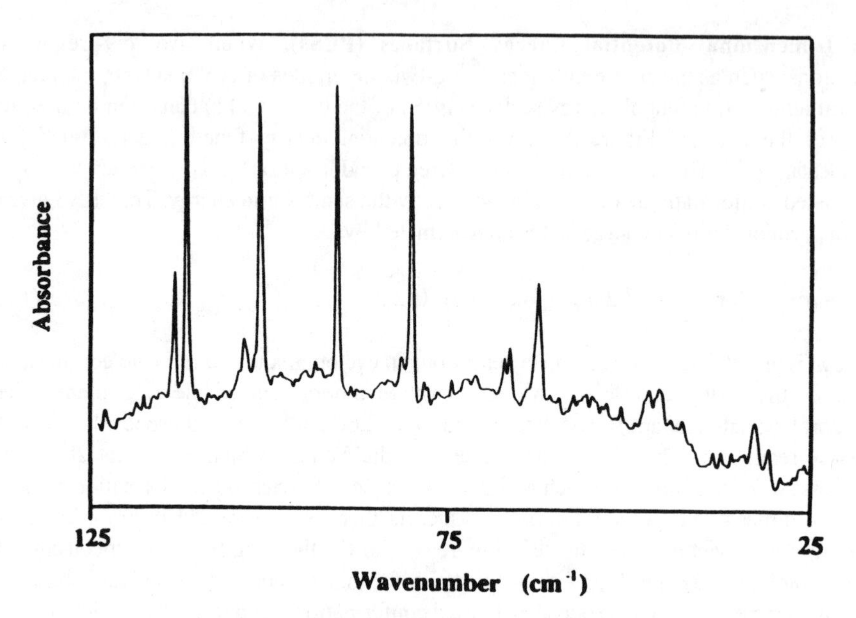

Fig. 6. Ring-bending spectrum of 1,3 oxathiolane.

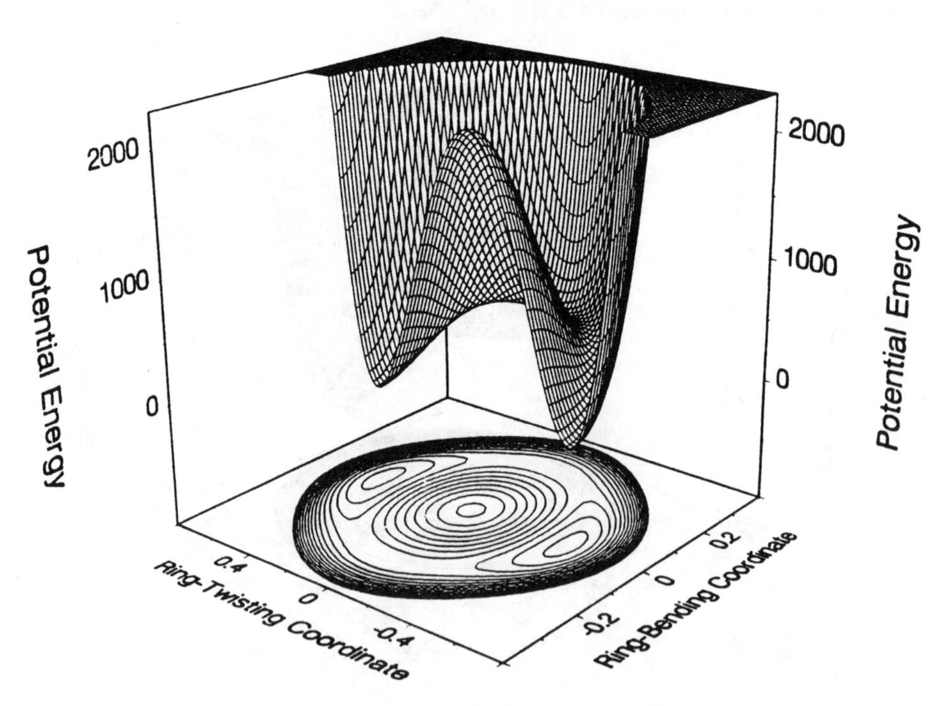

Fig. 7. Potential energy surface for 1,3-oxathiolane

3 Electronic Excited States

3.1 Experimental

In recent years, the combination of tunable high power lasers and pulsed supersonic jet techniques has provided the means of obtaining high resolution electronic spectra that are greatly simplified due to the sample cooling provided by the supersonic jet. The cooling results in the removal of hot bands and the narrowing of the bandwidths for the remaining bands. The reduced spectral congestion allows a much more accurate and thorough analysis of the low-frequency vibrations in electronic excited states than was previously possible using fluorescence or absorption techniques. Figure 8 shows the experimental arrangement used in our laboratory. A pulsed Nd:YAG laser operating at 1064 nm is frequency doubled or tripled and then used to drive a tunable dye laser. Typically, the dye laser output is again doubled to produce the desired frequency range. The pulsed molecular jet is synchronized with the laser system and electronics to produce efficient operation [10–12]. Fluorescence excitation spectra (FES) are detected using a phototube or using a time-of-flight mass spectrometer for ionization detection. The fluorescence can also be dispersed by a 0.64 m scanning monocromator and then detected using a phototube or a CCD. Sensitized phosphorescence excitation spectra (SPES) [35] using a liquid nitrogen cold tip can also be measured. Frequency calibration of the system is accomplished by recording the optogalvanic spectrum of neon in a Cr–Ne hollow cathode lamp; the frequency accuracy is ± 1 cm^{-1}. Spectra recorded using this system will be presented below.

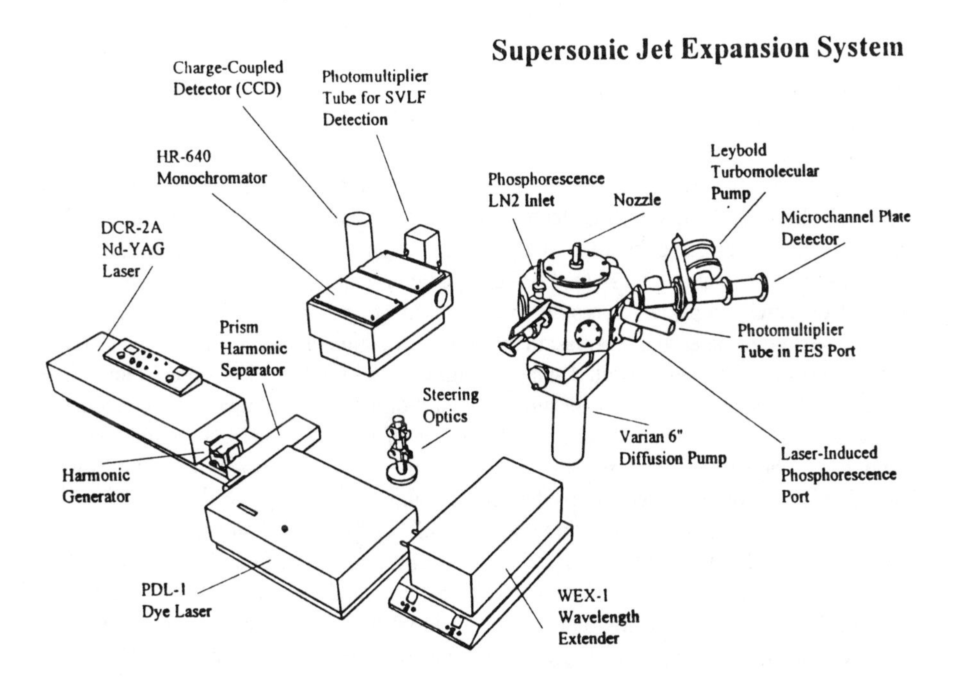

Fig. 8. Experimental arrangement for excited state studies

3.2 Fluorescence Excitation Spectra (FES) of Jet-Cooled Cyclic Ketones

A number of carbonyl compounds have been analyzed in their $S_1(n,\pi^*)$ states. In this electronic state, after the transfer of a non-bonding electron to an antibonding π^* state, a carbonyl compound typically distorts from a planar to a pyramidal configuration about the carbonyl carbon atom. This was predicted for formaldehyde in 1953 by Walsh [36] and verified spectroscopically by Brand in 1956 [37]. Formaldehyde's six vibrational degrees of freedom and its large rotational constants made the analysis of much of its vibronic spectra feasible even before high-resolution or jet spectroscopy techniques were available. The $S_1(n,\pi^*)$ state of acetaldehyde has been studied by fluorescence and absorption methods since 1954 as a prototype of larger carbonyl compounds [38–40]. However, its electronic absorption spectra are complex and ill-resolved, and little agreement existed on the interpretation. Only recently, with the vibrational and rotational cooling obtained in a supersonic jet, did it become possible to make accurate assignments [41–43]. Recently, our laboratory has recorded and analyzed the FES data for the cyclic ketones shown below, and the results for these molecules will be presented here.

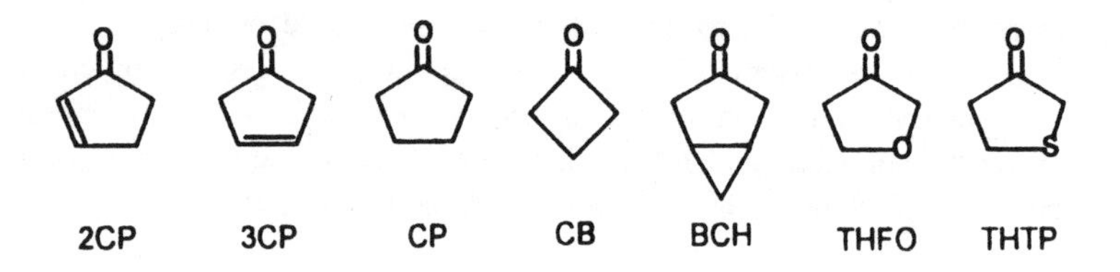

2-Cyclopenten-1-one (2CP). Figure 9 shows the low-resolution FES of 2CP; the data for the d_2 species have also been recorded [10]. The spectra were recorded in a region from below the electronic origin to about 1800 cm^{-1} beyond it. Figure 10 shows the expanded low energy region of the FES spectrum; the $S_1(n,\pi^*)$ excited state vibrational frequencies are indicated. The 2CP-d_0 and 2CP-d_2 electronic origins were observed at 27 210 and 27 203 cm^{-1}, respectively. In comparison to 3CP and CP, which will be considered later, the 2CP frequency is considerably lower. This is the expected result from the conjugation between the C=O and C=C groups, which results in a lower energy π^* orbital. The electronic origin for 2CP is extremely intense, in contrast to the $S_1(n,\pi^*)$ origins of similar molecules [43,44]. This is the result of a planar excited state structure and a high Franck–Condon factor. The vibrational frequencies of the relevant fundamental vibrations determined for the electronic ground state [45] and S_1 excited state [10] are given in Table 1. The FES assignments can be found elsewhere [10]. Thirteen of the thirty S_1 excited state vibrational frequencies were determined from the fluorescence excitation spectrum.

Figure 11 shows the ring-puckering energy levels for the ground and excited electronic states for thee d_0, d_1, and d_2 isotopomers. The ring-puckering series of bands at 67, 158, 256, and 360 cm^{-1} can be fit nicely with a one-dimensional potential energy function. Of the more than fifty fluorescence excitation bands observed for 2CP-d_0, thirty-four involve ν_{30} (the ring puckering), and twenty involve ν_{29} (ring twisting). At higher energies combinations with the C=0 in-plane (ν_{19}) and out-of-plane (ν_{28}) wags, which occur at 348 and 422 cm^{-1},

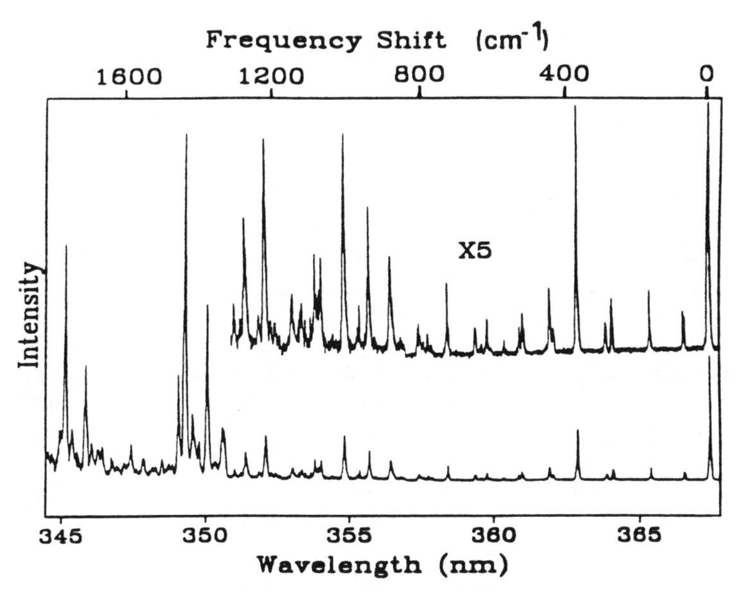

Fig. 9. FES of jet-cooled 2CP

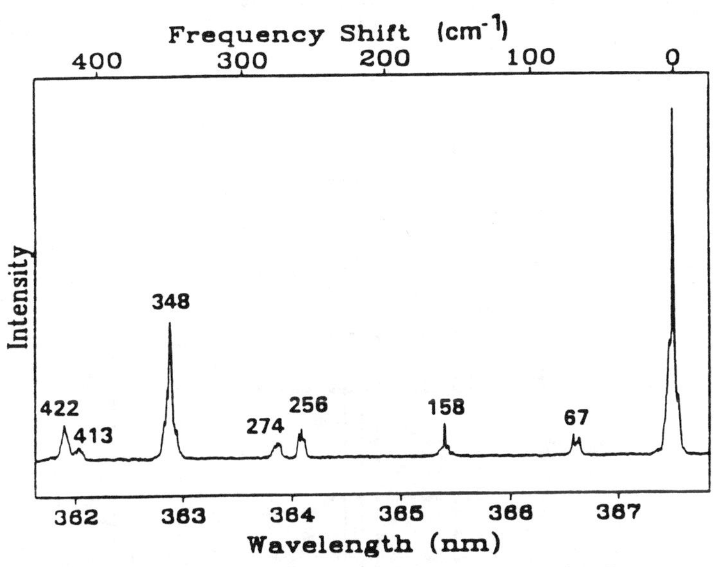

Fig. 10. Low-energy region of the FES spectrum of 2CP

are common. These modes are both considerably lower in frequency than in the electronic ground state reflecting the decrease in π character of the C=O bond. Intense bands at 1357 and 1418 cm^{-1} for 2CP are due to the C=O and C=C stretches, respectively. These two vibrations are Franck–Condon active due to the increased bond lengths for both bonds. Intense combination bands for each of these stretches were observed with the carbonyl in-plane wag, 19^1_0. All of the other ring mode transitions have also been observed for both the d_0 and d_2 species, and many of them were also found to be associated with combination bands. The only other excited state frequency determined from the FES data was for ν_{26}, a CH out-of-plane bending motion, at 768 cm^{-1}.

Table 1. Vibrational frequencies (cm^{-1}) for the ground and excited $S_1(n,\pi^*)$ states of 2CP-d_0 and 2CP-d_2

Approx. Description	2CP-d_0 Ground	2CP-d_0 Excited	2CP-d_2 Ground	2CP-d_2 Excited
ν_5 C=O stretch	1748	1357	1743	1360
ν_6 C=C stretch	1599	1418	1602	1421
ν_{13} Ring mode	1094	1037	1114	1037
ν_{14} Ring mode	999	974	957	960
ν_{15} Ring mode	912	906	851	843
ν_{16} Ring mode	822	849	810	814
ν_{17} Ring mode	753	746	746	727
ν_{18} Ring mode	630	587	630	580
ν_{19} C=O def (∥)	464	348	449	338
ν_{26} α-CH bend	750	768	737	762
ν_{28} C=O def (⊥)	537	422	?	403
ν_{29} C=C twist	287	274	281	267
ν_{30} Ring-puckering	94	67	85	59

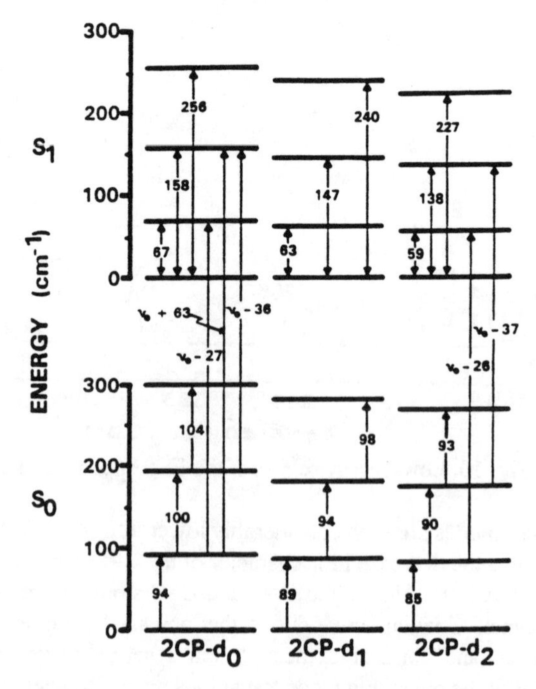

Fig. 11. Energy levels for the ring-puckering vibrations of 2CP and isotopomers.

The ring-puckering energy levels and observed transitions for the ground and excited electronic states are shown in Fig. 11. These were used to determine the coefficients a and b in (3). The potential function, which fit the data for all three isotopic species in the excited state was determined to be

$$V(\mathrm{cm}^{-1}) = 2.5\times10^{6}\,x^{4} + 1.8\times10^{3}\,x^{2}. \tag{14}$$

In the electronic ground state the coefficients a and b are 0.6×10^{6} and 2.6×10^{4} respectively. The electronic excitation has caused the potential energy function to change from one with a substantial quadratic contribution in the S_0 state to one that is almost purely quartic in the S_1 state. The dimensioned potential energy functions for both the ground and excited electronic states are shown together in Fig. 12. The result of the increased quartic nature of the potential is to flatten the curve near the minimum and to increase the slope of the walls of the potential at higher energies. Thus, 2CP has less resistance to ring-puckering in the $S_1(n,\pi^*)$ state than in the ground state.

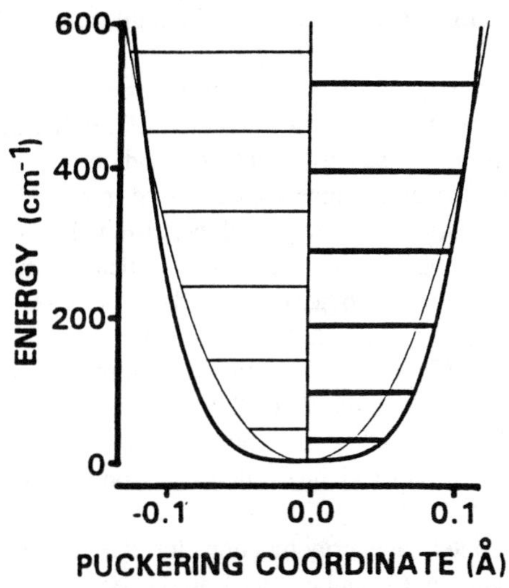

Fig. 12. Potential energy function and energy levels for the ring-puckering vibration of 2CP in its ground (thin lines) and S_1 excited (thick lines) states

Figure 13 shows a simplified molecular orbital energy diagram for the n and π orbitals of 2CP. Excitation to the $S_1(n,\pi^*)$ excited electronic state removes a non bonded electron from the carbonyl oxygen and transfers it to a π^* orbital, which is antibonding between carbon atoms 2 and 3 and also between carbon 1 and the oxygen atom. However, an increase in the bond order between carbons 1 and 2 is expected. A simple Hückel calculation [10] for the bond orders for the ground and excited states gave the following result:

1.8 1.5 1.9	1.5 1.6 1.7
O = C — C = C	O = C — C = C
Ground	Excited

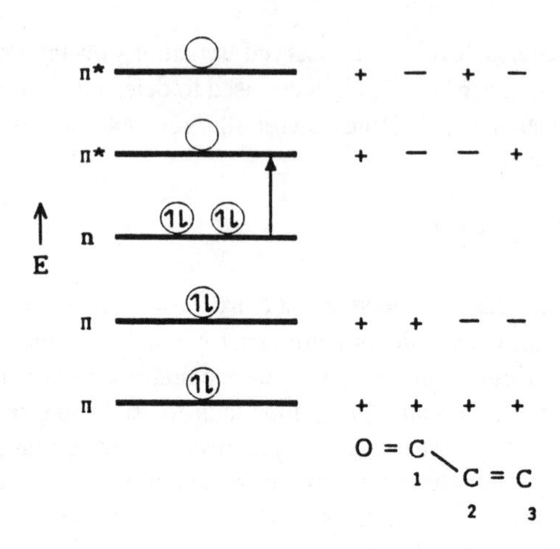

Fig. 13. Hückel molecular orbital diagram for 2CP

While the calculation is only approximate, it provides a semi-quantitative picture of the bonding changes expected following the excitation to the S_1 electronic state. As can be seen in Table 1, the decreases in the C=O and C=C bond orders manifest themselves in the ν_5 and ν_6 stretching frequencies, which are decreased from 1748 to 1357 cm^{-1} and from 1599 to 1418 cm^{-1}, respectively. The expected increase in the C(1)–C(2) bond stretching force constant has less effect on the other ring mode frequencies, since these vibrations also depend on the force constants of the other C–C bonds and C–C–C angles. Nonetheless, the observed frequency increase for ν_{16} from 822 to 849 cm^{-1} does appear to reflect this effect. Although the value of 1357 cm^{-1} for ν_5 in the electronic excited state has dropped considerably relative to the ground state, this decrease is smaller than that for 3CP (1227 cm^{-1}) and CP (1230 cm^{-1}). For these non-conjugated ring molecules, the π^* orbital is localized on the carbonyl group, and the excited state bond order is decreased to a greater extent.

The ring-puckering potential energy functions for 2CP in both the ground and excited states have the form given in (3) and are compared to each other in Fig. 12. As we have discussed above and previously shown [46,47] the potential energy coefficients a and b both have contributions from angle strain (a_s and b_s) and torsional (a_t and b_t) forces:

$$a = a_s + a_t, \qquad a_s >> a_t \tag{15}$$

$$b = b_s + b_t. \tag{16}$$

Furthermore, b_s depends both on the initial strain in the ring and on the angle bending force constants. For 2CP in the electronic ground state b_s is positive and greater in magnitude than b_t, which is negative. The latter contribution arises primarily from the CH_2–CH_2 torsion, and this effect is at a maximum for a planar ring. In the electronic excited state, b_t is expected to be little changed. However, the initial strain at the carbonyl carbon atom should be significantly reduced, and the magnitude of b_s should be lowered. This is confirmed by our experimental results here, which show that the sum of b_s and b_t has been reduced nearly to zero, and the quadratic constant has barely maintained a slight positive value. (The value of b has dropped from 2.6×10^4 to 0.2×10^4 cm^{-1}Å^{-2}.) Thus, the ring remains planar in the electronic excited state, but its rigidity has been substantially reduced.

3-Cyclopenten-1-one (3CP). Figure 14 shows the jet-cooled fluorescence excitation spectrum [48] of 3-cyclopenten-1-one. The band origin is observed at 30238 cm^{-1}. Each $v = 0$ (in the electronic ground state) to $v = n$ (in the A_2 electronic excited state) transition of the C=O wag has B_2 vibrational symmetry (assuming the molecule to lie in the xz plane) for n = odd, but has A_1 vibrational symmetry for n = even. Only transitions to the n = odd states can be observed. These show up as intense Type B bands arising from $A_2 \times B_2 = B_1$ symmetry. Figure 15 shows the low-frequency region of the spectrum in expanded form where the Type B bands are clearly evident.

The first six of these transitions are labelled in Fig. 14. It should be noted that since the $v = 0$ and $v = 1$ levels in the $S_1(n,\pi^*)$ state are near-degenerate, the band origin lies very close to the $0 \rightarrow 1$ frequency. The other bands in the spectrum include many combinations of the C=0 wag with the ring-puckering vibration and combinations of these with other fundamentals including the C=O stretch. Of particular note is the $3^1_0 29^4_0 30^1_0$ band at 32 211 cm^{-1}, shifted 1973 cm^{-1} from the band origin and 746 cm^{-1} from the C=0 stretching (υ_3) value at 1227 cm^{-1}. The 1973 cm^{-1} value corresponds to the sum of ν_3, of the $0 \rightarrow 4$ wagging transition (ν_{29}), and of the ring-puckering (ν_{30}) frequency of 127 cm^{-1}. This shows the $0 \rightarrow 4$ spacing to be 619 cm^{-1}, which is 21 cm^{-1} less than the $0 \rightarrow 5$ spacing.

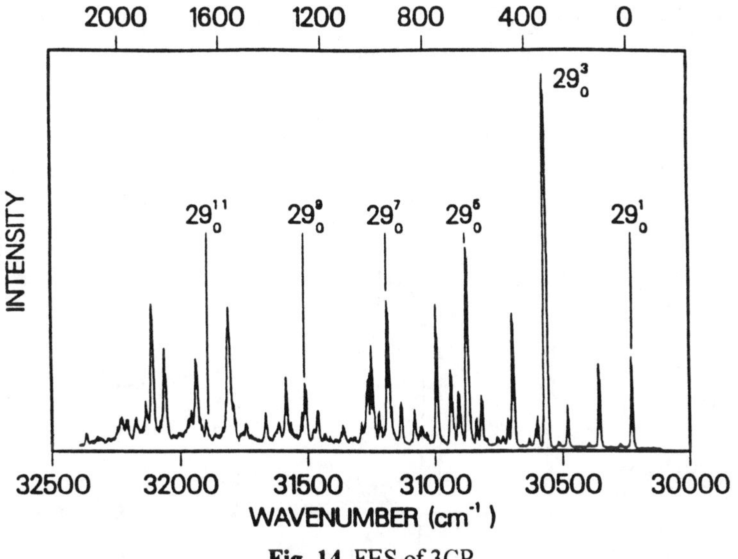

Fig. 14. FES of 3CP

In order to analyze the C=O wagging vibration in the electronic excited state, we have utilized our computer programs, described previously, for calculating the reduced masses [20,21] and energy levels [26,27] for the potential energy function given by (3). Here x is used for the C=O wagging coordinate given in terms of the wagging angle ϕ and the C=O bond distance R by [19]

$$x = R\phi \,. \tag{17}$$

The reciprocal reduced mass expansion g_{44} for this coordinate has the form given in (3).

For 3-cyclopenten-1-one, and several other ketones to be discussed below, the reduced mass and the carbonyl-wagging potential energy parameters which best fit the observed frequency separations are given in Table 2. The experimentally determined potential energy function is shown in Fig. 16 in both quartic/quadratic (solid line) and Gaussian barrier (dotted line) forms along with both the observed and calculated frequency separations. The minimum energy corresponds to a wagging angle of ±26°, and the barrier to inversion is 926 cm^{-1} (2.65 kcal/mole). For 3CP the ring-puckering frequency of 127 cm^{-1} in the S_1 state is considerably higher than the value of 83 cm^{-1} in the ground state.

Cyclopentanone (CP). The survey spectrum of cyclopentanone [49] is shown in Fig. 17. The band origin is at 30 276 cm^{-1}. In the ground state this molecule is twisted [50], and in the C_{2v} approximation, the vibrational ground state is nearly doubly degenerate, with symmetry species A_1 and A_2. The twisting conformation (and degeneracy) carries through to the electronic excited state, as demonstrated by the similarity in the ring-bending and twisting frequencies in the S_1 state. The purely electronic transition is again $^1A_2 \leftarrow {}^1A_1$, which is forbidden in the C_{2v} approximation. However, combination with odd quantum transitions of the B_2 C=O wagging results in Type B bands from B_1 symmetry. If either the ground or excited electronic state also is in combination with the near-degenerate A_2 twisting state, the even quanta C=O wagging transitions can also be observed as Type A (A_1) bands [$A_2 \times A_2 \times (B_2)^n = A_1$ for n = even]. Figure 17 shows that transitions for both even and odd quantum states of the C=O wag in the S_1 state are readily observed. The energy diagram for the

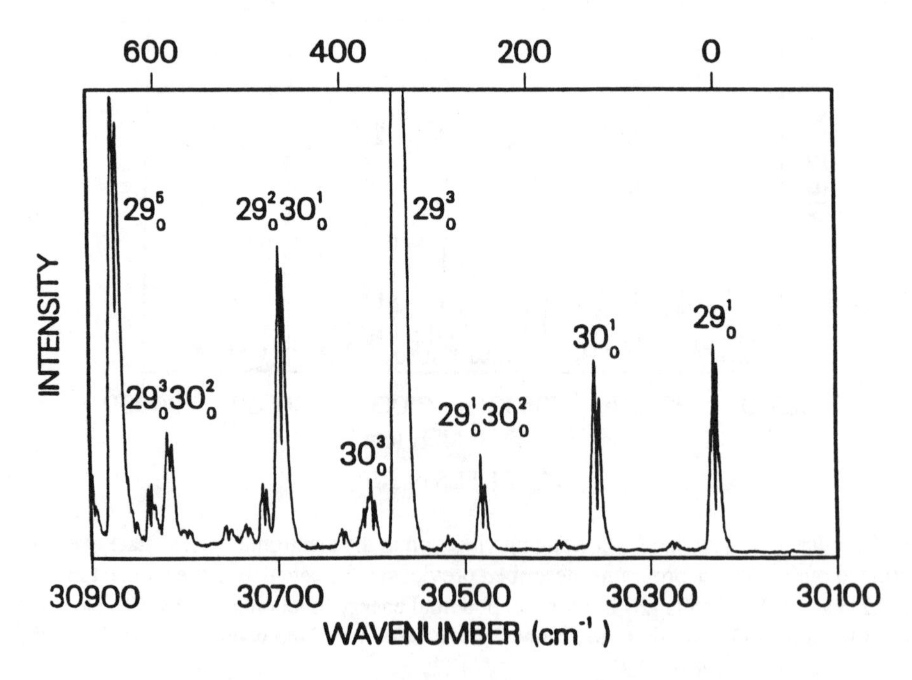

Fig. 15. Low-frequency region of the FES of 3CP

Table 2. Potential energy parameters and reduced masses for C=O wagging vibrations in the $S_1(n,\pi^*)$ electronic state

Molecule	μ (a u)	$V = ax^4 + bx^2$ a (cm^{-1}/Å^4)	 b(cm^{-1}/Å^2)	Barrier (cm^{-1})	ϕ_{min}
$CH_2CH_2CH_2CH_2C{=}O$ (CP)	5.57	10.49×10^3	-5.34×10^3	680	22°
$CH_2CH{=}CHCH_2C{=}O$ (3CP)	5.26	8.11×10^3	-5.48×10^3	926	26°
$CH_2CH_2CH_2C{=}O$ (CB)	4.24	2.47×10^3	-4.38×10^3	1940	41°
$CH_2CH_2OCH_2C{=}O$ (THFO)	5.21	8.51×10^3	-6.26×10^3	1152	23°
$CH_2CH_2SCH_2C{=}O$ (THTP)	3.57	13.4×10^3	-5.94×10^3	659	20°
bicyclohexanone (BCH)[a]	6.49	7.84×10^3	-5.18×10^3	873	23°

[a] The cubic term $0.097\times10^3x^3$ is also present in V.

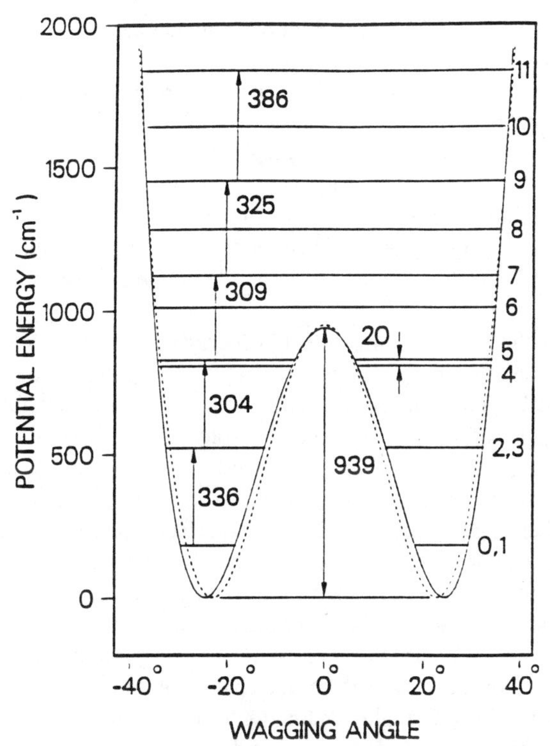

Fig. 16. Carbonyl wagging potential energy function for the excited state of 3CP

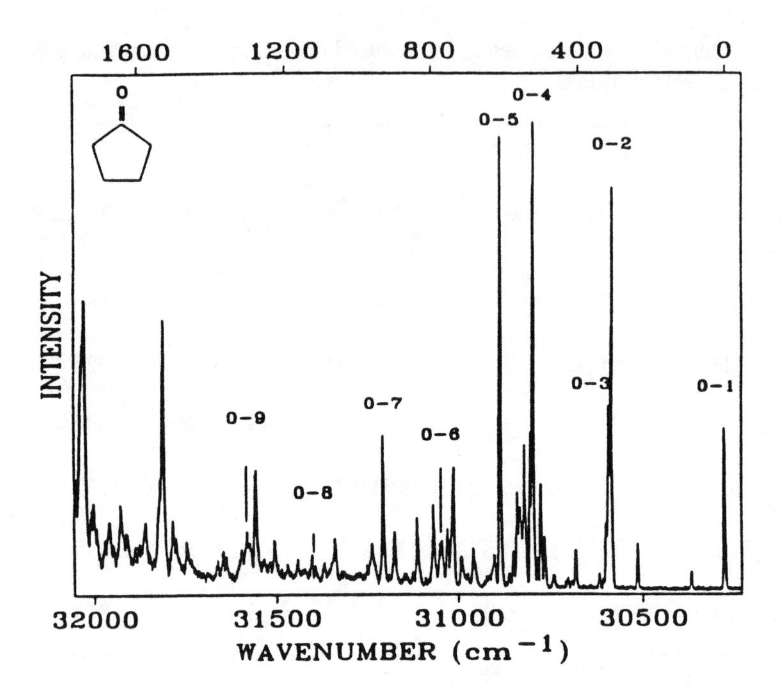

Fig. 17. FES for cyclopentanone

ground and excited states is shown in Fig. 18. The near-degeneracy of the carbonyl wagging in the S_1 excited state and the near degeneracy of the ring-twisting levels in both states greatly increase the complexity of the energy diagram. The twisting degeneracy allows the g even quantum states of the carbonyl wagging in the excited states to be directly examined. For 3CP and cyclobutanone (discussed below) only transitions to odd states can be observed. The spectral assignment for CP was aided by examination of the band contours for the Type A and B bands [49], and even overlapped bands were fit with the contour calculation.

Figure 19 shows the C=O wagging potential energy function for cyclopentanone with a barrier of 680 cm^{-1} and the energy minima at ±22°. The kinetic and potential energy terms are given in Table 2. The comparison between the S_0 and S_1 states for the fundamental vibrational frequencies of CP and several other molecules is given in Table 3.

The observed data for the ring-twisting and ring-bending motions of CP in the S_0 and S_1 states are quite similar, since the ring has a similar twisted ring conformation for each state. However, the ring bending/twisting potential energy surfaces for the S_0 and S_1 states, shown in Figs. 20 and 21, can be seen to be very different. In the ground state there is no saddle point in the PES, but in the S_1 state the bent form has dropped in energy, giving rise to a saddle point corresponding to a barrier to pseudorotation of about 550 cm^{-1}. This value is a measure of the energy difference between the lower energy twisted conformation and the bent structure. The decrease in energy of the bent conformation in the excited state is a result of a decrease in the C–C(O)–C angle bending force constant associated with the increased antibonding character of the carbon-oxygen bond.

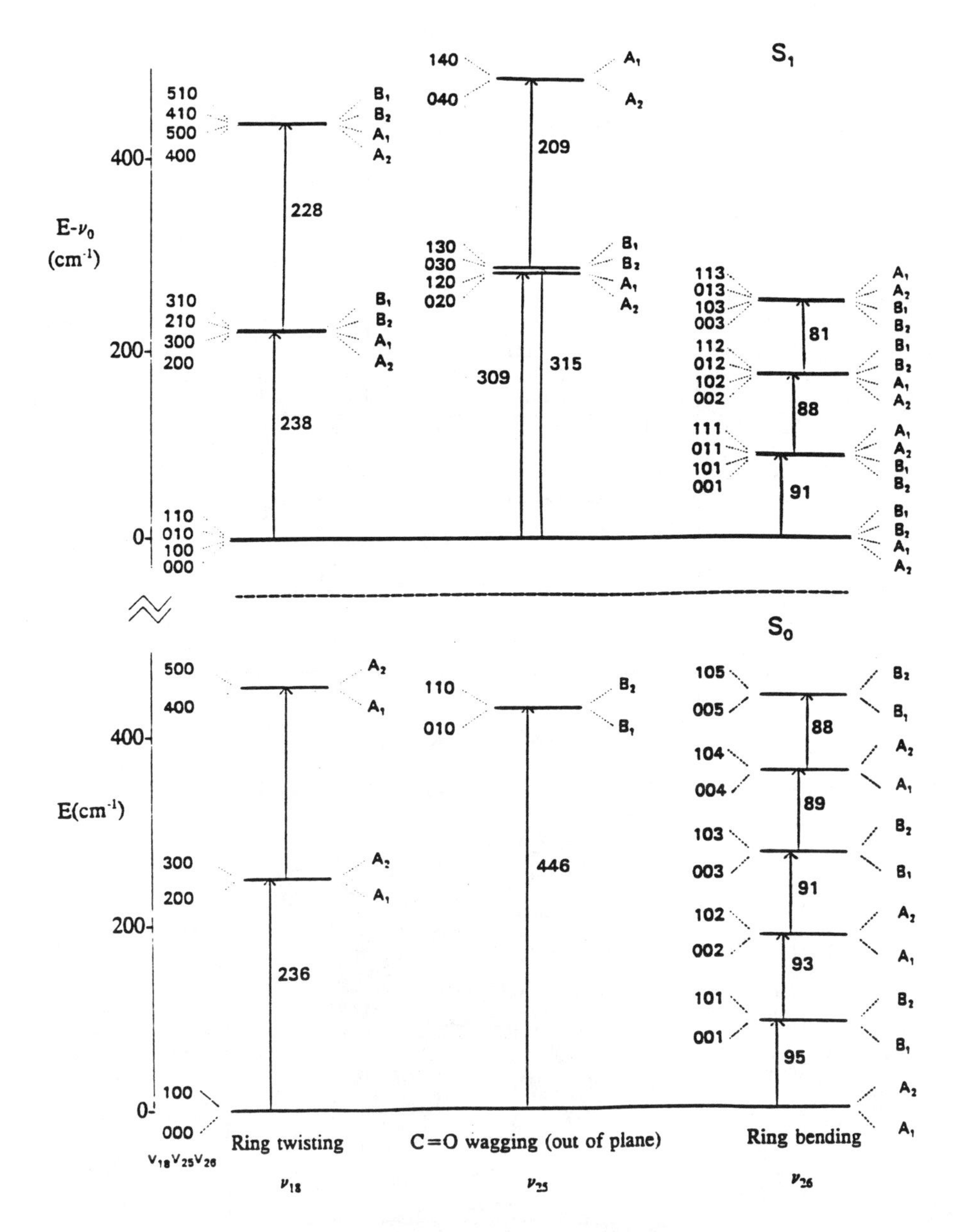

Fig. 18. Energy diagram for cyclopentanone

Cyclobutanone (CB). The fluorescence excitation spectrum of cyclobutanone [51] is shown in Fig. 22, with the band origin at 30 292 cm^{-1}. Only transitions involving the odd quantum levels of the wag are allowed as Type B bands ($A_2 \times B_2 = B_1$ gives the Type B band type). As it turns out, the barrier to inversion of cyclobutanone is sufficiently high that the lowest six pairs of levels are nearly doubly degenerate, with vibrational symmetry species A_1 and B_2. Even though transitions involving only the vibrational states of B_2

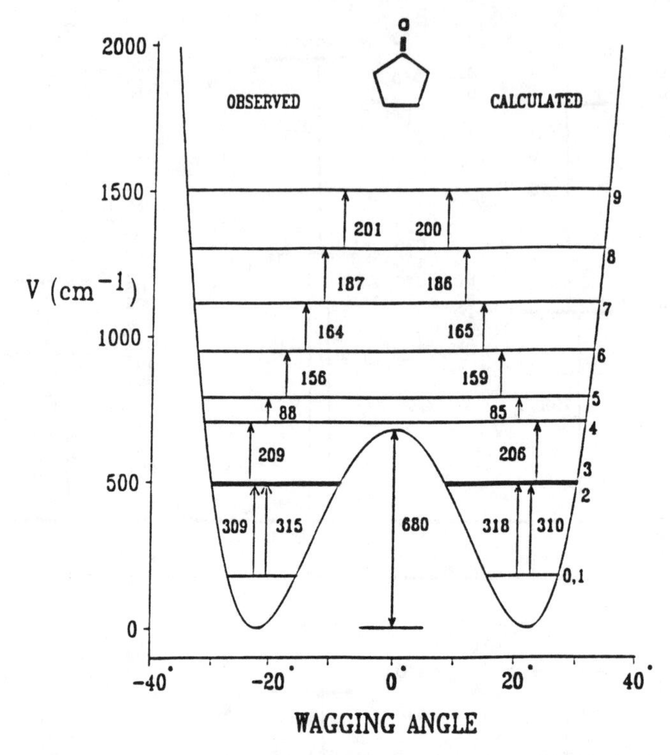

Fig. 19. Carbonyl wagging potential energy function of CP in its S_1 state

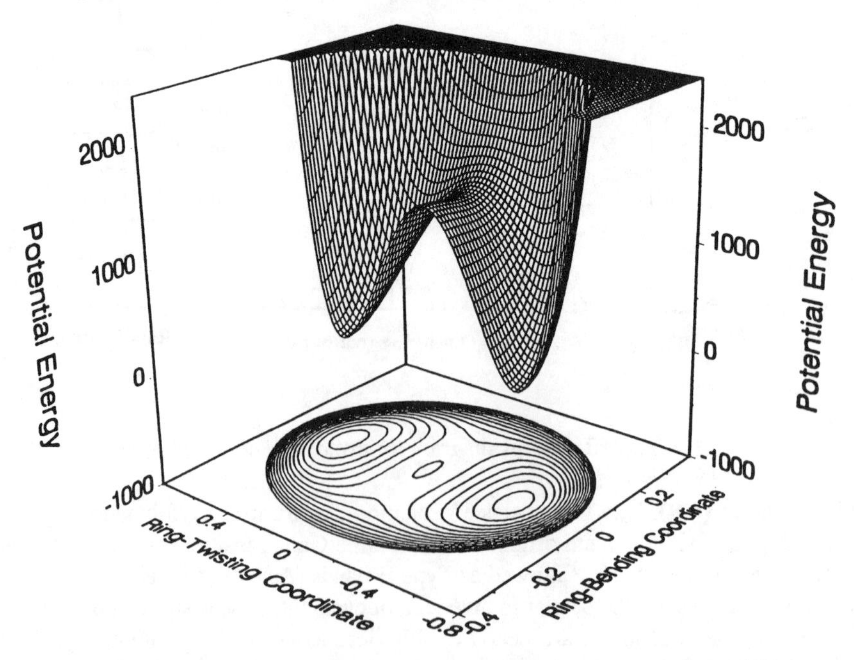

Fig. 20. PES for the out-of-plane ring modes in the ground state of CP

Table 3. Comparison of frequencies of several vibrations of cyclic ketones in the ground and $S_1(n,\pi^*)$ electronic excited states

Vibration	2CP		THTP		CP		BCH		3CP		THFO		CB	
	S_0	S_1	S_0	S_1	S_0	S_1	S_0	S_1	S_0	S_1	S_0	S_1	S_0	S_1
Ring puckering	94	67	67	58	95	91	86	134	83	127	114	82	36	106
Ring twisting	287	274	170	....	38	238	....	....	378	377	237	224	....	....
C=O wag o.p.	537	42	427	326	446	309	....	315	450	336	463	344	395	355
C=O wag i.p.	464	348	483	329	467	342	....	....	458	339	463	365	454	392
C=O stretch	1748	1357	1760	1240	1770	1230	....	....	1773	1227	1755	1232	1816	1251
Reference	10		53		50	49	52		10	48	53		51	

i.p. = in plane

o.p. = out of plane

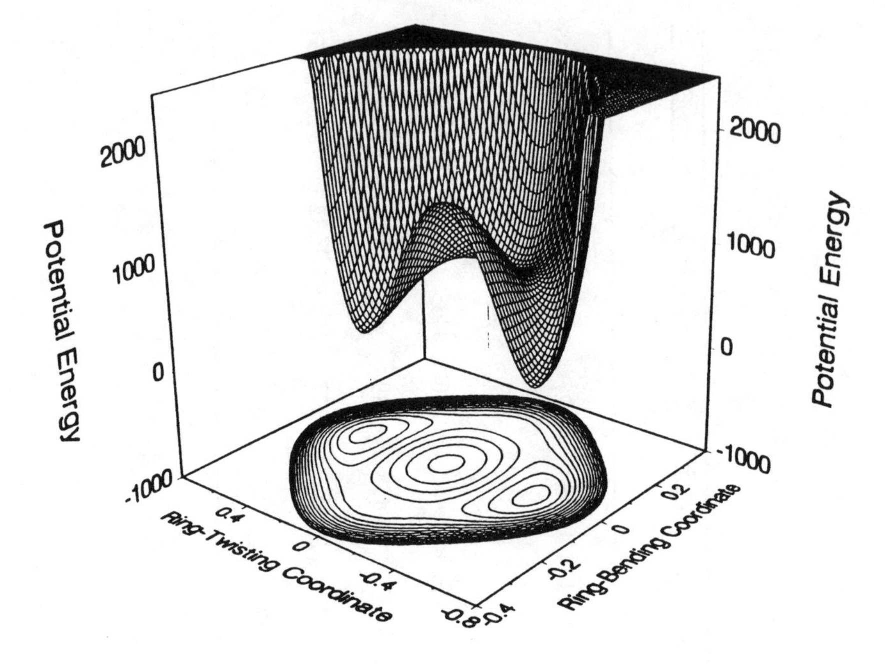

Fig. 21. PES for the out-of-plane ring modes for the S_1 state of CP

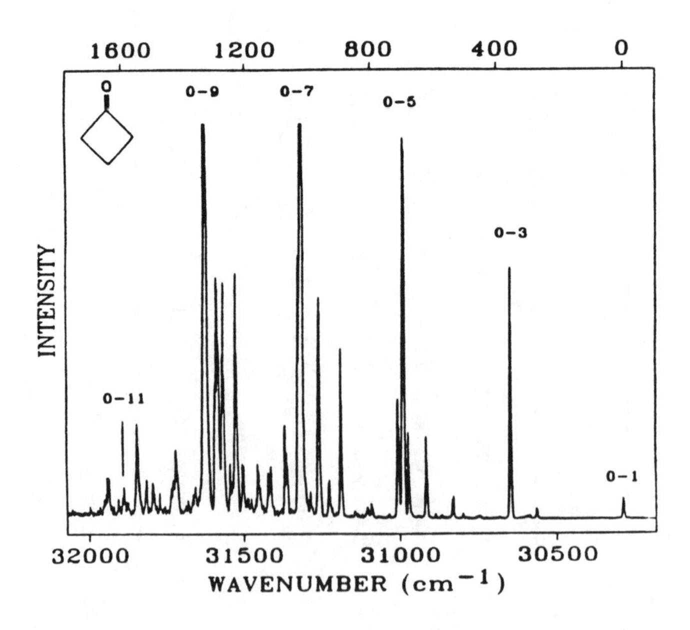

Fig. 22. FES of cyclobutanone

symmetry can be observed, these have essentially identical frequencies with those involving the A_1 states. Thus no significant information is lost. Figure 23 shows the C=0 wagging potential energy function with a barrier of 1940 cm^{-1} and energy minima at ±41°. The reduced mass and potential energy parameters for cyclobutanone are shown in Table 2. The ring-puckering frequency shows a dramatic increase in going from 37 cm^{-1} in the S_0 state to 106 cm^{-1} in the excited state. The high ring angle strain appears to be a factor here. A two-dimensional potential energy surface for the carbonyl-wagging and ring-puckering vibrations was also determined, and this is shown in Fig. 24. As shown below, the combination of these two motions can result in two different energy minima in the excited state, depending on whether the molecule puckers in the direction of the carbonyl wagging or not. Thus, the PES has some asymmetry resulting from this effect.

Other ketones. The fluorescence excitation spectra of three other ketones have also been recorded and analyzed. The results for bicyclo[3.1.0]hexan-3-one (BCH) [52], tetrahydrofuran-3-one (THFO) [53], and tetrahydrothiophen-3-one (THTP) [53] are also summarized in Table 2. The BCH molecule was of special interest in that the carbonyl wagging potential energy function is expected to be asymmetric as the C=O group wags toward or away from the cyclopropane ring. However, the asymmetry was found to be surprisingly small. This is shown in Fig. 25 where the energy difference between the two wagging minima can be seen to be less than 50 cm^{-1}. The THFO and THTP molecules have

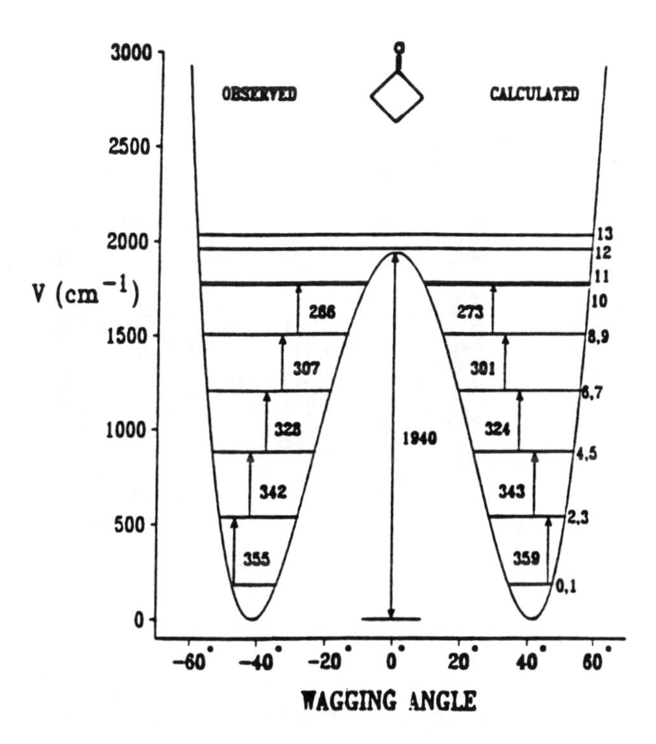

Fig. 23. Potential energy function for the carbonyl wagging of CB in its S_1 state

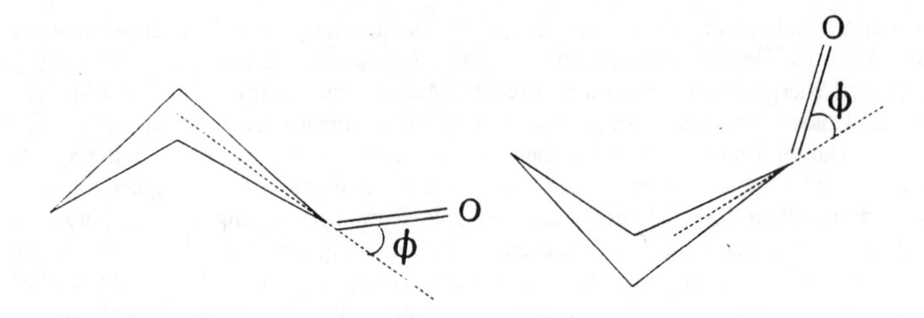

symmetric carbonyl-wagging functions with inversion barriers of 1152 and 659 cm^{-1}, respectively. The ring bending/twisting PES determined for the THFO excited state is similar to that of CP in that the bent conformation is relatively low in energy, resulting in saddle points on the surface.

Summary. The potential energy barriers for the cyclic ketones are summarized in Table 2. Examination of the results shows that the barriers to inversion increase as the internal ring angle at the carbonyl group becomes more strained. This is depicted in Fig. 26 where the barrier is plotted against the internal ring angle as calculated using molecular mechanics methods. The carbonyl oxygen thus is pulled more out of the plane as the C–C(O)–C angle is tightened.

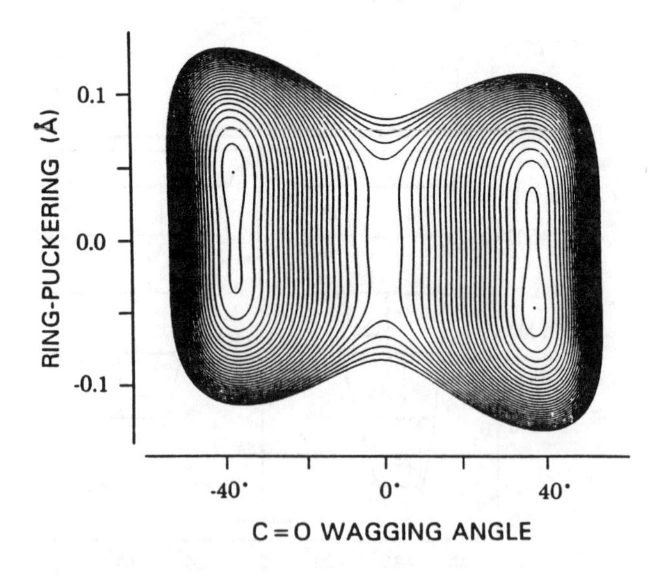

Fig. 24. Two-dimensional PES for CB in its S_1 state

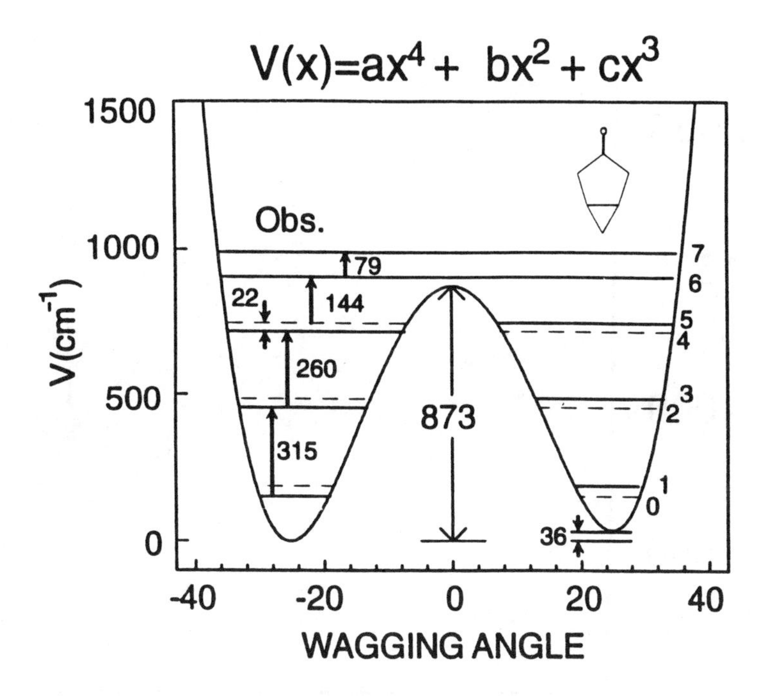

Fig. 25. Carbonyl wagging potential energy function for the BCH excited state

3.3 Fluorescence spectra of *trans*-stilbene and derivatives

***trans*-Stilbene.** For more than twenty years the photoisomerization of stilbene has been of great interest to both theoretical and experimental chemists [54–64]. It is generally believed that the isomerization is facilitated in the electronic excited state by the internal rotation (θ) about the C=C bond, which has a potential energy minimum corresponding to a perpendicular (twisted) configuration (analogous to the ethylene D_{2d} structure in its $S_1(\pi,\pi^*)$ state). A barrier of about 1200 cm^{-1} hindering the *trans* to twist internal rotation has been determined for the S_1 state by Syage, Felker, and Zewail [55, 56, 64]. Because the dynamics of the photoisomerization are strongly affected by the vibrational potential energy surfaces, there has been considerable interest in assigning the vibrational spectra of *trans*-stilbene, especially for the torsional modes, for both the ground (S_0) and excited (S_1) electronic states [54, 55, 57, 59–63]. However, there had not been general agreement on the vibrational frequencies associated with each mode. Several groups proposed frequency assignments [55, 57, 59, 61] and Waldeck [54] has reviewed this work.

We have recently reinvestigated [65] the fluorescence excitation spectra (Fig. 27) and dispersed fluorescence spectra (Fig. 28) and used these, together with vapor-phase Raman data [66], to reassign the eight low-frequency (below 300 cm^{-1}) vibrations of this molecule in its S_0 and $S_1(\pi,\pi^*)$ states. Figure 29 shows the energy levels for four of these low frequency modes in the ground and excited states. We then carried out kinetic energy (reciprocal reduced mass) calculations for the two phenyl torsions, ν_{37} (ϕ_1 - ϕ_2) and

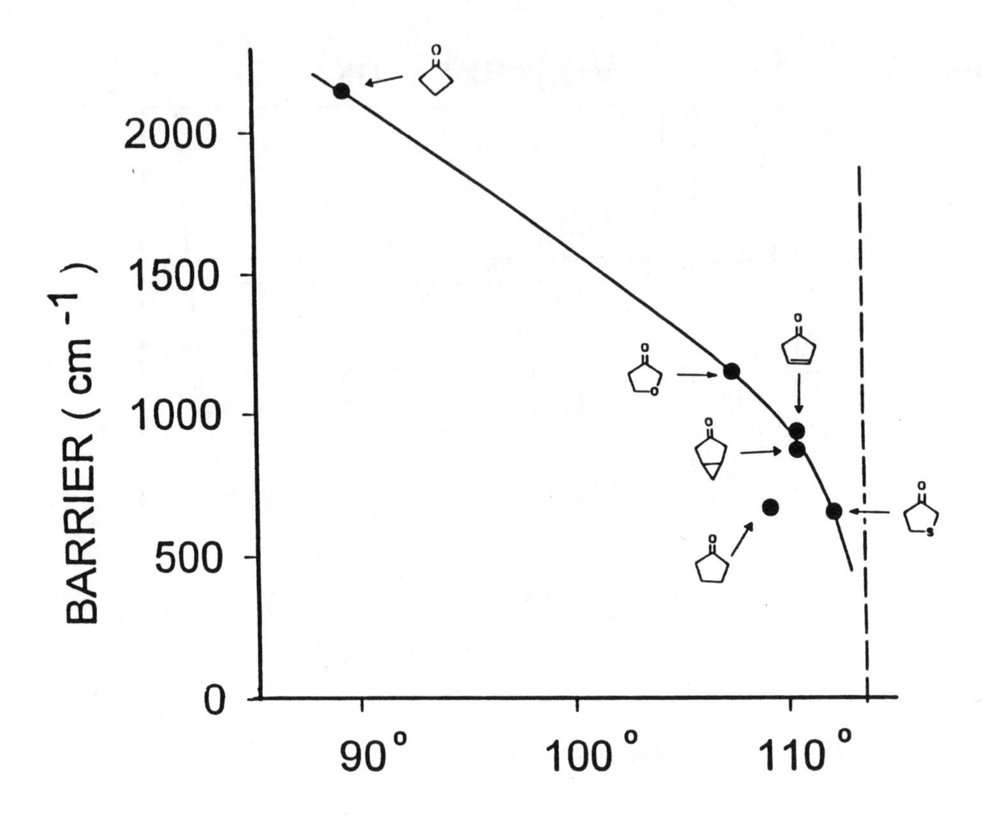

Fig. 26. Barriers to inversion in the excited states vs. interior C–C(O)–C angle

ν_{48} (ϕ_1 + ϕ_2), where ϕ_1 and ϕ_2 are defined in Fig. 30. These results were utilized along with the experimental data to determine the two-dimensional potential energy surfaces for these modes in both the S_0 and S_1 states. The function

$$V(\phi_1, \phi_2) = \tfrac{1}{2}V_2(2 + \cos 2\phi_1 + \cos 2\phi_2) + V_{12}'\cos 2\phi_1 \cos 2\phi_2 + V_{12} \sin 2\phi_1 \sin 2\phi_2 \quad (18)$$

with V_2 = 1550 cm^{-1}, V_{12}' = 337.5 cm^{-1}, and V_{12}= 402.5 cm^{-1} for the S_0 state and with V_2= 1500 cm^{-1}, V_{12} = -85 cm^{-1}, and V_{12}= -55 cm^{-1} for the $S_1(\pi,\pi^*)$ state fits the observed data (nine frequencies for S_0, six for S_1) extremely well. The barriers to simultaneous internal rotation of both phenyl groups are given by twice the V_2 values. The fundamental frequencies for these torsions are ν_{37} = 9 cm^{-1} and ν_{48} = 118 cm^{-1} for the S_0 state and ν_{37} = 35 cm^{-1} and ν_{48} = 110 cm^{-1} for the S_1 excited state. Figure 31 shows the potential energy surface for the phenyl torsions in the S_1 state. The third torsion, ν_{35}, which is the internal rotation about the C=C bond, was assigned at 101 cm^{-1} for the S_0 state based on a series of overtone frequencies (202 cm^{-1}, 404 cm^{-1}, etc.). For S_1 ν_{35} = 99 cm^{-1} based on observed frequencies at 198, 396 cm^{-1}, etc. Kinetic energy calculations were also carried out for this mode, and a one-dimensional potential energy function of the form

$$V(\theta) = \tfrac{1}{2}V_1(1 - \cos\theta) + \tfrac{1}{2}V_2(1 - \cos 2\theta) + \tfrac{1}{2}V_4(1 - \cos 4\theta) \quad (19)$$

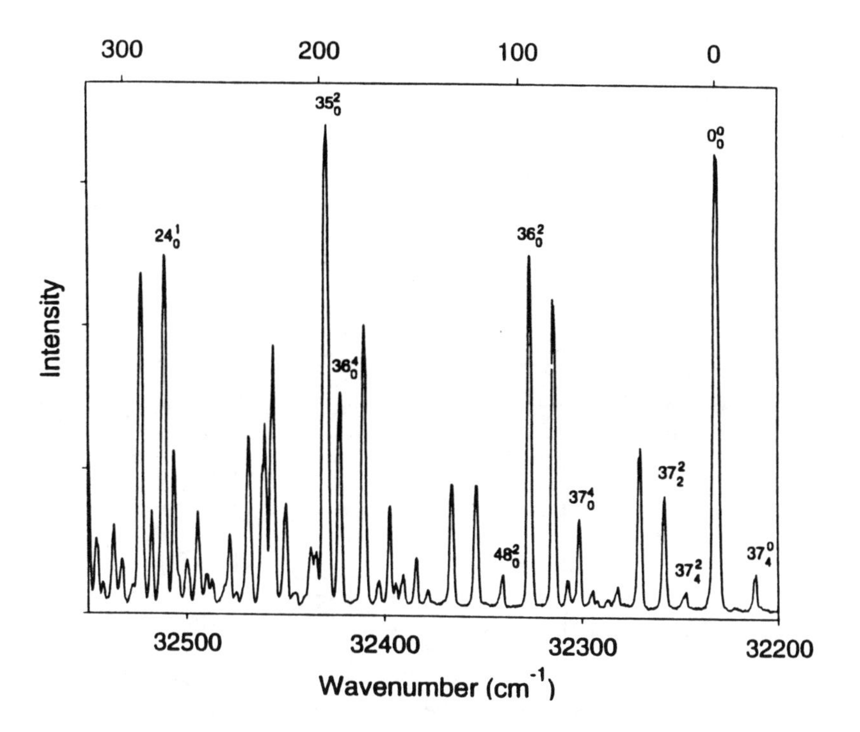

Fig. 27. FES of *trans*-stilbene

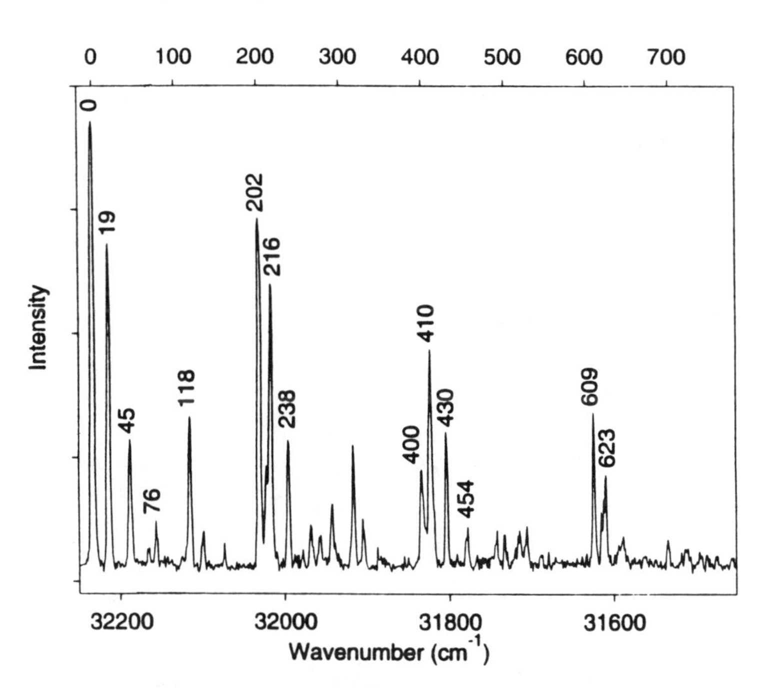

Fig. 28. Dispersed fluorescence spectra of *trans*-stilbene

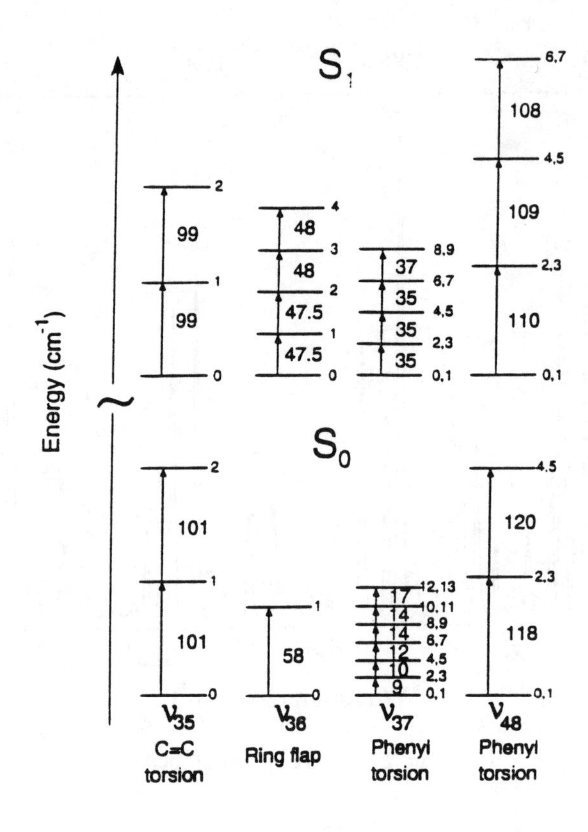

Fig. 29. Energy diagram for some low-frequency modes of *trans*-stilbene

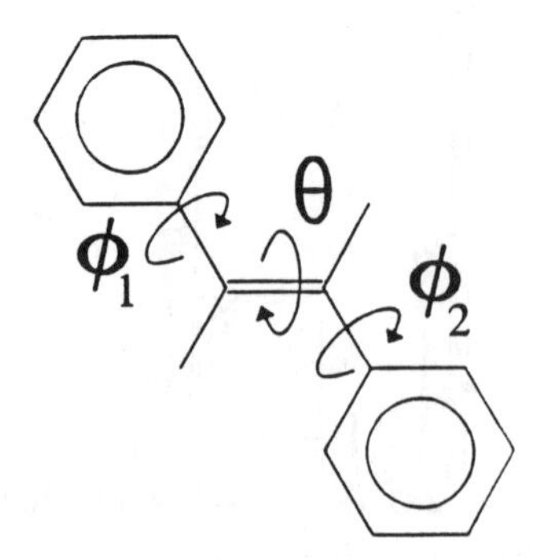

Fig. 30. Torsional coordinates for *trans*-stilbene

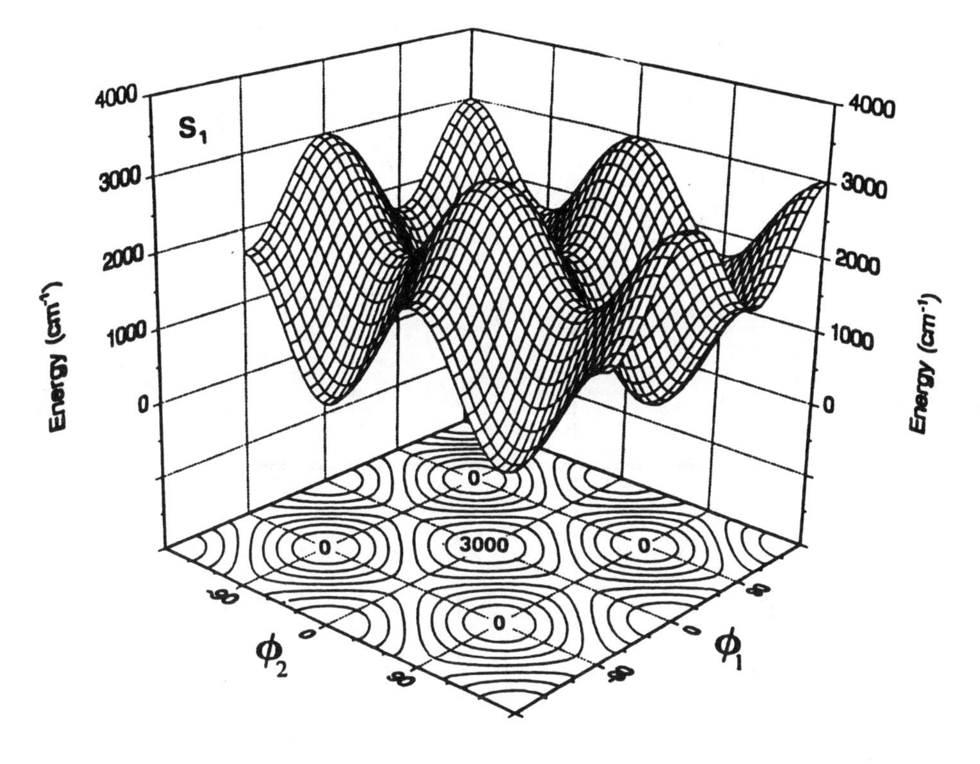

Fig. 31. PES for the phenyl torsions of *trans*-stilbene for the S_1 state

was utilized to reproduce the frequencies for the ground state. Since the observed vibrational levels extend to less than 3 kcal/mole beyond the potential energy minimum, they can only be used to approximate the torsional barrier, which was calculated to be 50 ± 20 kcal/mole. This is in general agreement with the accepted value of 48 kcal/mole [54, 58]. For the excited state an additional V_8 term was added to (19) in order to fit the data for the *trans* potential energy well. The data indicate that the *trans* → twist barrier for the S_1 state is 2000 ± 600 cm^{-1}, somewhat higher than estimated from dynamics studies [55, 56, 64]. Figure 32 shows the potential functions for the C=C torsion in the one-dimensional approximation for both the ground and excited states.

4-Methoxy-*trans*-stilbene. Both the FES and Raman spectra for this methoxy derivative of *trans*-stilbene are very similar to those of the parent compound [67]. Table 4 compares the vibrational frequencies of the two molecules for both the ground and excited states. The phenyl torsion frequencies can be seen to be at 123 and 8 cm^{-1} for the methoxy compound in the ground state and at 118 and 9 cm^{-1} for the unsubstituted stilbene. In the excited state the corresponding values are at 109 and 28 cm^{-1} and 110 and 35 cm^{-1}, respectively. Analysis

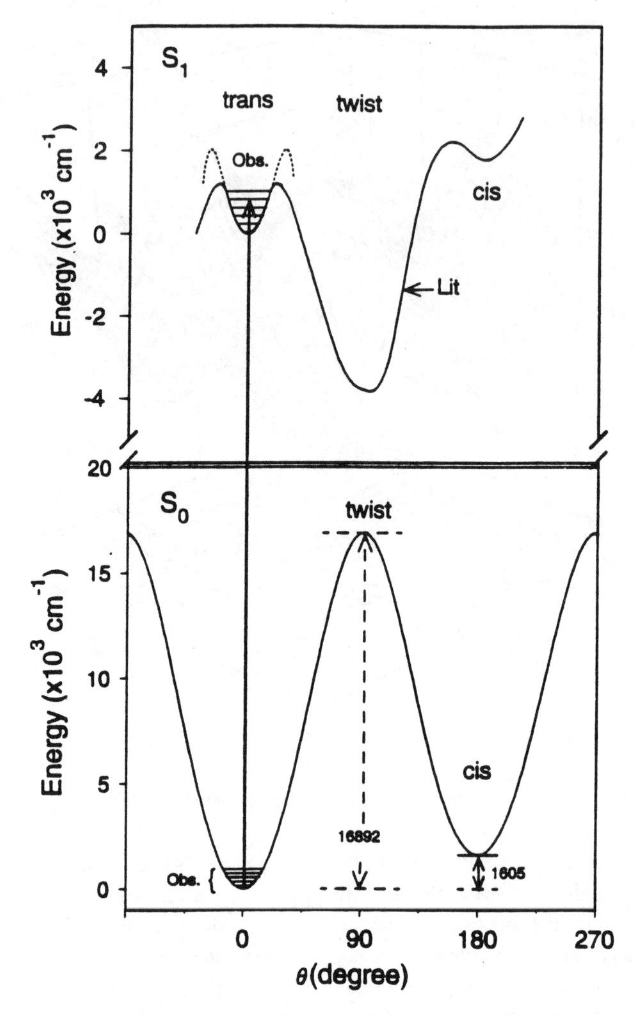

Fig. 32. Potential energy functions for the C=C torsion of *trans*-stilbene

of the data shows that the barrier to simultaneous phenyl torsion is 2860 cm^{-1} for both ground and excited states, or about 5% lower than for *trans*-stilbene. For the C=C torsion the substitution increases the barrier from 49 to about 52 kcal/mole in the ground state and by about 250 cm^{-1} in the excited state. The increase in the ethylinic torsion barrier is consistent with an increase observed in dynamics studies.

4 Summary

The FES spectra of seven cyclic ketones have been analyzed and all but 2-cyclopenten-1-one, which retains some C=C/C=O conjugation in its excited state, have been found to have double minimum potential energy functions for the carbonyl wagging

Table 4. Vibrational assignments (cm^{-1}) for the low-frequency vibrations of 4-methoxy-*trans*-stilbene for the S_0 and $S_1(\pi,\pi^*)$ electronic states.

Description	*trans*-stilbene				4-methoxy- *trans*-stilbene			
	C_{2h}		S_0	S_1	C_s		S_0	S_1
C_e-Ph bend	A_g	ν_{24}	273	280	A′	ν_{54}	234	279
phenyl wag (i.p.)		ν_{25}	152	150		ν_{55}	144	-
phenyl wag (i.p.)	B_u	ν_{72}	76	70		ν_{56}	68	-
phenyl flap	B_g	ν_{47}	211	200	A″	ν_{80}	179	-
phenyl torsion		ν_{48}	118	110		ν_{81}	123	109
C=C torsion	A_u	ν_{35}	101	99		ν_{82}	86	86
phenyl flap		ν_{36}	58	48		ν_{83}	59	46
phenyl torsion		ν_{37}	9	35		ν_{84}	8	28

vibration in the $S_1(n,\pi^*)$ electronic excited state. The inversion barrier and the wagging angle were found to increase with angle strain of the –C–C(O)–C– angle. Thus, cyclobutanone has the largest barrier of 1940 cm^{-1} and the largest wagging angle for the excited state. The data also allow the out-of-plane ring motions to be studied in the excited states. For the five-membered rings, the bent conformations become energetically less unfavorable in the excited states as the excitation of an electron to an antibonding orbital decreases the angle strain which opposes ring bending. In addition to these vibrations, the expected decrease in the C=O stretching frequency was also observed for all compounds.

The fluorescence data for *trans*-stilbene and its methoxy derivative have enabled us to rigorously determine the two-dimensional phenyl torsional potential energy surfaces for each molecule. The barriers are much higher than previously estimated. The potential energy function for the C=C torsion, which governs the photoisomerization, was also calculated for both molecules for both the ground and excited states, and the results have been correlated to dynamics studies.

Much remains to be learned about electronic excited states. The determination of potential energy surfaces which govern conformational or dynamic processes has proven to be of great value in elucidating the nature of these states.

Acknowledgments

Financial support from the National Science Foundation, the Robert A. Welch Foundation, and the Texas ARP program is gratefully acknowledged. The excellent work of the graduate students listed in the references has been greatly appreciated.

REFERENCES

1 J. Laane, Quart. Rev. **25**, 533 (1971).
2 C. S. Blackwell and R. C. Lord, *Vibrational Spectra and Structure* **1**, 1 (1972).
3 J. Laane, *Vibrational Spectra and Structure* **1**, 25 (1972).
4 L. A. Carreira, R. C. Lord, and T. B. Malloy, Jr., *Topics in Current Chemistry*, Springer-Verlag, Berlin **82**, 3 (1979).
5 T. B. Malloy, Jr., L. E. Bauman, L. A. Carreira, *Topics in Stereochemistry* **14**, 97 (1979).
6 A. C. Legon, Chem. Rev. **80**, 231 (1980).
7 J. Laane, Pure & Appl. Chem. **59**, 1307 (1987).
8 J. Laane, M. Dakkouri, B. van der Veken, and H. Oberhammer (eds.), *Structures and Conformations of Non-Rigid Molecules*, Kluwer Publishing, Amsterdam, p. 65 1993.
9 J. Laane, Ann. Rev. Phys. Chem. **45**, 179 (1994).
10 C. M. Cheatham and J. Laane, J. Chem. Phys. **94**, 7743 (1991).
11 C. M. Cheatham, M. Huang, N. Meinander, M. B. Kelly, K. Haller, W.-Y. Chiang, and J. Laane, J. Mol. Struct. **377**, 81 (1996).
12 C. M. Cheatham, M.-H. Huang, J. Laane, J. Mol. Struct. **377**, 93 (1996).
13 M. A. Harthcock and J. Laane, J. Mol. Spectrosc. **94**, 461 (1982).
14 J. R. Villarreal, L. E. Bauman, and J. Laane, J. Chem. Phys. **80**, 1172 (1976).
15 L. E. Bauman, P. M. Killough, J. M. Cooke, J. R. Villarreal, and J. Laane, J. Phys. Chem. **82**, 2000 (1982).
16 P. M. Killough, R. M. Irwin, and J. Laane, J. Chem. Phys. **76**, 3890 (1982).
17 P. W. Jagodzinski, L. W. Richardson, M. A. Harthcock, and J. Laane, J. Chem. Phys. **73**, 5556 (1980).
18 M. A. Harthcock and J. Laane, J. Chem. Phys. **79**, 2103 (1983).
19 J. Laane, M. A. Harthcock, P. M. Killough, L. E. Bauman, J. M. Cooke, J. Mol. Spectrosc. **91**, 286 (1982).
20 M. A. Harthcock and J. Laane, J. Mol. Spectrosc. **91**, 300 (1982).
21 R. W. Schmude, M. A. Harthcock, M. B. Kelly, and Laane, J. J. Mol. Spectrosc. **124**, 369 (1987).
22 M. M. Tecklenburg and J. Laane, J. Mol. Spectrosc. **137**, 65 (1989).
23 M. M. Strube and J. Laane, J. Mol. Spectrosc. **129**, 126 (1988).
24 R. P. Bell., Proc. Roy. Soc. **A183**, 328 (1945).
25 J. Laane, J. Phys. Chem. **95**, 9246 (1991).
26 J. Laane, Appl. Spectrosc., **24**, 73 (1970).
27 M. A. Harthcock and J. Laane, J. Phys. Chem. **89**, 4231 (1985).
28 E. Cortez and J. Laane, J. Amer. Chem. Soc. **115**, 12132 (1993).
29 I. Tvaroska, T. Bleha, R. S. Tyson, D. Horton, (eds.), *Anomeric and Exo-Anomeric Effects in Carbohydrate Chemistry*, Academic Press, Inc. **47**, 45 (1989).
30 L. Norskov-Lauritsen and N. L. Allinger, J. Computational Chem. **4**, 326 (1984).
31 J. Choo, J. Laane, R. Majors, and J. R. Villarreal, Amer. Chem. Soc. **115,** 8396 (1993).
32 L. E. Bauman and J. Laane, J. Phys. Chem. **92**, 1040 (1988).
33 J. E. Kilpatrick, K. S. Pitzer., and R. Spilzer, J. Am. Chem. Soc. **64**, 2483 (1947).

34 S. J. Leibowitz, J. R. Villarreal, and J. Laane, J. Chem. Phys. **96**, 7298 (1992); S. J. Leibowitz and J. Laane, J. Chem. Phys. **101**, 2740 (1994).
35 M. Ito, J. Phys. Chem. **91**, 517 (1987).
36 A. D. Walsh, J. Chem. Soc. 2306 (1953).
37 J. C. D. Brand, J. Chem. Soc. 858 (1956).
38 V. R. Rao and I. A. Rao, Indian J. Phys. **28**, 491 (1954).
39 K. K. Innes and L. E. Giddings, J. Mol. Spectrosc. **7**, 435 (1961).
40 L. M. Hubbard, D. F. Bocian, and R. R. Birge, J. Am. Chem. Soc. **103**, 3313 (1981).
41 M. Noble and E. K. C. Lee, J. Chem. Phys. **81**, 1632 (1984).
42 M. Baba, I. Hanazaki, and U. Nagashima, J. Chem. Phys. **82**, 3938 (1985).
43 M. Noble, E. C. Apel, and E. K. C. Lee, J. Chem. Phys. **78**, 2219 (1983).
44 N. Mikami, A. Hiraya, I. Fujiwara, and M. Ito, Chem. Phys. Lett. **74**, 531 (1980).
45 C. M. Cheatham and J. Laane, J. Chem. Phys. **94**, 5394 (1991).
46 J. D. Lewis and J. Laane, J. Mol. Spectrosc. **53**, 417 (1974).
47 J. Laane, J. Chem. Phys. **50**, 776 (1969).
48 P. Sagear and J. Laane, J. Chem. Phys. **102**, 7789 (1995)
49 J. Zhang, W.-Y. Chiang, and J. Laane, J. Chem. Phys. **98**, 6129 (1993)
50 J. Choo and J. Laane, J. Chem. Phys. **101**, 2772 (1994).
51 J. Zhang, W.-Y Chiang, and J. Laane, J. Chem. Phys. **100**, 3455 (1994).
52 W.-Y. Chiang and J. Laane, J. Phys. Chem. **99**, 11640 (1995).
53 P. Sagear, S. N. Lee, and J. Laane, J. Chem. Phys. **106**, 3876 (1997).
54 D. H. Waldeck, Chem. Rev. **91**, 415 (1991).
55 J. A. Syage, P. M. Felker, and A. H. Zewail, J. Chem. Phys. **81**, 4685 (1984).
56 J. A. Syage, P. M. Felker, and A. H. Zewail, J. Chem. Phys. **81**, 4706 (1984).
57 L. H. Spangler, R. Zee, and S. Zwier, J. Phys. Chem. **91**, 2782 (1987).
58 J. Saltiel, S. Ganapathy, and C. Werking, J. Phys. Chem. **91**, 2755 (1987).
59 T. Suzuki, N. Mikami, and M. Ito, J. Phys. Chem. **93**, 5124 (1989).
60 T. Urano, H. Hamaguchi, M. Tasumi, K. Yamanouchi, S. Tsuchiya, and T. L. Gustafson, J. Chem. Phys. **91**, 3884 (1989).
61 T. Urano, M. Maegawa, K. Yamanouchi, and S. Tsuchiya, J. Phys. Chem. **89**, 3459 (1989).
62 F. Negri, G. Orlandi, and F. Zerbetto, J. Phys. Chem. **93**, 5124 (1987).
63 K. Palmo, Spectochim. Acta **44A**, 341 (1988).
64 L. Bañares, A. A. Heikal, and A. H. Zewail, J. Phys. Chem., **96**, 4127 (1992).
65 W.-Y. Chiang and J. Laane, J. Chem. Phys. **100**, 8755 (1994).
66 K. Haller, W.-Y. Chiang, A. del Rosario, and J. Laane, J. Mol. Struct. **379**, 19 (1996).
67 W.-Y. Chiang and J. Laane, J. Phys. Chem. **99**, 11823 (1995).

Structures and Dynamics of Excited Electronic States and Ionic Radicals Studied by Time-Resolved Raman Spectroscopy

Hiroaki Takahashi
Department of Chemistry, School of Science and Engineering
Waseda University, Tokyo 169, Japan,
E-mail:takahash@mn.waseda.ac.jp

Abstract. Structures and dynamics of excited electronic states and ionic radicals have been investigated for several compounds with photochemical interest. Picosecond and nanosecond time-resolved Raman spectroscopy was employed with the aid of laser flash photolysis and *ab initio* molecular orbital calculations. It was observed that the multiple bonds such as C≡C, C=C, C=O, C=N and phenyl rings were weakened in both the excited electronic states and ionic radicals; the weakening is much larger in the former than in the latter. Information on the dynamics of these transient species provided the basis for understanding photochemical reaction mechanisms.

1 Introduction

In order to understand the mechanism of photochemical reactions, it is of utmost importance to clarify the structures and dynamics of photolytically generated transient molecular species, including excited electronic states, such as S_1 and T_1, and ionic radicals, $R^{+\bullet}$ and $R^{-\bullet}$, which play key roles in the physical and chemical processes involved. Due to the recent progress in pulsed laser techniques and optical multichannel detectors, it is now instrumentally feasible to measure resonance Raman spectra of excited electronic states and ionic radicals having picosecond or even subpicosecond lifetimes with the use of time-resolved Raman pump-probe techniques.

Time-resolved Raman spectroscopy is an ideal tool particularly for the purpose of obtaining structural information on the excited electronic states and ionic radicals, when the absorption maxima of these transients are known in advance, and resonance enhancement can be exploited. As the concentrations of excited electronic states and ionic radicals are inevitably very low, resonance enhancement of Raman signals is absolutely necessary to obtain their Raman spectra. In this respect measurements of transient absorption spectra of excited electronic states and ionic radicals should be performed along with Raman measurements when their absorption maxima are not available in the literature. In addition, in order to extract structural information from

Raman spectra it is necessary to know the assignment of observed Raman bands. Measurements of isotopic frequency shifts of the Raman bands and calculations of normal vibrational frequencies using force constants obtained by the *ab initio* molecular orbital method should also be carried out for making reliable vibrational assignments. For this reason results from time-resolved absorption spectroscopy and *ab initio* MO calculations are also presented in this article.

A number of studies have been made since 1975 on the structures and dynamics of excited electronic states and anion and cation radicals of various kinds of compounds by utilizing time-resolved resonance Raman spectroscopy. Since earlier investigations on this subject have been thoroughly reviewed and collected by many workers [1–11], emphasis is placed in the present review on the recent results obtained in our laboratory.

2 Experimental Setup

A pump-probe method is commonly employed for the measurement of time-resolved Raman spectroscopy. Since our experimental setup for nanosecond time-resolved Raman spectroscopy has been described previously [12], only a brief description is presented here. An excimer-laser-pumped dye laser (Lambda Physik, LPX120I) is used as a light source for pumping, and another excimer-laser-pumped dye laser is used as a light source for Raman probing. The delay time between the pump and probe laser pulses is electrically varied using a delay pulse generator (Stanford Research, DG535). Raman signals are collected on the slit of a polychromator (Jobin Yvon, HR320) and are detected on a multichannel analyzer (SMA, D/SIDA 700G).

As an example of the experimental setup for picosecond time-resolved Raman spectroscopy, our arrangement is schematically shown in Figs. 1 and 2.

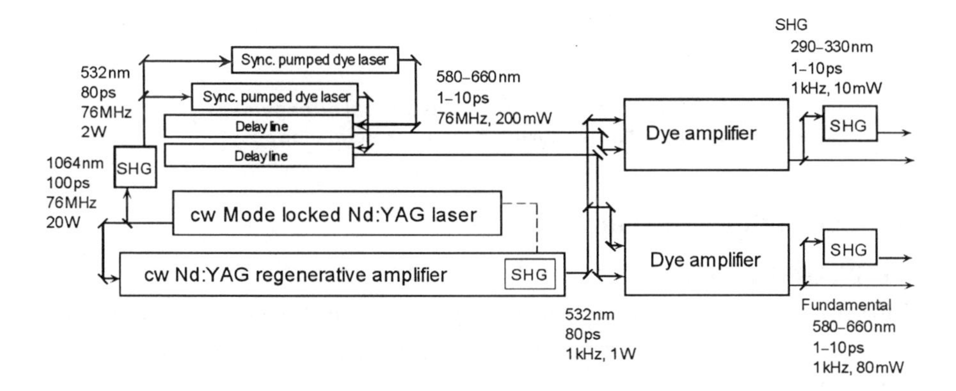

Fig. 1 Schematic presentation of the laser system for picosecond time-resolved Raman and absorption spectroscopy

The laser system of this arrangement consists of a cw mode-locked Nd:YAG laser (Coherent, Antares 76-S) with two synchronously pumped dye lasers (Coherent CR702-2) and a cw Nd:YAG regenerative amplifier (Quantronix, 4316E-RG) with two dye amplifiers (Leonix, LAMP-002) (Fig. 1). This laser system is normally operated with one mode-locked synchronously pumped dye laser and one dye amplifier. In this case each amplified laser pulse is divided into two pulses, one of which is used to generate second harmonics for pumping, and the other is used for Raman probing (Fig. 2). Two mode-locked synchronously pumped dye lasers and two dye amplifiers are simultaneously used when different wavelengths other than the second harmonics of the Raman probing light are needed for the pumping. The delay time of the probe pulse from the pump pulse is varied by an optical delay system, which can range from −10 ps to 8 ns with a step of 66 ± 6 fs. The shortest pulse width is approximately 1 ps, and the pulse repetition rate is 1 kHz.

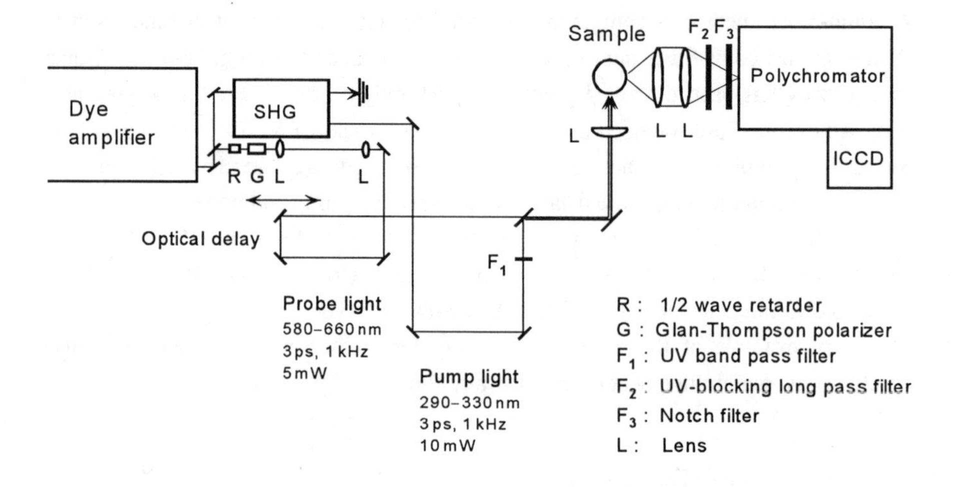

Fig. 2 Schematic presentation of the optical arrangement for picosecond time-resolved Raman Spectroscopy

3 Results and Discussion

3.1 Diphenylacetylene

Diphenylacetylene (DPA) provides an interesting example of the study of the structure and dynamics of excited electronic states and ionic radicals. In the ground state S_0, DPA is known to be in a linear planar structure having D_{2h} symmetry in the crystalline state [13, 14] and most probably also in the gaseous state [15] and in solution [16]. However, not much work has been carried out on the structure of its excited electronic states and ionic radicals.

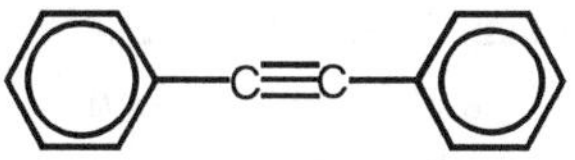

Diphenylacetylene

The structure of the lowest excited triplet state T_1 was investigated by ESR spectra [17], and it was suggested that the central C≡C triple bond of T_1 has a bond length close to that of a C=C double bond. Polarized T–T absorption spectra combined with calculations of transition energies and oscillation strengths by the modified PPP method [18] suggested that T_1 (absorption peak, 418 nm in glycerol) takes a *trans* bent structure having C_{2h} symmetry. Structures of the cation radical $R^{+\bullet}$ and anion radical $R^{-\bullet}$ of DPA were investigated by absorption spectra [19], and it was suggested that both ionic radicals (absorption peak: $R^{+\bullet}$, 424 nm in s-butyl chloride; $R^{-\bullet}$, 445 nm in 2-methyltetrahydrofuran) take planar structures similar to the structure of the S_0 state.

We have measured resonance Raman spectra of T_1, $R^{+\bullet}$, and $R^{-\bullet}$ of DPA and its isotopically substituted analogues, DPA-d_5 (deuteration of a phenyl group), DPA-d_{10} (fully deuterated), and DPA-^{13}C (^{13}C-substitution of a carbon atom of the C≡C bond) by utilizing nanosecond time-resolved Raman spectroscopy [20]. Figure 3 compares Raman spectra of S_0, $R^{+\bullet}$, $R^{-\bullet}$, and T_1 of DPA in solutions. It was observed that the stretching frequency of the C≡C triple bond exhibited large downshifts on going from S_0 to $R^{+\bullet}$ and further to $R^{-\bullet}$ and T_1 in the order S_0 (2217 cm^{-1}), $R^{+\bullet}$ (2142 cm^{-1}), $R^{-\bullet}$ (2091 cm^{-1}), and T_1 (1972 cm^{-1}). The assignment of these Raman bands to the C≡C stretch was supported by the frequency shifts on the ^{13}C-substitution: S_0, −34 cm^{-1}; $R^{+\bullet}$, −39 cm^{-1}; $R^{-\bullet}$, −38 cm^{-1}; and T_1, −32 cm^{-1} (Table 1). Because of its high frequency, the C≡C stretch is an almost localized vibrational mode, having the major contribution of the C≡C bond vibration and the minor contribution of the C–Ph bond vibration. Therefore, these frequency shifts can safely be correlated to the weakening of the C≡C triple bond. The extent of the downshifts of the C≡C stretch in $R^{+\bullet}$, $R^{-\bullet}$, and T_1 imply that although the C≡C bond is weakened drastically in $R^{+\bullet}$, $R^{-\bullet}$, and T_1, it still retains the triple-bond character. Also, the spectral changes on going from S_0 to $R^{+\bullet}$, $R^{-\bullet}$, and T_1 strongly suggest that the molecule retains the D_{2h} symmetry in all these transients. Our *ab initio* calculations of optimized geometries (Table 2) and normal frequencies of S_0, $R^{+\bullet}$, and $R^{-\bullet}$ supported the above conclusions [21].

Picosecond time-resolved absorption spectra of DPA indicated that the S_1 state exhibited peaks at 437 and 700 nm that decayed with a lifetime of about 200 ps, the S_2 state at 500 nm with a lifetime of about 8 ps, and the T_1 state at 415 nm [22]. Recently, Ishibashi and Hamaguchi [23] measured picosecond time-resolved CARS (coherent anti-Stokes Raman scattering) spectra of DPA and DPA-^{13}C and observed transient bands at

2099 cm^{-1} with a lifetime shorter than 20 ps and 1557 cm^{-1} with a lifetime of about 200 ps. They assigned the bands at 1557 and 2099 cm^{-1} to the C≡C stretches of S_1 and S_2, respectively, based on the frequency shifts on ^{13}C-substitution: −22 cm^{-1} for the 1557 cm^{-1} band, and −39 cm^{-1} for the 2099 cm^{-1} band. The extremely low frequency of the C≡C stretch of the S_1 state suggests that the C≡C bond no longer retains the triple-bond character and should be considered to be almost a double bond, implying that S_1 is in a *trans* bent structure having C_{2h} symmetry. This is in good accord with the well-known *trans* bent structure of the S_1 state of acetylene [24]. On the other hand, the 2099 cm^{-1} of the C≡C stretch indicates that although the C≡C bond is markedly weakened in the S_2 state, it still retains the triple-bond character, implying that the S_2 state is in a linear planar structure having D_{2h} symmetry.

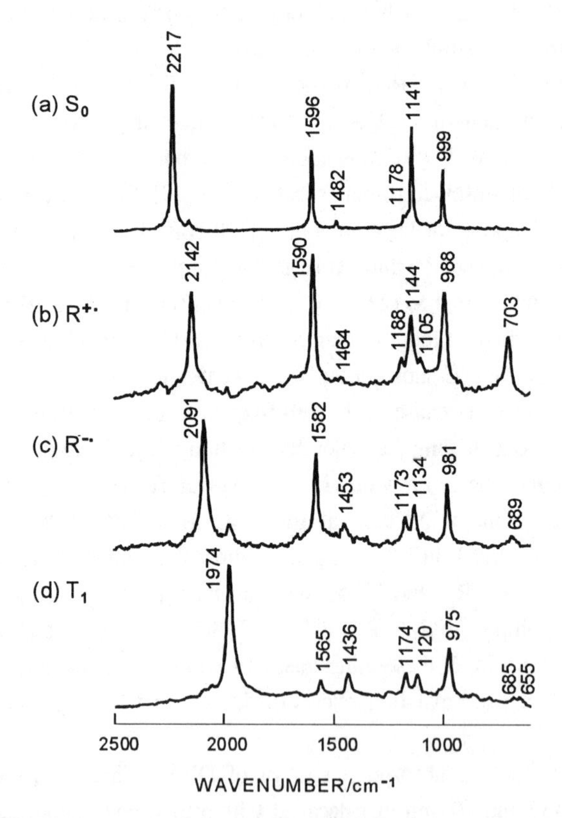

Fig. 3 Raman spectrum of S_0 and resonance Raman spectra of $R^{+\cdot}$, $R^{-\cdot}$, and T_1 of DPA. (a) S_0 in CCl_4, 420 nm probe light; (b) $R^{+\cdot}$ in acetonitrile, 420 nm probe light; (c) $R^{-\cdot}$ in dimethylformamide, 460 nm probe light; (d) T_1 in cyclohexane, 420 nm probe light. Pump wavelength, 308 nm.

In Table 1 are listed Raman frequencies of some of the structurally important vibrational modes of S_0, $R^{+\bullet}$, $R^{-\bullet}$, T_1, S_1, and S_2 of DPA in solutions. We see that the phenyl ring stretch, 8a and 8b modes [25], also show downshifts on going from S_0 to $R^{+\bullet}$, $R^{-\bullet}$, T_1, and S_1. Since the downshifts of the 8a and 8b modes are not very large compared to these of the C≡C stretch, it may be concluded that the electronic excitations and electron elimination/addition are strongly localized on the C≡C bond. The bond-length change of C–Ph is not clear. Although the frequencies assigned to the C–Ph stretch show a tendency to decrease slightly on going from S_0 to $R^{+\bullet}$, $R^{-\bullet}$, and T_1, this vibrational mode is not localized on the C–Ph bonds alone but involves also phenyl-ring stretches and C–H in-plane bends. Therefore, the downshifts of the C–Ph stretch may not be directly correlated with the weakening of the C–Ph bond.

Table 1. Comparison of Raman frequencies (cm^{-1}) of structurally important vibrational modes of the ground state, electronic excited states, and ionic radicals of DPA in solutions. The assignment is based on isotopic frequency shifts (cm^{-1}) and *ab initio* normal coordinate calculations [21]. Values in parentheses are the frequency shifts on the ^{13}C-substitution of a carbon atom of the C≡C bond. The 8a mode of S_2 and 19a mode and C–Ph stretch of S_1 and S_2 were not observed.

Vibrational modes	Electronic excited states and ionic radicals					
	S_0	$R^{+\bullet}$	$R^{-\bullet}$	T_1	S_1	S_2
C≡C str.	2217 (–34)	2142 (–39)	2091 (–38)	1974 (–32)	1557 (–22)	2099 (–39)
8a*	1596	1590	1582	1565	1577 (–4)	
19a*	1482	1464	1453	1436		
C–Ph str.	1141 (–1)	1144 (–4)	1134	1120		

* according to the notations by E. B. Wilson [25].

In order to know the detailed structure of the excited electronic states and ionic radicals, it is necessary to calculate optimized geometries and normal frequencies of these transients by using the *ab initio* MO method. Our calculated optimized geometries of S_0, $R^{+\bullet}$, $R^{-\bullet}$, T_1 using the 4-31G basis set at the ROHF version of the SCF level [21] are given in Table 2 along with the observed values of S_0. It is seen that the C≡C bond lengthens considerably in $R^{+\bullet}$ and $R^{-\bullet}$, and drastically in T_1. This is in good accord with the downshifts of observed Raman frequencies of the C≡C stretch shown in Table 1, except that while the observed C≡C stretching frequencies of $R^{+\bullet}$ and $R^{-\bullet}$ are considerably different, the calculated C≡C bond lengths of these ionic radicals are the same. Since the *ab initio* force constant (diagonal element of the force constant matrix) of the C≡C stretch was calculated to be almost the same for $R^{+\bullet}$ and $R^{-\bullet}$ (12.52 and 12.51 md/Å, respectively) [21], the intrinsic frequency of the C≡C stretch of the two ionic radicals must be almost the same, and the observed frequency difference should be attributed to the difference in the vibrational coupling. The calculated C–Ph bond length shows a tendency to decrease on going from S_0 to $R^{+\bullet}$, $R^{-\bullet}$, and T_1, implying that the C–Ph bond is strengthened in these transients. However, the observed Raman bands assigned to the C–Ph stretch show the opposite tendency, that their frequencies decrease on going from S_0 to $R^{+\bullet}$, $R^{-\bullet}$, and T_1. The reason for this disagreement may again be attributable to vibrational coupling; the vibrational modes of these bands involves the contribution of not only the C–Ph stretch but also the phenyl ring stretches and C–H in-plane bends.

Table 2. Optimized geometries of S_0, $R^{+\bullet}$, $R^{-\bullet}$, and T_1 of DPA calculated by using the 4-31G basis set at the RHF version (for S_0) and ROHF version (for $R^{+\bullet}$, $R^{-\bullet}$, and T_1) of the SCF level. Values in parentheses are observed values [21].

	S_0	$R^{+\bullet}$	$R^{-\bullet}$	T_1
	Bond length (Å)			
C≡C	1.194 (1.198)	1.224	1.224	1.257
C-Ph	1.431 (1.434)	1.384	1.387	1.341
C_1-C_2	1.392 (1.386)	1.413	1.424	1.443
C_2-C_3	1.381 (1.376)	1.370	1.372	1.359
C_3-C_4	1.384 (1.368)	1.392	1.394	1.402
	Bond angle (deg.)			
C_6-C_1-C_2	119.0 (119.1)	119.6	116.1	116.9
C_1-C_2-C_3	120.4 (120.3)	119.9	121.4	120.8
C_2-C_3-C_4	120.2 (120.0)	119.7	121.6	121.0
C_3-C_4-C_5	119.8 (120.1)	121.2	117.9	119.5

C_1 denotes the phenyl carbon atom attached to the –C≡C– group.

The ionization to generate the cation radical of DPA is a biphotonic process. The intensity of the Raman band at 2142 cm^{-1} of $R^{+\bullet}$ increases quadratically with the laser power of UV (308 nm) light used for pumping when the pump power is not extremely large. The biphotonic ionization of DPA with 308 nm light is considered to occur both via S_1 and via T_1. Transient resonance Raman spectra of DPA were measured using 360 nm light as the pump light and benzophenone as the triplet sensitizer [20]. The wavelength 360 nm (3.44 eV) is not energetically large enough to pump DPA into the S_1 state, which is located about 4.0 eV above the S_0 state [26], but is sufficiently large to pump benzophenone into the S_1 state, which is located 3.23 eV above the S_0 state [27]. Since the energy of the T_1 state of benzophenone (3.00 eV [27]) is higher than that of DPA (2.71 eV [27]), the production of T_1 of DPA without generating S_1 is considered to be possible using benzophenone as a sensitizer. As expected, neither T_1 nor $R^{\bullet +}$ was observed by the 360 nm pumping in the absence of benzophenone, but addition of benzophenone into the same solution gave rise to not only T_1 but also $R^{\bullet+}$. This indicates that $R^{\bullet+}$ was generated from T_1 by absorbing a photon of 360 nm light.

3.2 5-Dibenzosuberenol

5-Dibenzosuberenol (5H-Dibenzo[a,d]cyclohepten-5-ol; DBCH-5-ol) is an interesting compound in the sense that two phenyl groups attached to the ethylenic –C=C– bond are fixed in an almost *cis* configuration by the –CH(OH)– bridge and may provide information for understanding the photochemistry of *cis*-stilbene. In addition, this compound may take two configurations with respect to the position of the OH group: pseudo-equatorial and pseudo-axial positions.

HO H

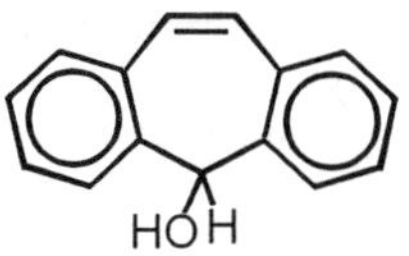

5-Dibenzosuberenol

Although the structure of this compound is not known, the structures of closely related compounds, dibenz[b,f]azepine and dibenz[b,f]oxepin, in which the –CH(OH)– bridge of DBCH-5-ol is replaced by an –NH– bridge in the former and by an –O– bridge in the latter compound, were determined by x-ray diffraction [28, 29] to have a "butterfly-like" structure (saddle-shaped) with the central 7-membered ring in a boat conformation and its "wings" (two closely planar phenyl rings) bent backwards (the dihedral angle between the two planes of the phenyl rings being 144.4° and 134°, respectively). It can reasonably be assumed that DBCH-5-ol also takes the "butterfly-like" structure, and therefore, the OH group attached to the central ring can take pseudo-equatorial and pseudo-axial positions. It is of interest to investigate which of the two configurations (OH-equatorial and OH-axial configurations) is more stable in the excited electronic states and ionic radicals.

Another interesting problem of this compound is that of photoionization. Laser flash-photolysis studies by Johnston et al. [30] have shown that the transient phenomena in benzosuberenyl compounds are dominated by the productions of the T_1 state and cation radical $R^{+\bullet}$ generated by photoejection of an electron. They suggested that the photoionization of this compound requires more than one photon and is dependent on the excitation wavelength.

We have measured laser-flash photolysis spectra of DBCH-5-ol in both picosecond and nanosecond time regions (Fig. 4). In the picosecond time region an absorption band was observed at 607 nm with a lifetime of about 2 ns. This band is attributable to the S_1 state. In the nanosecond time region a band was observed at 423 nm that was quenched by oxygen. This band is reasonably assigned to the T_1 state. With increasing intensity of the UV laser light for pumping, a doublet band with peaks at 428 and 470 nm and a weak band around 800 nm were observed . These absorption peaks can be assigned to $R^{+\bullet}$.

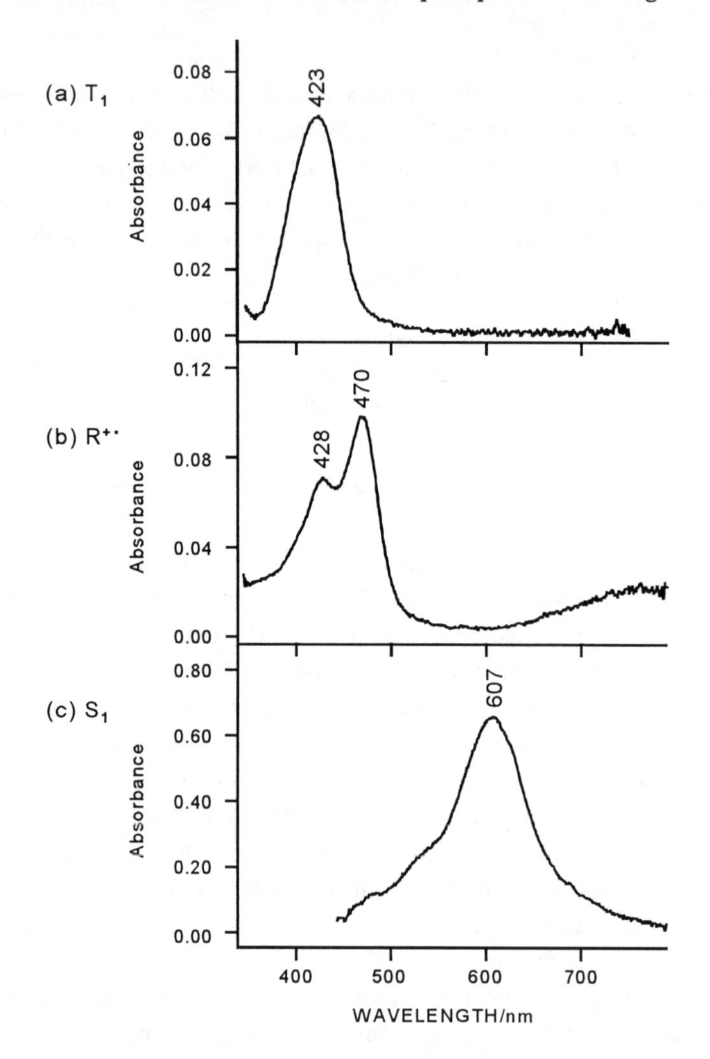

Fig. 4 Transient absorption spectra of T_1, $R^{+\bullet}$, and S_1 of DBCH-5-ol in acetonitrile. Concentration, 3.0×10^{-4} mol dm^{-3}. Pump wavelength, 308 nm for (a) and (b), and 300 nm for (c).

In Fig. 5 are compared the Raman spectrum of S_0 and the resonance Raman spectra of $R^{+\bullet}$, S_1, and T_1 of DBCH-5-ol [31]. The bands at 1621 cm^{-1} of S_0, 1548 cm^{-1} of $R^{+\bullet}$, 1532 cm^{-1} of S_1, and 1503 cm^{-1} of T_1 can be assigned to the ethylenic C=C stretch, because these Raman bands exhibit downshifts of −21, ∼−30, −10, and −11 cm^{-1}, respectively, on the ^{13}C substitution of a carbon atom of the C=C bond. Since the downshifts are considerably less than the theoretical value of ~−30 cm^{-1}, particularly for S_1 and T_1, a more explicit description of the vibrational mode for these Raman bands is the one that involves the largest contribution of the C=C stretch with substantial contribution of the phenyl 8b vibration. The assignments of some of the structurally important Raman bands of S_0, $R^{+\bullet}$, S_1, and T_1 are given in Table 3 along with the isotopic frequency shifts on the ^{13}C substitution.

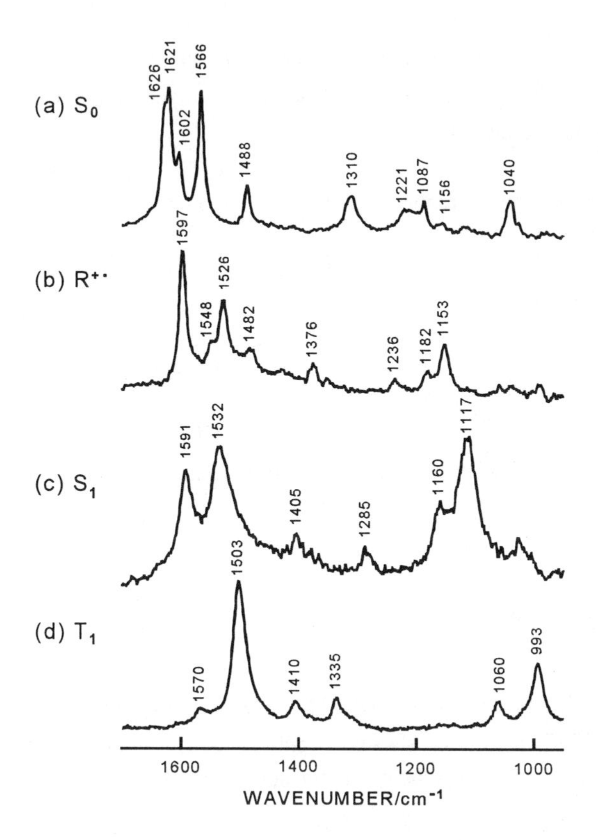

Fig. 5 Comparison of the Raman spectrum of S_0 and the resonance Raman spectra of $R^{+\cdot}$, S_1, and T_1 of DBCH-5-ol in acetonitrile. Probe wavelength: (a) S_0, 420 nm; (b) $R^{+\cdot}$, 465 nm, (c) S_1, 600 nm, (d) T_1, 420 nm. Pump wavelength: (b), (d) 308 nm; (c) 300 nm. Concentration: (a) nearly saturated; (b), (c), (d) 3.0×10^{-3} mol dm^{-3}.

It is seen that the C=C stretch is downshifted in the order S_0, 1621 cm^{-1}; $R^{+\bullet}$, 1548 cm^{-1}; S_1, 1532 cm^{-1}; T_1, 1503 cm^{-1}. This indicates that the C=C bond weakens in this order. 'Although the downshift from S_0 to T_1 is remarkably large, the C=C stretch frequency 1503 cm^{-1} of T_1 suggests that the C=C bond still retains the double-bond nature in T_1. The phenyl 8a and 8b modes also show the downshift in the same order as the C=C stretch, suggesting that the phenyl rings weaken in this order. The weakening of the phenyl rings, however, is much smaller than that of the C=C bond. In the case of S_0, the band at 1621 cm^{-1} has a shoulder at 1626 cm^{-1}. The temperature-dependence of the relative intensities of this doublet and that of the ^{1}H-NMR spectrum (400 MHz) indicated that DBCH-5-ol exists as two mutually invertible isomers; one having a pseudo-equatorial OH group and the other a pseudo-axial OH group attached to the central nonplanar cycloheptatriene ring, with the former being stable by approximately 0.5 kcal mol^{-1}. No indication of the existence of this type of isomer was detected for $R^{+\bullet}$, S_1, and T_1, probably due to the large energy difference between the equatorial and axial OH configurations of these transients.

Table 3. Comparison of frequencies (cm^{-1}) of the alkene C=C stretch, phenyl 8a and 8b modes, and C–Ph stretch of S_0, S_1, T_1, and $R^{+\bullet}$ of DBCH-5-ol. Values in parentheses are the frequency shifts (cm^{-1}) on the ^{13}C-substitution of a carbon atom of the alkene C=C bond.

Vibrational mode	Electronic excited states and $R^{+\bullet}$			
	S_0	$R^{+\bullet}$	S_1*	T_1
C=C stretch	1621 (−21)	1548 (~−30)	1532 (−10)	1503 (−11)
Phenyl 8a	1602	1597 (−1)	1591 (−2)	1570
Phenyl 8b	1566 (−8)	1526 (−11)		1410 (−10)
C–Ph stretch	1156 (−3)	1153 (−6)	1117 (−7)	1060 (−5)

*Reference [32].

The photoionization mechanism of DBCH-5-ol was investigated. The intensity of the Raman band at 1597 cm^{-1} of $R^{+\bullet}$ showed a positive second-order dependence on the pump laser power, indicating that the photoionization of this compound occurred through a biphotonic process.

The dependence of the yield of $R^{+\bullet}$ and T_1 on the wavelength of the pump laser was investigated. It was observed that with 308 nm pumping both $R^{+\bullet}$ and T_1 were simultaneously detected, while with 340 nm pumping only $R^{+\bullet}$ was detectable. However, using benzophenone as a triplet sensitizer, both $R^{+\bullet}$ and T_1 were simultaneously detectable even under the 340 nm pumping. On the basis of the above results the following conclusions were obtained.

(1) The biphotonic ionization of DBCH-5-ol occurs through S_1. The possibility that the biphotonic ionization occurs through T_1 still remains.

(2) From high vibrational levels of S_1, the intersystem crossing to T_1 takes place very rapidly with high quantum yield, comparable to that of the $S_1 \rightarrow S_n$ photoexcitation, while from low vibrational levels, the intersystem crossing to T_1 is much slower than the

$S_1 \rightarrow S_n$ photoexcitation. The very slow intersystem crossing to T_1 with very low quantum yield may be attributable to the low density of states in the vicinity of low vibrational levels.
(3) The energy of the triplet state of DBCH-5-ol is lower than that of benzophenone, which is 69.2 kcal mol^{-1}.

3.3 N,N,N',N'-Tetramethyl-*p*-phenylenediamine

The ionization potential of N,N,N',N'-tetramethyl-*p*-phenylenediamine (TMPD) is 6.6 eV in the gaseous state [33], which corresponds to the energy of a photon of 188 nm wavelength light. Due to the remarkably low ionization potential for organic compounds, TMPD readily undergoes photoionization in polar solvents to produce $R^{+\bullet}$ on irradiation with UV light. There was, however, some confusion with regard to the photoionization in nonpolar solvents. Since the absorption spectrum of $R^{+\bullet}$ very much resembles that of T_1 both in the wavelength and doublet-band shape, the absorption band of T_1 was sometimes incorrectly assigned to $R^{+\bullet}$ or vice versa.

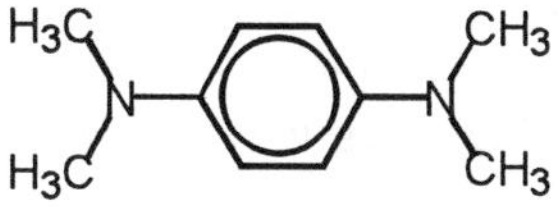

TMPD

It is now generally accepted that photoejection of electrons from TMPD in polar solvents proceeds monophotonically via S_1 through a geminate electron–cation pair [34–36], whereas in nonpolar solvents monophotonic ionization does not occur on irradiation with UV light of wavelength longer than 300 nm [37–39]. We have demonstrated using nanosecond time-resolved Raman spectroscopy that TMPD undergoes monophotonic ionization in polar solvents and CCl_4 on irradiation with UV light, while in aliphatic hydrocarbons monophotonic ionization does not occur and only T_1 is generated [40]. Since the Raman spectrum of $R^{+\bullet}$ is quite different from that of T_1, misassignment should never occur. The unusual monophotonic ionization of TMPD in nonpolar CCl_4 was interpreted in terms of CCl_4 abstracting an electron from T_1 of TMPD to form ion pair consisting of $R^{+\bullet}$ and Cl^- and thus stabilizing $R^{+\bullet}$.

Hirata and Mataga have shown using picosecond absorption spectroscopy that $R^{+\bullet}$ (absorption shoulders at 575 and 610 nm) is produced even in nonpolar hexane by a biphotonic process [41]. The lifetime of $R^{+\bullet}$ (more precisely, geminate electron–ion pair) in hexane, however, is extremely short and is converted to S_1 (absorption peaks at 670 and 740 nm) by recombination in less than 200 ps.

In Fig. 6 are shown time-resolved absorption spectra of TMPD in cyclohexane. The peaks at 555 and 597 nm were quenched by oxygen and were assigned to T_1. The bands at 666 and 728 nm disappeared almost completely at 30 ns after excitation. These two peaks were attributable to S_1.

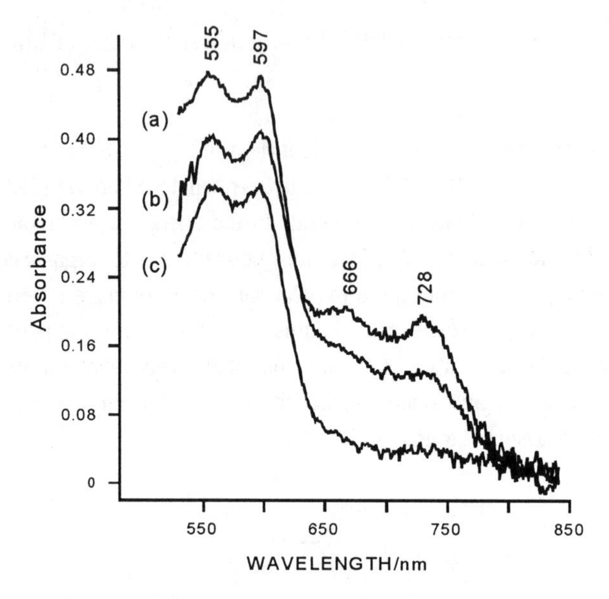

Fig. 6 Time-resolved absorption spectra of TMPD in cyclohexane: (a) 0 ns; (b) 10 ns; (c) 30 ns after pumping with 308 nm light.

Table 4. Comparison of the frequency shifts (cm^{-1}) on isotopic (D, ^{15}N) substitutions of S_0, $R^{+\bullet}$, T_1, and S_1 of TMPD.

Species	Raman band (cm^{-1})	Frequency shift (cm^{-1})			Assignment
		Isotopic substitution			
		d_4	^{15}N	d_{12}	
S_0	1621	−24	0	0	8a
	1340	−3	−13	−13	N–Ph sym. Str.
$R^{+\bullet}$	1629	−27	−2	−3	8a
	1505	0	−9	−43	N–Ph sym. Str.+ CH_3 sym. def.
T_1	1530	−14	−13	−17	8a + N–Ph sym. Str.
S_1	1521	−20	−3	−5	8a

The Raman spectrum of S_0 and the resonance Raman spectra of $R^{+\bullet}$, T_1 [40], and S_1 [42] of TMPD are shown in Fig. 7. The Raman spectrum of S_0 and resonance Raman spectra of $R^{+\bullet}$, T_1, and S_1 of isotopically substituted analogues, TMPD-$^{15}N_2$ (^{15}N substitution of the two nitrogen atoms), TMPD-d_4 (deuteration of the phenyl group), TMPD-d_{12} (deuteration of all four methyl groups), were also measured and vibrational assignments were made based on the frequency shifts on the isotopic substitutions [43]. Assignments of structurally important vibrational modes are listed in Table 4.

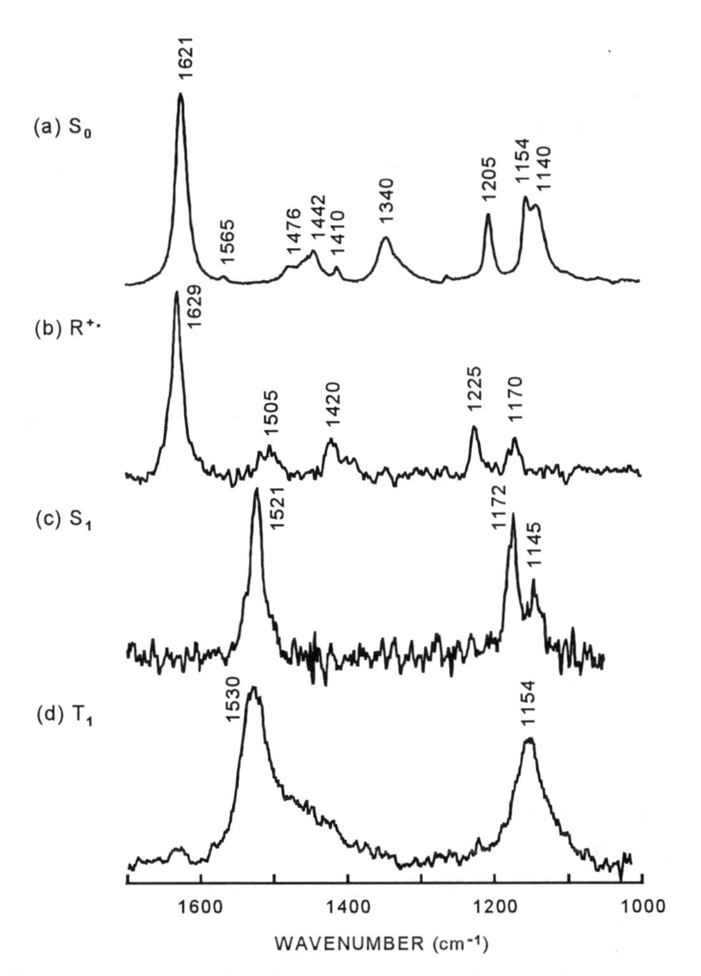

Fig. 7 Comparison of the Raman spectrum of S_0 and the resonance Raman spectra of $R^{+\bullet}$, S_1, and T_1 of TMPD in cyclohexane. (a) S_0, 532 nm probe light, nearly saturated; (b) $R^{+\bullet}$, 620 nm probe light, 2.0×10^{-3} mol dm^{-3}; (c) S_1, 720 nm probe light, 2.0×10^{-3} mol dm^{-3}; (d) T_1, 620 nm probe light, 2.0×10^{-3} mol dm^{-3}. Pump wavelength, 308 nm.

The band at 1621 cm^{-1} of S_0 and the band at 1629 cm^{-1} of $R^{+\bullet}$ are undoubtedly assigned to the phenyl 8a mode on the basis of the frequency shifts on the phenyl deuteration. The upshifts of the phenyl 8a mode and the N–Ph stretch from 1340 to 1505 cm^{-1} on going from S_0 to $R^{+\bullet}$ is in accord with the quinoid-like structure of $R^{+\bullet}$ in the crystalline state determined by x-ray diffraction [44]. The band at 1521 cm^{-1} of S_1 can also be assigned to the phenyl 8a mode, implying that the phenyl group is considerably weakened in the S_1 state. It is seen in Fig. 7 that the spectrum of T_1 is very similar to that of S_1. However, as can be seen in Table 4, the isotopic frequency shifts of the band at 1530 cm^{-1} of T_1 is quite different from those of the band at 1521 cm^{-1} of S_1. This suggests that the structure of T_1 is considerably different from that of S_1, although the spectral features of these two excited states look strikingly alike.

Optimized geometries of S_0, $R^{+\bullet}$, T_1, and S_1 of TMPD calculated using the *ab initio* STO-3G approximation [45] are given in Table 5.

Table 5. Optimized geometries of S_0, $R^{+\bullet}$, T_1, and S_1 of TMPD calculated by *ab initio* STO-3G approximation.

Species		Bond length			
		C_1–C_2	C_2–C_3	N–Ph	N–CH_3
S_0	Obs.[46]	1.397	1.390	1.414	1.438
	Calc.	1.397	1.381	1.425	1.465
$R^{+\bullet}$	Obs.[44]	1.426	1.365	1.359	1.463
	Calc.	1.437	1.352	1.368	1.489
T_1	Calc.	1.428	1.419	1.384	1.477
S_1	Calc.*	1.421	1.415	1.400	1.476

C_1 denotes the phenyl carbon atom attached to the –$N(CH_3)_2$ group.
* unpublished results.

Comparing the optimized geometries of T_1 and S_1, it is readily recognized that although the phenyl group is weakened both in T_1 and S_1, the former is more quinoid-like; the bond length difference between C_1–C_2 and C_2–C_3 is more pronounced in T_1 than in S_1 and the N–Ph bond length is shorter in T_1 than in S_1. This structural difference well accounts for the difference in the isotopic frequency shifts between T_1 and S_1; because of the quinoid-like structure, the phenyl 8a and N–Ph stretch modes mix with each other in T_1, which is the reason for the 1530 cm^{-1} band showing frequency shifts on both the phenyl deuteration and ^{15}N substitution.

3.4 Benzil

Photochemistry of benzil (diphenylethanedione) is of interest with regard to the configurations about the central C–C bond in its excited electronic states and anion radical. The structure of the S_0 state is reported to be in a twisted conformation, with the two benzoyl planes making an angle of 111°36′ in the crystalline state [47] and 98° in solutions [48].

Benzil

Fluorescence and phosphorescence studies of benzil have suggested both skewed and trans-planar structures for S_1 and T_1. The fluorescence peaks at 440 and 500 nm were assigned to the skewed and trans-planar structures of S_1, respectively [49], and the phosphorescence peaks at 525 and 565 nm were assigned to the skewed and trans-planar structures of T_1, respectively [49–51]. Picosecond time-resolved absorption study of benzil showed that S_1 exhibits an absorption peak at 525 nm and T_1 at 490 nm in cyclohexane, while in the microcrystalline state the absorption peaks were observed at 560 nm for S_1 and at 500–510 nm for T_1. The geometries in cyclohexane and in the microcrystalline state were considered to be different from each other [51].

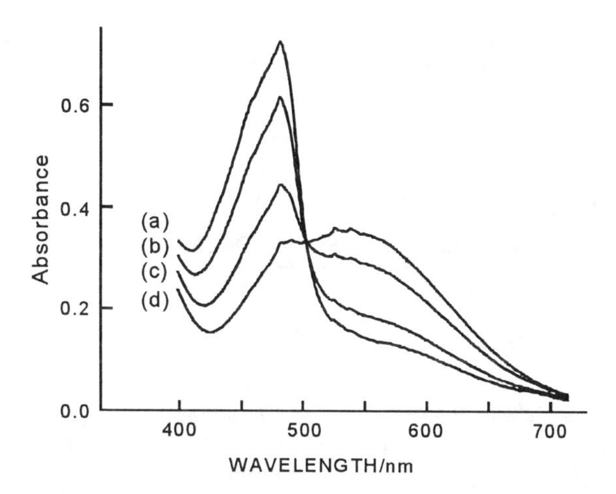

Fig. 8 Time-resolved absorption spectra of benzil in methanol with addition of a small amount of triethylamine: (a) 100 ns; (b) 200 ns; (c) 500 ns; (d) 1 μs after pumping with 308 nm light.

A time-resolved ESR study [52] showed that in the presence of triethylamine, $R^{-\bullet}$ is produced from T_1 through a one-photon process. The structure of $R^{-\bullet}$ has not been studied in any detail. Although the transient absorption peak at 620 nm in acetonitrile-triethylamine has been assigned to the skewed conformer, the evidence for this assignment is not very convincing.

Time-resolved absorption spectra of benzil observed in deoxygenated methanol-triethylamine [53] are shown in Fig. 8. The spectra were highly sensitive to the concentration of triethylamine. When triethylamine is not present, only the 480 nm band was observed. However, when an excess amount of triethylamine is added, only the 545 nm band was observable in the nanosecond time region. Since the presence of electron-donating species such as triethylamine 1,4-diazabicyclo[2,2,2]octane (DABCO) quench T_1 by transferring an electron to generate $R^{-\bullet}$, the band at 480 nm can be assigned to T_1 and the band at 545 nm to $R^{-\bullet}$.

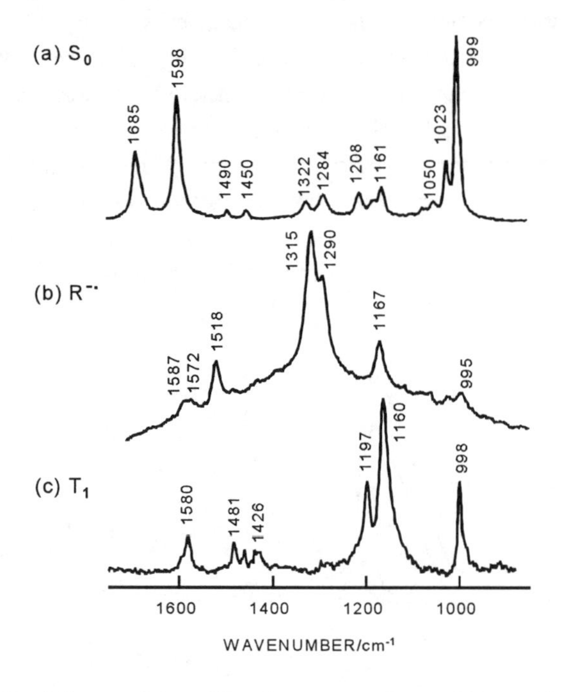

Fig. 9 Comparison of the Raman spectrum of S_0 and the resonance Raman spectra of $R^{-\bullet}$ and T_1 of benzil: (a) S_0 in CCl_4 with 640 nm probing; (b) $R^{-\bullet}$ and in DMSO with 640 nm probing; (c) T_1 in cyclohexane with 480 nm probing.

The Raman spectrum of S_0 and the resonance Raman spectra of $R^{-\bullet}$ and T_1 [53] are compared in Fig. 9. It is seen that the C=O symmetric stretch is downshifted from 1685 to 1518 cm^{-1} on going from S_0 to $R^{-\bullet}$ and further to 1426 cm^{-1} in T_1. The vibrational assignments were made based on the frequency shifts on isotopic substitutions: Benzil-$^{13}C_2$ (^{13}C substitution of the central two carbon atoms), Benzil-$^{18}O_2$ (^{18}O substitution of the two oxygen atoms), and Benzil-d_{10} (deuteration of the two phenyl groups). The normal modes of the Raman bands at 1518 cm^{-1} of $R^{-\bullet}$ and at 1426 cm^{-1} of T_1 are not the pure C=O stretch, because the downshifts of these bands on ^{18}O substitution are not sufficiently large. However, it is certain that the C=O stretch makes the largest contribution to the normal modes of these Raman bands. It is interesting to note that the central C=C stretch is upshifted from 1050 cm^{-1} to 1315 cm^{-1} on going from S_0 to $R^{-\bullet}$ and to 1160 cm^{-1} in T_1. Simple HOMO-LUMO consideration suggests that since the HOMO (highest occupied molecular orbital) has antibonding nature and the LUMO (lowest unoccupied molecular orbital) has bonding nature with respect to the π-bonding of C–C single bonds, the elevation of an electron from the HOMO to LUMO to generate T_1 would cause a much larger increase in the bond order of the central C–C bond than the addition of an electron to the LUMO to generate $R^{-\bullet}$. Actually this is not the case. This discrepancy may be ascribed to different contributions of other vibrations in $R^{-\bullet}$ and T_1. It is also probable that the discrepancy arises from the difference in molecular structure; it is conceivable that $R^{-\bullet}$ takes a trans-planar structure, whereas T_1 takes a skewed structure like S_0.

Table 6. Comparison of frequencies (cm^{-1}) and isotopic frequency shifts (cm^{-1}) of some vibrational modes of S_0, $R^{-\bullet}$, and T_1 of benzil. Values in parentheses are frequency shifts on the ^{13}C substitution of the two carbon atoms of the central C–C bond, ^{18}O substitution of the two C=O bonds, and deuteration of the two phenyl groups, respectively.

Vibrational mode	Frequencies and frequency shifts on isotopic substitutions (^{13}C substitution, ^{18}O substitution, deuteration)		
	S_0	$R^{-\bullet}$	T_1
C=O sym. Str.	1685 (–40, –38, 0)	1518 (–34, –11, –15)	1426 (–10, –11, 4)
8a	1598 (0, –1, –32)	1587 (0, 0, –50)	1580 (0, –1, –34)
8b	1583 (0, 0, –38)	1572 (0, 1, n)	n
C–Ph sym. Str.	1284 (–19, 0, –36)	1290 (n, 0, n)	n
C–C str.	1050 (n, 0, –34)	1315 (–44, –1, –47)	1160 (–8, –1, –9)

n: not detected due to overlapping by nearby bands.

The absorption peak of $R^{-\bullet}$ exhibits a remarkably large hypsochromic shift in hydrogen-bond-donor solvents; the band at 620 nm in acetonitrile is shifted to 540 nm in methanol. The absorption band in the 540–650 nm spectral region of benzil $R^{-\bullet}$ may be attributable to the transition associated with the intramolecular charge transfer induced by the charge migration from the C=O groups to the phenyl rings, and the transition from the SOMO (singly occupied molecular orbital) π^* to LUMO π^{**} makes the major contribution to the intensity of the band [54]. It is generally believed that the absorption bands arising from π–π^* transitions exhibit bathochromic shifts in hydrogen-bond-donor solvents. Therefore, the hypsochromic shifts of the band having the π^*–π^{**} character of benzil $R^{-\bullet}$ in hydrogen-bond-donor solvents are quite unusual.

The effect of hydrogen-bond-donor solvents is also seen in the stretching frequency of the central C–C bond; the frequency is upshifted from 1314 cm^{-1} in DMSO to 1317 cm^{-1} in acetonitrile, 1327 cm^{-1} in t-butyl alcohol, 1339 cm^{-1} in 2-propanol, 1342 cm^{-1} in ethanol, 1346 cm^{-1} in methanol, and 1366 cm^{-1} in trifluoroethanol. A linear relationship was obtained between the C–C stretching frequency and the α-value of the Taft–Kamlet equation [55] of the solvent. Since the Taft–Kamlet α is a measure of the strength of hydrogen-bond-donor acidity of the solvent, this linear relationship indicates that the unusually large upshift of the C–C stretch on going from DMSO solution to methanol solution is ascribable to the hydrogen bond formation between $R^{-\bullet}$ and the solvent. The strengthening of the central C–C bond implies that the double-bond nature of the C–C bond, and hence the planarity of the molecule, increases as the hydrogen-bond-donor acidity of the solvent increases.

3.5 Coumarin

Coumarin (CM) and its derivatives are known to exhibit photosensitizing properties. Particularly, psoralen and its derivatives are important drugs used in the photochemotherapy of skin diseases such as psoriasis [56] and vitiligo [57]. On the other hand, they are known to photoinduce skin erythema [58, 59] and skin cancer in mice [60]. The biological activity of psoralens has been correlated with their photoreactivity towards the pyrimidine bases of DNA to form a crosslink between the two separate strands [61, 62], which inhibits the overactive DNA synthesis.

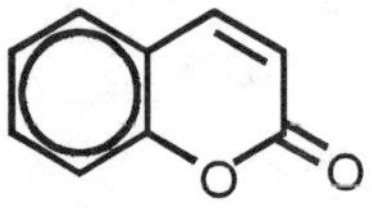

O O O O O

Coumarin 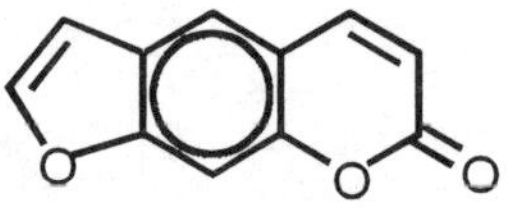 Psoralen

In order to understand the photosensitizing mechanism of psoralen and its derivatives, it is important to obtain information on the structures and dynamics of the excited states and photolytically generated transient species of coumarin. Due to its structural similarities with psoralen, coumarin serves as a useful model for this purpose.

Time-resolved absorption spectra of coumarin in deoxygenated ethanol are shown in Fig. 10. Three well-resolved peaks at 400, 425, and 450 nm and a shoulder at about 378 nm are seen at 100 ns after the pumping. It was observed that the peaks at 400, 425, and 450 nm were quenched by the presence of oxygen [63]. These peaks can be assigned to T_1. The spectrum of T_1 is in good agreement with that reported in the literature [64–66]. We observed that when an excess amount of DABCO is added, the band at 378 nm with a shoulder at 360 nm became much stronger than the bands of T_1. Since DABCO is a strong electron-donating agent and the bands of T_1 at 400, 425, and 450 nm were quenched by the addition of DABCO, the 378 nm band can be assigned to $R^{-\bullet}$ generated from T_1 by obtaining an electron from DABCO or the solvent ethanol.

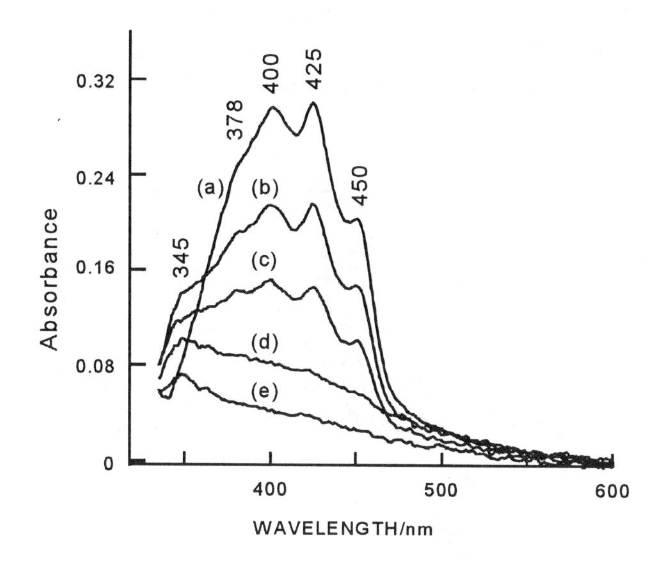

Fig. 10 Time-resolved absorption spectra of coumarin in ethanol: (a) 100 ns; (b) 200 ns; (c) 500 ns; (d) 2 μs; (e) 3 μs after pumping with 308 nm light. Concentration, 5.0×10^{-4} mol dm^{-3}.

The Raman spectrum of S_0 and the transient resonance Raman spectra of T_1 and $R^{-\bullet}$ are compared in Fig. 11. The bands at 1732 cm^{-1} of S_0, 1629 cm^{-1} of $R^{-\bullet}$, and 1552 cm^{-1} of T_1 can be assigned to the C=O stretch based on the frequency shifts on isotopic substitutions [63] (Table 7).

The downshifts of the bands at 1629 cm^{-1} of $R^{-\bullet}$ and 1552 cm^{-1} of T_1 on the ^{18}O substitution are not sufficiently large, and the former band also exhibits comparable downshifts on the ^{13}C substitution, while the later band shows an almost equal amount of downshifts on the phenyl deuteration. These isotopic frequency shifts indicate that these bands are not the pure C=O stretch, but are a mixed mode with the C(3)=C(4) or C(3)–C(2) stretch for the 1629 cm^{-1} band of $R^{-\bullet}$ and that with the phenyl 8a or 8b mode for the 1522 cm^{-1} band of T_1. The moderate frequency decrease of the C=O stretch on going from S_0 to T_1 implies that the T_1 of coumarin is of a π–π^* character [67, 53]; the frequency decrease of a C=O stretch of the T_1 state having n–π^* character is expected to be much larger [68]. The band at 1510 cm^{-1} can be assigned to the phenyl 8a and 8b modes of T_1 based on the frequency shift on phenyl deuteration. The vèry large downshift of the 8a and 8b modes from 1610 and 1567 cm^{-1} to 1510 cm^{-1} on going from S_0 to T_1 indicates that the involvement of the phenyl groups in the excitation is quite substantial. It is interesting to note that no band that is assignable to the C(3)=C(4) stretch was observed in the frequency region of double-bond stretches; only the bands at 1002 and 985 cm^{-1} exhibited frequency shifts to 999 and 982 cm^{-1}, respectively. These results suggests that the bond order of the C(3)=C(4) bond is decreased to that of a single bond in T_1, and therefore, the C(3)=C(4) stretch is no longer a localized mode and is distributed over several vibrational modes in the frequency region of single-bond stretches.

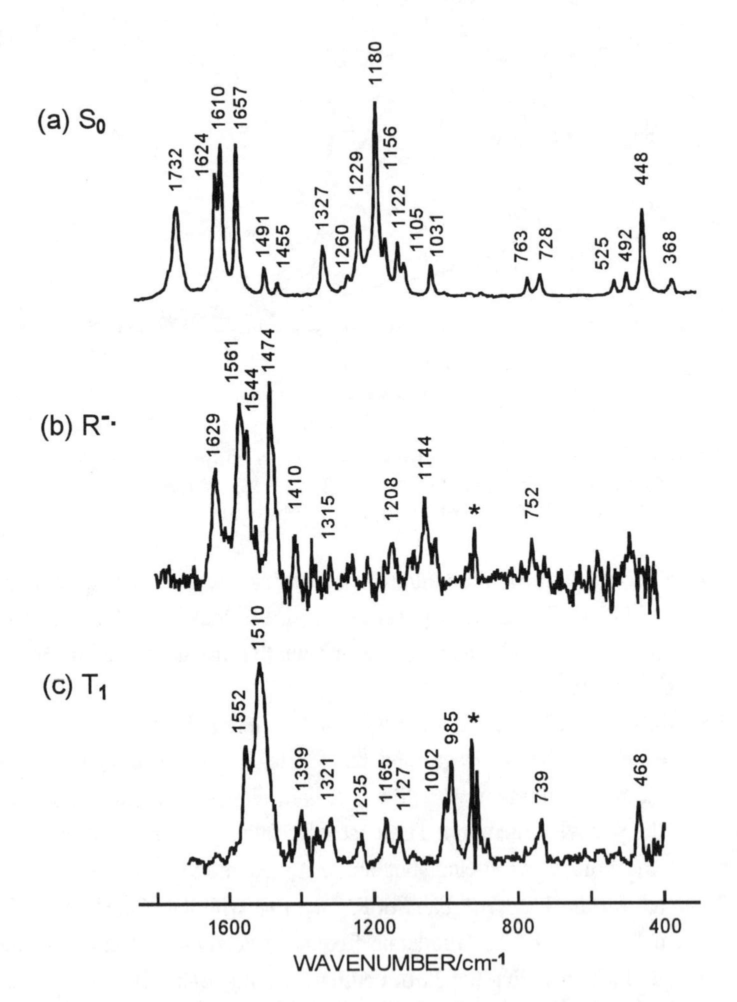

Fig.11 Comparison of the Raman spectrum of S_0 and Resonance Raman spectra of $R^{-\cdot}$ and T_1 of coumarin in acetonitrile: (a) S_0, nearly saturated, 532 nm probe light; (b) $R^{-\cdot}$, coumarin 3.0×10^{-3} mol dm^{-3}, DABCO 6.0×10^{-2} mol dm^{-3}, 383 nm probe light; (c) T_1, coumarin 3.0×10^{-3} mol dm^{-3}, 420 nm probe light. * due to subtraction of solvent bands.

Table 7. Comparison of frequencies (cm^{-1}) and frequency shifts (cm^{-1}) on isotopic substitutions of some vibrational modes of S_0, $R^{-\bullet}$, and T_1 of coumarin. Values in parentheses are frequency shifts on ^{18}O substitution of the C=O group, ^{13}C substitution of the carbon atom at the 3-position of the C=C bond in the pyrone ring, and deuteration of the phenyl group.

Vibrational mode	Frequencies and frequency shifts on isotopic substitutions (^{18}O substitution, ^{13}C substitution, deuteration)		
	S_0	$R^{-\bullet}$	T_1
C=O str.	1732 (−26, 0, −2)	1629 (−11, −11, 0)	1552 (−15, 0, −14)
C(3)=C(4) str.	1624 (0, −14, −4)	1050 (−2, 5, −1)	
8b	1610 (−1, 0, −24)	1561 (1, 1, −26)	1510 (0, 0, −19)
8a	1569 (0, −8, −22)	1544 (1, 1, −21)	

Ab initio MO calculations of the optimized structures (Fig. 12) and normal frequencies of S_0 and $R^{-\bullet}$ were carried out at the RHF (restricted Hartree–Fock) level for S_0 and ROHF (restricted open-shell HF) level for $R^{-\bullet}$ using the 4-31G (O*) basis set, which includes polarization functions for the oxygen atoms only.

The optimized geometry of S_0 is quite consistent with the one expected from the Raman spectrum. The C=O bond length was calculated to be 1.188 Å, a value that indicates that the C=O bond is localized and is consistent with the Raman frequency 1732 cm^{-1}. The calculated C(3)=C(4) bond length 1.329 Å is indicative of an almost localized C=C double bond and is consistent with the Raman frequency 1624 cm^{-1}.

The optimized geometry of $R^{-\bullet}$ is quite interesting and suggestive. The C=O bond was calculated to be 1.218 Å which is longer by 0.03 Å than the C=O bond length of S_0. This value may be considered to be in good accord with the relatively small downshift (−103 cm^{-1}) of the C=O stretch on going from S_0 to $R^{-\bullet}$. The calculated C(3)=C(4) bond length of 1.406 Å indicates that the bond order of the C(3)=C(4) bond is greatly decreased and is in good accord with the experimental result that no band assignable to the C(3)=C(4) stretch is observed in the double-bond-stretch region. The C(2)–C(3) and C(4)–C(4a) bond lengths were calculated to be 1.399 and 1.412 Å, respectively, indicating that the bond orders of these two bonds are greatly increased. The decrease of the bond order of the C(3)=C(4) double bond and the increase of the bond orders of the two C(2)–C(3) and C(4)–C(4a) single bonds indicate that bond-order reversal is occurring in the pyrone ring of $R^{-\bullet}$.

Although optimized geometry of T_1 was not obtained with the use of UHF level due to spin contamination, more pronounced bond-order reversal is expected in T_1 than in $R^{-\bullet}$. This prediction is in good accord with the experimental results that no band assignable to the C(3)=C(4) stretch is detectable in the double-bond-stretch region and the contribution of this stretch is distributed over the vibrational modes in the single-bond-stretch region.

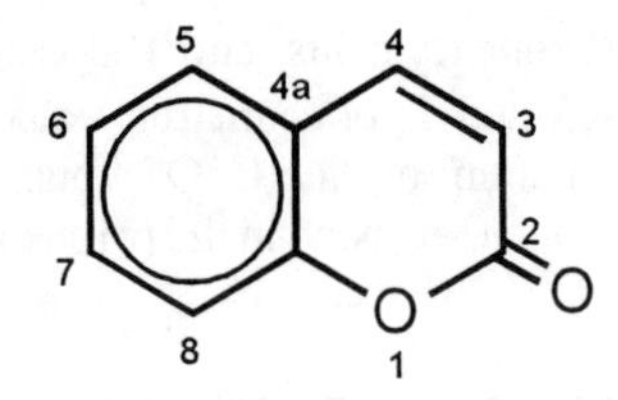

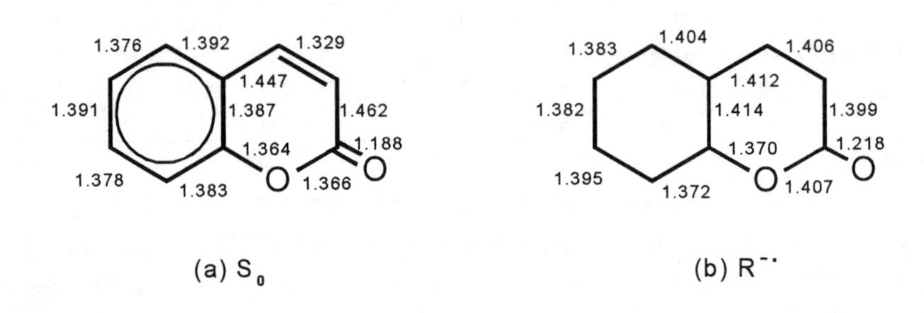

Fig. 12 Calculated bond lengths (Å) of S_0 and $R^{-\cdot}$ of coumarin obtained by *ab initio* MO method using the 4-31G(O*) basis set: (a) S_0, calculation at the RHF level; (b) $R^{-\cdot}$, calculation at the ROHF level. (O*) represents the inclusion of polarization functions for oxygen atoms only.

3.6 Phenothiazine

Phenothiazine tranquilizers, particularly chlorpromazine, have long been used for the treatment of psycotic disorders [69]. However, they are known to cause both phototoxic and photoallergic reactions in the skin [70] and eyes [71, 72] of patients receiving these drugs. Although the detailed mechanism of the phototoxicity and photoallergy of phenothiazine tranquilizers have not yet been clarified, it is without doubt that photolytically generated excited electronic states, ionic or neutral radicals of the phenothiazine derivatives, or the compounds derived from them play key roles in the mechanisms.

Phenothiazine

Chlorpromazine

With this view the photochemistry of phenothiazine (PTZ) was investigated by time-resolved absorption and time-resolved resonance Raman spectroscopies in order to obtain structures and dynamics of the excited electronic states and neutral and cation radicals that are considered to appear in the photochemistry of these compounds.

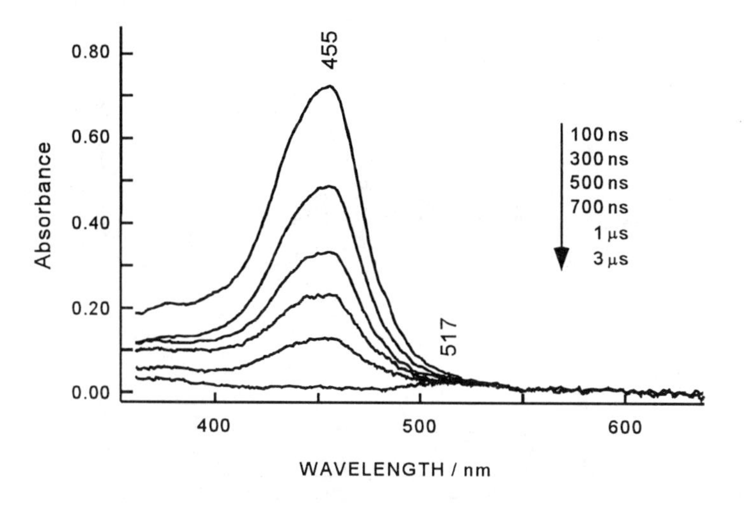

Fig. 13 Time-resolved absorption spectra of phenothiazine in methanol. Concentration, 1.0×10^{-3} moldm^{-3}. Pump wavelength, 308 nm.

As shown in Fig. 13, time-resolved absorption spectra of PTZ in deoxygenated methanol reveal that two transients are involved in the PTZ of this compound: a transient showing a strong absorption band at 455 nm with a lifetime of about 800 ns and another transient exhibiting a weak peak at 517 nm with a much longer lifetime [73]. In the presence of oxygen the lifetime of the 455 nm band was shortened markedly, and a new band appeared at 387 nm. The 455 nm band can be assigned to T_1, and the new band at 387nm may be attributable to a complex generated from T_1 and O_2. The intensity of the 517 nm band increased quadratically with increasing power of the pump UV (308 nm) light. This band is assignable to $R^{-\bullet}$. These results are in good accord with the reports in the literature [74–77].

The Raman spectrum of S_0 and the resonance Raman spectra of T_1 and $R^{-\bullet}$ of PTZ are compared in Fig. 14. It is seen that the phenyl ring stretch mode 8a and 8b at 1605 and 1575 cm^{-1} in the S_0 spectrum are both drastically downshifted to 1507 cm^{-1} in the spectrum of T_1, while they do not exhibit appreciable shifting in the spectrum of $R^{-\bullet}$. Usually the downshifts of 8a and 8b modes on going from S_0 to T_1 are around 20–30 cm^{-1}. The unusually large downshift 96 cm^{-1} of the 8a and 8b modes suggests that the excitation is strongly localized on the phenyl rings and that the phenyl rings of PTZ are drastically weakened in T_1.

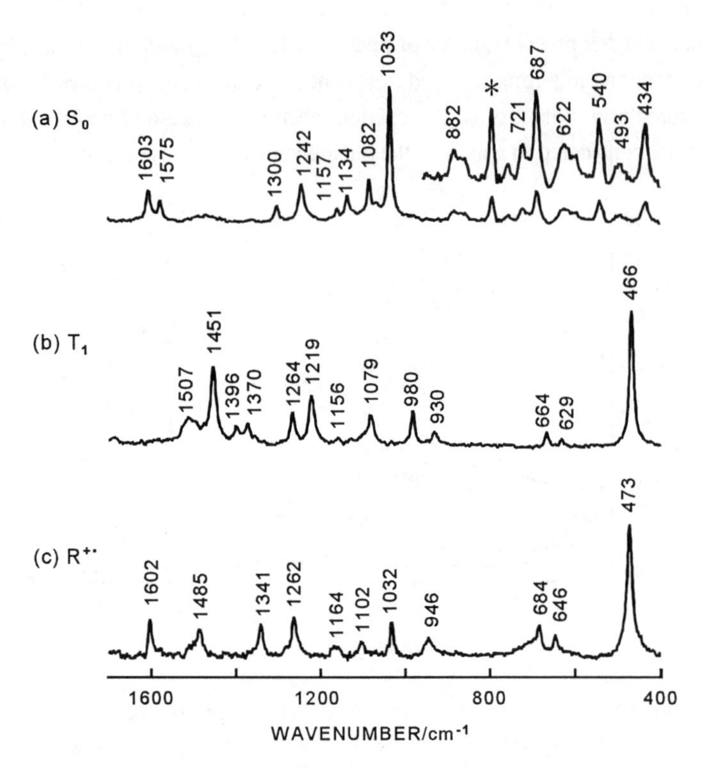

Fig. 14 Comparison of the Raman spectrum of S_0 and resonance Raman spectra of T_1 and $R^{+\bullet}$ of phenothiazine: (a) S_0 in acetone, nearly saturated. Probe wavelength, 581 nm; (b) T_1 in methanol, 2.0×10^{-3} moldm^{-3}. Probe wavelength, 456 nm. Pump wavelength, 308 nm; $R^{+\bullet}$ in methanol, 2.0×10^{-3} moldm^{-3}. Probe wavelength, 517 nm. Pump wavelength, 308 nm. The asterisk (*) denotes solvent bands.

In order to obtain more detailed structural information, the Raman spectrum of S_0 and the resonance Raman spectra of T_1 and $R^{-\bullet}$ of isotopically substituted PTZ, viz. ^{34}S substituted analogue (PTZ-^{34}S) and N–D substituted analogue (PTZ-d_1), were also measured. It was observed that in the spectrum of S_0 the band at 434 cm^{-1} of PTZ was downshifted to 430 cm^{-1} on the ^{34}S substitution, and the band at 1242 cm^{-1} of PTZ was downshifted to 1238 cm^{-1} on the N–D substitution (Table 8). The 434 cm^{-1} band of S_0 can be assigned to a vibrational mode involving the largest contribution of the C–S symmetric stretch, and the1242 cm^{-1} band of S_0 is assignable to a mode in which the contribution of the C–N symmetric stretch is the largest. The frequency 434 cm^{-1} may appear to be considerably lower than the usual C–S stretching frequencies, which are expected to lie in the range 580–750 cm^{-1} [78]. The lower frequency of the C–S symmetric stretch of PTH is ascribed to the coupling with skeletal deformations of the ring. In the spectrum of T_1 the band at 466 cm^{-1} can be assigned to the C–S symmetric stretch, and the band at 930 cm^{-1} is assignable to the first overtone of the 466 cm^{-1} band. This assignment is supported by the downshift of −13 cm^{-1} on ^{34}S substitution, which is about twice as large compared to the downshift of −6 cm^{-1} of the 466 cm^{-1} band. In the spectrum of $R^{-\bullet}$ the bands at 473 and 946 cm^{-1} can be assigned to the C–S symmetric stretch and its first overtone, respectively, based on the frequency shifts on the ^{34}S substitution.

Table 8. Frequencies (cm^{-1}) and isotopic frequency shifts (cm^{-1}) of some of the vibrational modes of S_0, T_1, and $R^{-\bullet}$ of phenothiazine. Values in parentheses are frequency shifts on the ^{34}S substitution and deuteration of the N–H group, respectively.

Vibrational mode	Frequencies and frequency shifts on isotopic substitutions (^{34}S substitution, N–D substitution)		
	S_0	T_1	$R^{-\bullet}$
8a	1603 (0, −1)	1507 (−1, −1)	1602 (0, 0)
8b	1575 (0, −1)		
C–N sym. str.	1242 (0, −4)	1219 (0, −4)	1262 (0, −5)
2×CS sym. str.		930 (−13, −4)	946 (−14, −5)
C–S sym. str.	434 (−4, 0)	466 (−6, −2)	473 (−7, −3)

As seen above, time-resolved resonance Raman spectroscopy has shown that the phenyl 8a and 8b modes are drastically downshifted in the T_1 state, whereas they do not exhibit appreciable frequency shift in $R^{-\bullet}$. In contrast, the C–S symmetric stretch is upshifted both in T_1 and $R^{-\bullet}$. These observations indicate that the phenyl rings of PTZ are drastically weakened, and therefore the C–C bonds of the phenyl rings are lengthened in T_1. However, these structural changes do not occur in $R^{-\bullet}$, and the C–S bonds are shortened both in the T_1 and $R^{-\bullet}$.

3.7 Flavin Mononucleotide

Photochemical reactions of flavins may be regarded as model reactions for understanding the enzymic reaction mechanisms of flavoproteins. Flavins in their excited electronic state are known to sensitize photooxidation of a wide range of biologically important substrates such as amino acids, proteins, and DNA and its nucleotides [79–81]. In this photoreaction process flavins are reduced first to semiquinones and then further reduced to hydroquinones. Several studies have shown that the T_1 state of flavins is the reactive intermediate in the photooxidation reactions [82, 83].

R: $-CH_2(CHOH)_3CH_2OPO_3{}^{2-}$

Flavin mononucleotide

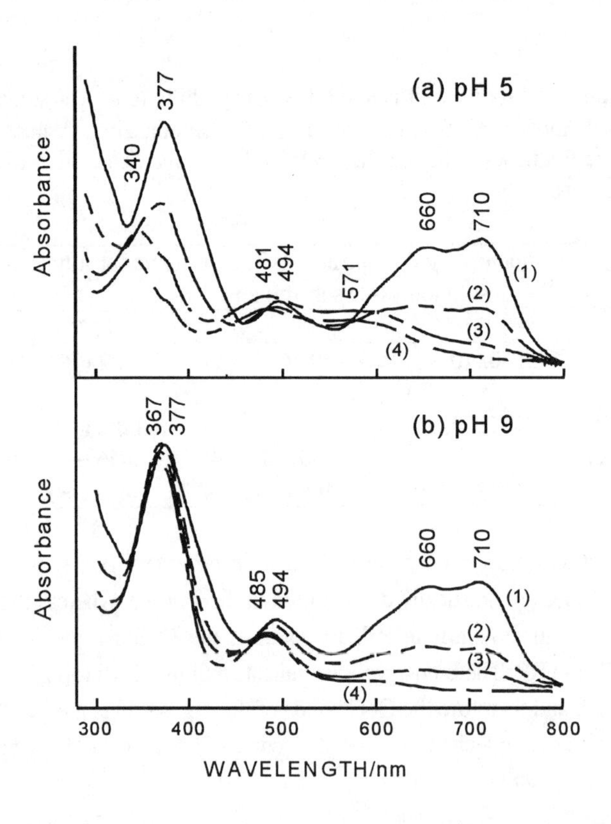

Fig. 15 Time-resolved absorption spectra of FMN in deoxygenated acidic and basic buffer solutions with an exess amount of EDTA: (a) acidic solution pH 5, (1) 60 ns, (2) 1 μs, (3) 3 μs, (4) 10 μs; (b) basic solution pH 9, (1) 60 ns, (2) 600 ns, (3) 1 μs, (4) 3 μs after the pumping with 456 nm light. Concentration: FMN, 5.0×10^{-4} moldm^{-3}; EDTA, 1.5×10^{-3} moldm^{-3}.

Time-resolved absorption spectra of flavin mononucleotide (FMN) with excess amount of EDTA in deoxygenated buffer solutions, pH 5 and 9, are shown in Fig. 15 [84]. The bands at 377, 494, 660, and 710 nm were quenched by oxygen. These bands are attributable to T_1. Since these bands were observed in both pH 5 and 9 solutions, T_1 is not protonated in the acidic solution pH 5. We observed that protonated T_1 of FMN in a solution pH 3.2 exhibited absorption bands at 397, 485, and 673 nm.

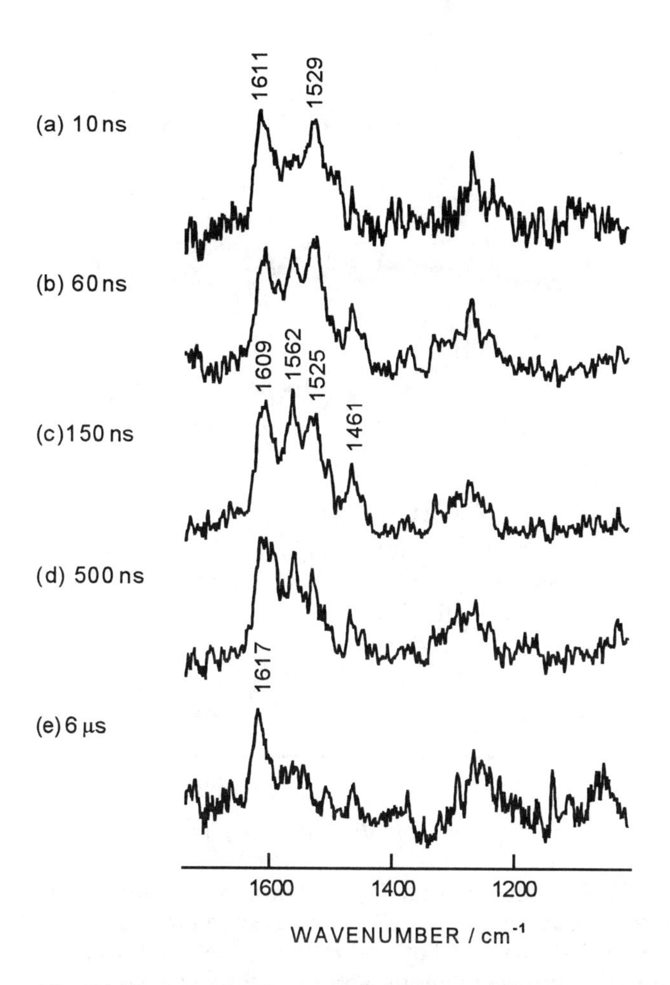

Fig. 16 Time-resolved resonance Raman spectra of FMN with an excess amount of EDTA in deoxygenated acidic solution pH 5 measured at several delay times after pumping. Pump wavelength, 456 nm. Probe wavelength, 390 nm. Concentraion, FMN, 2.0×10^{-3} moldm^{-3}; EDTA, 1.0×10^{-2} moldm^{-2}.

In acidic solution pH 5, new bands become prominent at 340, 481, and 571 nm at around 3 μs delay time. These bands are attributable to the semiquinone neutral radical, $RH^{\bullet}$, usually called the "blue form" of the semiquinone. In the basic solution pH 9, new bands are observed at 367 and 485 nm at 3 μs delay time. These bands can be assigned to the semiquinone anion radical $R^{-\bullet}$, usually called the "red form" of the semiquinone.

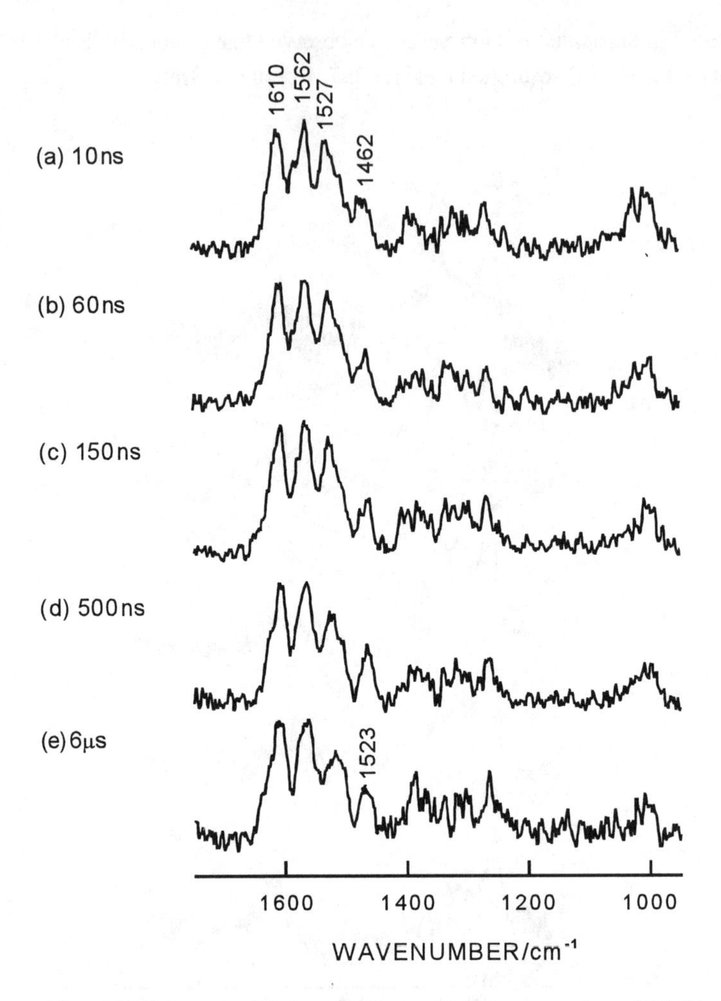

Fig. 17 Time-resolved resonance Raman spectra of FMN with an excess amount of EDTA in deoxygenated basic solution pH 9 measured at several delay times after the pumping. Pump wavelength, 456 nm. Probe wavelength, 390 nm. Concentration: FMN, $2.0x10^{-3}$ mol dm^{-3}; EDTA, $1.0x10^{-2}$ mol dm^{-3}.

We notice that the absorption peaks of $R^{-\bullet}$ may well be hidden under the peaks of T_1 or $RH^{\bullet}$, and the peaks of $RH^{\bullet}$ can be buried under the peaks of T_1. Therefore, it is not clear in Fig. 15(a) whether or not $R^{-\bullet}$ is generated in the acidic solution pH 5 at early delay times, nor is it clear in Fig. 15(b) whether $RH^{\bullet}$ is involved in the basic solution pH 9. The spectral changes in Fig. 15 appear to indicate that in the acidic solution pH 5, T_1 is directly converted into $RH^{\bullet}$ by abstracting a hydrogen atom from EDTA, while in the basic solution pH 9, T_1 is directly converted into $R^{-\bullet}$ by obtaining an electron from EDTA.

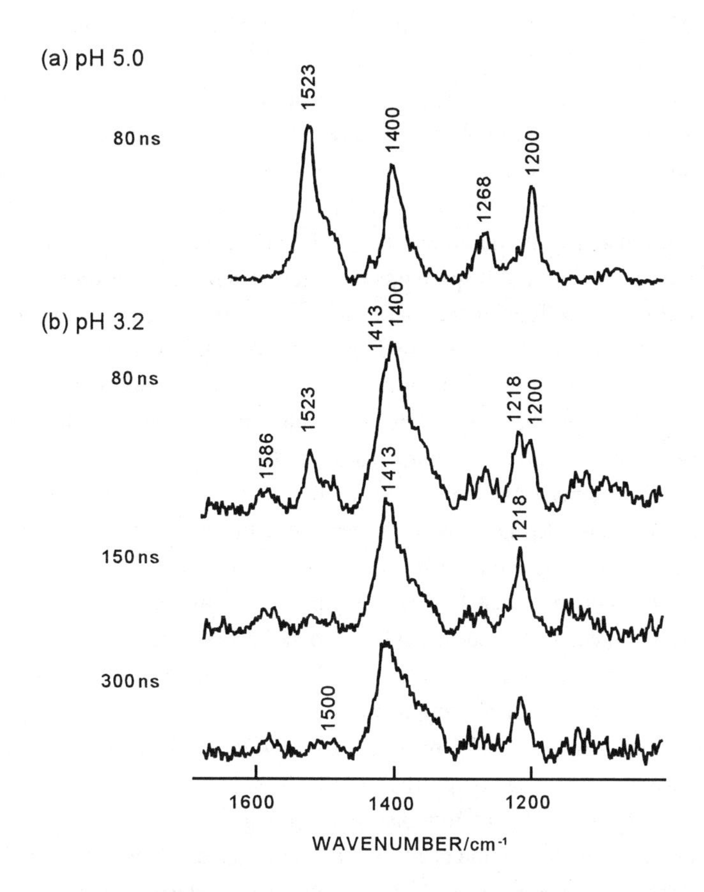

Fig. 18 Protonation of T_1 of FMN in acidic solution pH 3.2 observed by time-resolved resonance Raman spectroscopy: (a) transient resonance Raman spectrum of T_1 in acidic solution pH 5 measured at 80 ns delay time; (b) time-resolved resonance Raman spectra of FMN in acidic solution pH 3.2 at different delay times. Probe wavelength, 690 nm. Pump wavelength, 456 nm. Concentration, 2.0×10^{-3} mol dm^{-3}.

Time-resolved resonance Raman spectra of FMN in deoxygenated acidic solution pH 5 with an excess amount of EDTA are shown in Fig. 16. At 10 ns after the pumping, two bands are observed at 1611 and 1529 cm^{-1}. These bands are assigned to T_1 based on the marked decrease in intensity in the presence of oxygen. At 60 ns delay time two more bands appear at 1562 and 1461 cm^{-1}, and at 150 ns the band at 1562 cm^{-1} becomes stronger, while the bands at 1611 and 1529 cm^{-1} appear to be slightly shifted to lower frequencies. The apparent downshifts are due to the appearance of new bands at

1609 and 1525 cm^{-1}. The bands at 1609, 1562, 1525, and 1461 cm^{-1} are attributable to $R^{-\bullet}$. At 500 ns delay time the bands at 1562, 1525, and 1461 cm^{-1} decrease in intensity, and the band at 1609 cm^{-1} is upshifted to 1617 cm^{-1}. At 6 μs delay time only the band at 1617 cm^{-1} is observed. The band at 1617 cm^{-1} is attributable to $RH^{\bullet}$.

From the time evolution of the resonance Raman spectra it became clear that $RH^{\bullet}$ is produced via $R^{-\bullet}$ in acidic solution pH 5. This means that what is abstracted by T_1 from EDTA in the acidic solution is not a hydrogen atom to generate $RH^{\bullet}$ but is an electron to produce $R^{-\bullet}$, which is then converted into $RH^{\bullet}$ by combining with H^+. The rate of this protonation is considered to be dependent on the H^+ concentration, and in the acidic solution pH 5 the rise time of $RH^{\bullet}$ is approximately a few hundred nanoseconds.

Time-resolved resonance Raman spectra of FMN in deoxygenated basic solution pH 9 with an excess amount of EDTA are shown in Fig. 17. At 10 ns delay time four bands are observed at 1610, 1562, 1527, and 1462 cm^{-1}. These bands should be attributed to both T_1 and $R^{-\bullet}$. Although the spectral changes with time are not very distinct, the spectrum measured at 6 μs can be attributed solely to $R^{-\bullet}$. It is noteworthy that in the basic solution $R^{-\bullet}$ is generated much earlier than 10 ns delay time, whereas in the acidic solution it is not yet generated at 10 ns. This means that the abstraction of an electron from EDTA by T_1 occurs faster in basic solutions than in acidic solutions. Another important point to mention is that $RH^{\bullet}$ is not produced in the basic solution pH 9, implying that the protonation of $R^{-\bullet}$ does not occur in the basic solution.

In Fig. 18 are shown time-resolved resonance Raman spectra of FMN in the acidic solution pH 3.2 measured with 690 nm probe light. The transient resonance Raman spectrum of T_1 in acidic solution pH 5 measured with 690 nm probe light at 80 ns delay time is also shown for comparison purposes (Fig. 18a). Due to the difference in the probe wavelength, this T_1 spectrum differs from that shown in Fig. 16. At 80 ns delay time two transients are observed in the solution of pH 3.2. It is clear that the bands at 1523, 1400, 1268, and 1200 cm^{-1} can be assigned to T_1. These bands rapidly decrease in intensity and at 150 ns they disappear almost completely. Since the remaining bands at 1586, 1413, and 1218 cm^{-1} were quenched by oxyen, they were assigned to the protonated species of T_1, namely T_1($FMNH^+$). We note that the protonation of T_1 occurs rather slowly in the nanosecond time range in the acidic solution pH 3.2.

Based on the above results, the photoreduction process of FMN in acidic and basic solutions can be represented schematically as shown in Fig. 19. In aqueous solutions with pH range from 0 to 10, FNN exists as an oxidized form [85]. When irradiated with UV or blue light in the presence of substrate such as EDTA or amino acids, FMN is reduced via T_1 to the semiquinone, which takes either the anion radical form $R^{-\bullet}$ or the neutral radical form $RH^{\bullet}$ depending on the H^+ concentration of the solution. Since the pK_a of this semiquinone is about 8.4 [85], the semiquinone is considered to take the $R^{-\bullet}$ form when the pH is larger than 8.4, whereas it is in the form of $RH^{\bullet}$ when the pH is smaller than 8.4. The photoreduction proceeds as follows: abstraction of an electron from the substrates by T_1 to generate $R^{-\bullet}$, which is then converted into $RH^\bullet$ by protonation when the pH is smaller than 8.4. When the pH is smaller than 4.4, which is the pK_a value of T_1 [84], the protonation of T_1 occurs to produce T_1 ($FMNH^+$), which is then converted into $RH^\bullet$ by abstracting an electron from the substrates. Any indication of the abstraction of a hydrogen atom by T_1 to produce $RH^\bullet$ directly was not obtained.

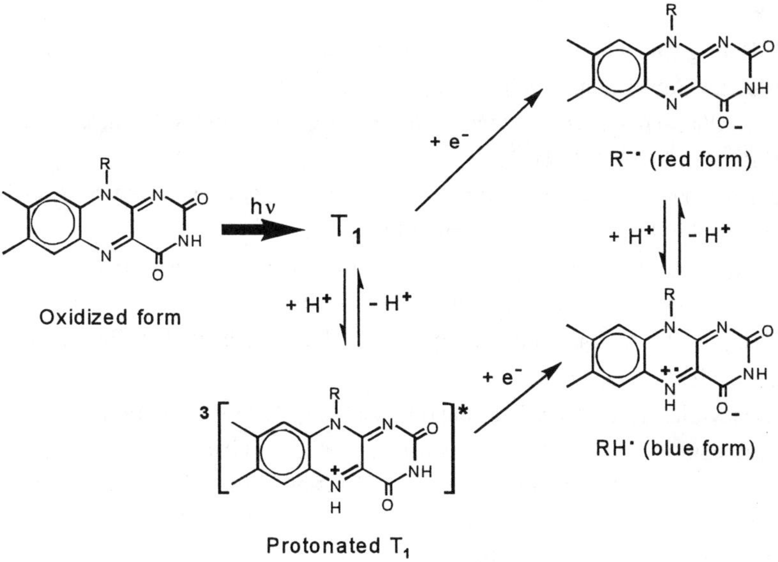

Fig. 19 Photoreduction mechanism of FMN

Acknowledgments

The author thanks M. Sakai and M. Mizuno for their help in preparing the figures.

References

1 G. H. Atkinson, in *Advances in Laser Spectroscopy*, Vol. 1, ed. by B. A. Garetz and J. R. Lombardi (Heyden; London, 1982).

2 G. H. Atkinson, in *Advances in Infrared and Raman Spectroscopy*, Vol. 9, ed. by R. J. H. Clark and R. E. Hester (Heyden; London, 1982).

3 G. H. Atkinson (ed.), *Time-resolved Vibrational Spectroscopy* (Academic Press; New York, 1983).

4 G. H. Atkinson (ed.), *Time-Resolved Vibrational Spectroscopy* (Gordon Breach;

New York, 1985).
5 J. Terner and M. A. El-Sayed, Acc. Chem. Res. **18**, 331 (1985).
6 A. Laubereau and M. Stockburger (eds.), *Time-resolved Vibrational Spectroscopy* (Springer; Berlin, Heidelberg, 1985).
7 M. Stockburger, T. Alshuth, D. Oesterhelt, and W. Gärtner, in *Spectroscopy of Biological Systems,* ed. by R. J. H. Clark and R. E. Hester (Wiley; New York, 1986).
8 H. Hamaguch, in *Vibrational Spectra and Structure,* Vol. 16, ed. by J. R. Durig (Elsevier; Amsterdam, 1987).
9 H. Takahashi (ed.), *Time-Resolved Vibrational Spectroscopy* V (Springer; Berlin, Heidelberg, 1991).
10 Y. Koyama and Y. Mukai, in Biomolecular Spectroscopy, Part B, ed. by R. J. H. Clark and R. E. Hester (Wiley; New York, 1993).
11 A. Lau, F. Siebert, and W. Werncke (eds.), *Time-Resolved Vibrational Spectroscopy* VI (Springer; Berlin, Heidelberg, 1994).
12 S. Hashimoto, A. Shimojima, T. Yuzawa, H. Hiura, J. Abe, and H. Takahashi, J. Mol. Struct. **242**, 1 (1991).
13 A. Mavridis and I. Moustakali-Mavridis, Acta Cryst. **B33**, 3612 (1977).
14 A. A. Espiritu and J. G. White, Z. Kristallog. **147**, 177 (1977).
15 K. Okuyama, T. Hasegawa, M. Ito, and N. Mikami, J. Phys. Chem. **88**, 1711 (1984).
16 H. Suzuki, Bull. Chem. Soc. Jpn. **33**, 389 (1960).
17 J. Higuchi, T. Ito, and O. Kanehisa, Chem. Phys. Lett. **23**, 440 (1973).
18 T. Hoshi, K. Ota, J. Yoshino, K. Morofushi, and Y. Tanizaki, Chem. Lett. 357 (1977).
19 H. Suzuki, K. Koyano, T. Shida and A. Kida, Bull. Chem. Soc. Jpn. 55, 3690 (1982).
20 H. Hiura and H. Takahashi, J. Phys. Chem. **96**, 1909 (1992).
21 A. Shimojima and H. Takahashi, J. Phys. Chem. **97**, 9103 (1993).
22 Y. Hirata, T. Okada, N. Mataga, and T. Nomoto, J. Phys. Chem. **96**, 6559 (1992).
23 T. Ishibashi and H. Hamaguchi, Chem. Phys. Lett. **264**, 551 (1977).
24 C. K. Ingold and G. W. King, J. Chem. Soc. 2275 (1953).
25 E. B. Wilson, Jr., Phys. Rev. **45**, 706 (1934).
26 A. I. Kruppa, T. L. Leshina, and R. Z. Sagdeev, Chem. Phys. Lett. **121**, 386 (1985)
27 S. L. Murov, *Handbook of Photochemistry* (Marcel Dekker; New York, 1973).
28 J. P. Reboul, B. Cristau, and J. C. Soyfer, Acta Cryst. **B36**, 2683 (1980).
29 J. A. G. Drake and D. W. Jones, Acta Cryst. **B38**, 200 (1982).
30 L. J. Johnston, J. Lobaugh, and V. Wintgens, J. Phys. Chem. **93**, 7370 (1989).
31 J. Abe, N. Kunimatsu, A. Shimojima, and H. Takahashi, Chem. Phys. Lett. **178**,

547 (1991).
32 M. Sakai, M. Mizuno, and H. Takahashi, to be published in J. Raman Spectrosc.
33 G. Briegleb and J. Czekalla, Z. Electrochem. **63**, 6 (1956).
34 T. Imura, N. Yamamoto, H. Tsubomura, and K. Kawabe, Bull. Chem. Soc. Jpn. **44**, 3185(1971).
35 Y. Hirata and N. Mataga, J. Phys. Chem. **87**, 1680; 3190 (1983).
36 S. Nakamura, N. Kanamaru, S. Nohara, H. Nakamura, Y. Saito, J. Tanaka, M. Sumitani, N. Nakashima, and K. Yoshihara, Bull. Chem. Soc. Jpn. **57**, 145 (1984).
37 N. Yamamoto, Y. Nakato, and H. Tsubomura, Bull. Chem. Soc. Jpn. **39**, 2603 (1966).
38 Y. Nakato, N. Yamamoto, and H. Tsubomura, Bull. Chem. Soc. Jpn. **40**, 2480 (1967).
39 J. T. Richards and J. K. Thomas, Trans. Faraday Soc. **66**, 621 (1970).
40 H. Isaka, S. Suzuki, T. Ohzeki, Y. Sakaino, and H. Takahashi, J. Photochem. **38**, 167 (1987).
41 Y. Hirata, A. Nogata, and N. Mataga, Chem. Phys. Lett. **189**, 159 (1992).
42 T. Kuroda, H. Hiura, and H. Takahashi,
43 H. Isaka, J. Abe, T. Ohzeki, Y. Sakaino, and H. Takahashi, J. Mol. Struct. **178**, 101 (1988).
44 J. L. de Boer and A. Vos, Acta Cryst. **B28**, 835 (1968).
45 J. Abe, T. Miyazaki, and H. Takahashi, J. Chem. Phys. **90**, 2317 (1989).
46 I. Ikemoto, G. Katagiri, S. Nishimura, K. Yakushi, and H. Kuroda, Acta Cryst. **B35**, 2264 (1979).
47 C. J. Brown and R. Sadanaga, Acta Cryst. **18**, 158 (1965).
48 C. W. N. Cumper and A. P. Thurston, Perkin Trans. II, 106 (1972).
49 T. S. Fang, R. E. Brown, and L. A. Singer, Chem. Phys. Lett. **60**, 117 (1978).
50 D. S. Roy, K. Bhattacharyya, S. C. Bera, and M. Chowdhury, Chem. Phys. Lett. **69**, 134 (1980).
51 N. Ikeda, M. Koshioka, H. Masuhara, and K. Yoshihara, Chem. Phys. Lett. **150**, 452 (1988).
52 M. Mukai, S. Yamaguchi, and N. Noboru, J. Phys. Chem. **93**, 4411 (1989).
53 K. Ebihara, H. Hiura, and H. Takahashi, J. Phys. Chem. **96**, 9120 (1992).
54 T. Shida, S. Iwata, and S. Imamura, J. Phys. Chem. **78**, 741 (1974).
55 M. Kamlet, J. L. M. Abboud, M. H. Abraham, and W. Taft, J. Org. Chem. 48, 2877 (1983).
56 J. A. Parrish, T. B. Fitzpatrick, L. Tannenbaum, and M. A. Patak, New Engl. J. Med. **291**, 1207 (1974).
57 T. B. Fitzpatrick, J. A. Parrish, and M. A. Pathak, in *Sunlight and Man*, eds. M. A. Pathak, C. Harber, M. Seiji, and A. Kurita (University of Tokyo Press; Tokyo, 1974), p.783.

58 M. A. Pathak, J. H. Fellman, and K. D. Kaufman, J. Invest. Dermatol. **35**, 165 (1960).
59 L. Musajo, Ann. Ist. Super. Sanita **5**, 376 (1969).
60 A. C. Giese, Photophysiol. **6**, 77 (1971).
61 F. Dall'Acqua, S. Marciani, L. Ciavatta, and G. Z. Rodighiero, Naturforsch. **26b** 561 (1971).
62 L. Musajo and G. Rodighiero, Photophysiol. 7, 115 (1972).
63 Y. Uesugi, M. Mizuno, A. Shimojima, and H. Takahashi, J. Phys. Chem. A **101**, 268 (1997).
64 B. R. Henry and R. V. Hunt, J. Mol. Spectrosc. **39**, 466 (1971).
65 E. J. Land and T. G. Truscott, Photochem. Photobiol. **29**, 861 (1979).
66 P.-T. Chou, M. L. Martinez, and S. L. Studer, Chem. Phys. Lett. 188, **49** (1992).
67 T. Tahara, H. Hamaguchi, and M. Tasumi, J. Phys. Chem. **94**, 170 (1990).
68 T. Tahara, H. Hamaguchi, and M. Tasumi, J. Phys. Chem. **91**, 5875 (1987); Chem. Phys. Lett. **152**, 135 (1988).
69 K. H. Schulz, A. Wiskemann, and K.Wulf, Arch. Klin. Exp. Dermatol. **202**, 285 (1956).
70 L. W. C. Massey, Canad. Med. Assoc. J. **92**, 186 (1965).
71 B. A. Kirshbaum and H. Beerman, Am. J. Med. Soc. **248**, 445 (1964).
72 W. Bruinsma, Dermatologica, **145**, 377 (1972).
73 G. Sarata, Y. Noda, M. Sakai, and H. Takahashi, J. Mol. Struct. **413-414**, 49 (1997).
74 B. R. Henry and M. Kasha, J. Chem. Phys. **47**, 3319 (1967).
75 T. Iwaoka, H. Kokubun, and M. Koizumi, Bull. Chem. Soc. Jpn. **44**, 341 (1971); **45**, 73 (1972).
76 H. J. Shine and E.E. Mach, J. Org. Chem. **30**, 2130 (1965).
77 S. A. Alkaitis, G. Beck, and M. Gratzel, J. Am. Chem. Soc. **97**, 5723 (1975).
78 F. R. Dollish, W. G. Fateley, and F. F. Bentley, *Characteristic Raman Frequencies of Organic Compounds* (John Wiley; New York, 1974).
79 M. B. Taylor and G. K. Radda, Methods Enzymol. **18B**, 496 (1971).
80 A. Knowles and G. N. Mautner, Photochem. Photobiol. **15**, 199 (1972).
81 K. Kuramori and Y. Kobayashi, FEBS Lett. **72**, 295 (1976).
82 A. Knowles and E. M. F. Roe, Photochem. Photobiol. 7, 421 (1968).
83 P. F. Heelis, B. J. Persons, G. O. Phillips, and J. F. McKellar, Photochem. Photobiol. **28**, 169 (1978).
84 M. Sakai and H. Takahashi, J. Mol. Struct. **379**, 9 (1996).
85 J. T. Mc Farland, in *Biological Applications of Raman Spectroscopy*, Vol 2, ed. by T. Spiro (Wiley; New York, 1987).

Solvation of Radicals in Small Clusters

J. A. Fernández , J. Yao, J. A. Bray and E. R. Bernstein

Department of Chemistry, Colorado State University
Fort Collins, CO 80523, USA.

Abstract. The spectra and solvation behavior of several radical systems, cooled in a supersonic jet expansion, are studied using laser induced fluorescence and time-of-flight mass spectroscopic techniques. The systems included in this study are small radicals, CH_3O and NCO, and larger aromatic radicals, picolyl, lutidyl, cyclopentadienyl, and its derivatives. Intensive *ab initio* and Leonard-Jones potential energy surface calculations are performed to understand the spectra, the geometries, and the van der Waals interactions of these radicals and their clusters. Inter- and intra-molecular dynamics of excited state species, possible chemical reactions, as well as substituent effects on cluster chemistry are discussed.

1 Introduction

Gas phase radicals are open-shell, highly reactive species. These two characteristics make their interactions with other molecules and radicals difficult to model [1]. On the other hand, a good understanding of their interactions is necessary to rationalize the chemistry and behavior of systems such as flames or the atmosphere, in which radicals play important roles as intermediate species [2]. Different approaches to the problem have been tried with more or less success: pure kinetic studies [3], highly expensive computational methods [4], dynamic approaches [5], etc. One of the newer techniques considers the study of radical solvation at very low temperatures (a few Kelvin) produced in a supersonic jet expansion. At these low temperatures, van der Waals (vdW) interactions become important, and radicals can form clusters with various closed shell atoms and molecules.

The study of vdW clusters gives important information about intra- and intermolecular dynamics [6], energy transfer [7], unimolecular dissociation [8], solvation [9], chemical reactions [10], potential energy surfaces [11], and collisional processes [12] in stable molecules. Recently some laboratories have started to apply these studies to small radicals (OH [13], CH [14], NH [15]...) for which intensive computational algorithms can be performed to explore energy transfer in collisional processes.

Our laboratory is involved in the study of radical reactivity and dynamics and in the identification of new species. vdW cluster studies have been very helpful in solving problems dealing with radical reactivity and dynamics. For example, we reported that while benzyl radical forms stable clusters with C_2H_4 in its ground state, upon electronic excitation of the radical, it undergoes a double bond addition reaction [16]. This fact, together with extensive *ab initio* calculations, enables modeling of the reaction pathway. We also predicted that the addition of CH_3 to C_2H_4 will occur in a similar manner. We are presently engaged in exploring this prediction for $CH_3 + C_2H_4$.

In this chapter, we present four pieces of our most recent work on vdW clusters containing radicals. Sec. 1 reports the study of CH_3O clusters with nonpolar solvents. We explain the nature and relative importance of the interactions that form the clusters: evidence of a cluster-formation controlled reaction for CH_3O in an excited electronic state with CH_4 is presented.

Sec. 2 deals with the difficult problem of NCO excited state dynamics. This linear radical experiences a series of effects [17] (Renner-Teller, Fermi resonances, and large spin-orbit coupling) that yields a rich, and yet not totally understood, spectroscopy. Using molecular jet expansions to cool the radicals we can obtain simplified spectra that are easier to assign and analyze than static gas spectra. The information obtained in the study contributes to the solution of the problem of NCO $B \leftarrow X$ transition spectrum assignment. The study of NCO vdW clusters provides important information about NCO excited state dynamics.

Sec. 3 presents studies of two new radicals: picolyl and lutidyl. The identification of a new species is always a complex problem that requires careful experimentation and strong *ab initio* calculations. We make use of vdW clusters to unravel the electronic spectrum of these new species.

Finally, in Sec. 4 we present preliminary results of a more extensive work, still in progress, about a family of radicals: cyclopentadienyl (cpd), methylcyclopentadienyl (mcpd), and cyanocyclopentadienyl (cncpd). We employ vdW cluster studies to elucidate how the addition of a functional group to an aromatic ring affects radical reactivity and the ring π electronic cloud.

2 CH_3O Clusters with Nonpolar Solvents

Determination of cluster structure is by no means a simple task; it requires complex experimental arrangements and extensive theoretical calculations. The task involves determination of cluster geometry (not only for the most stable Isomer, but also for all the isomers formed for a given stoichiometry), cluster binding energies in both ground and excited electronic states, and the type of interactions responsible for forming the cluster. In other words, determination of cluster structure entails elucidation of cluster potential energy surface for all involved electronic states.

The experimental setup has been described elsewhere [18]. Briefly, the experiments have been carried out in a stainless steel vacuum chamber at a working pressure of ca. 10^{-4} Torr. To produce the clusters, CH_3OH vapor is mixed with He carrier gas and the solvent. The mixture passes through a pulsed valve (General Valve 0.7 mm nozzle diameter) and expands in the chamber, producing both molecular cooling and cluster formation. Typical pressures and concentrations vary from one solvent to another, but are typically in the range of 150-500 psi for the expansion gas with a stagnation pressure concentration of 10-20% for the solvent gas. Cooling is one of the most critical parameters in this study. CH_3OH photolysis produces extremely hot radicals. The vibration-rotation-translation energy transfer processes that cool the CH_3O radical are not very efficient, due to the radical's high vibrational frequencies: many collisions are required to reach the low temperatures necessary for the cluster formation process to take place. Higher stagnation pressures are then required compared to those needed for the cooling and clustering of stable molecules. Both cooling and photolysis are aided by a 1.0 id × 15 mm quartz tube fixed at the nozzle exit.

An ArF excimer laser (193 nm), with output power of ≈80 mJ/pulse, aligned collinearly with the supersonic expansion axis and focused with a 15 cm lens is used to photolyze the methanol precursor and produce the radicals. The excimer laser output is aligned to be co-axial with the molecular jet from the nozzle. The excimer beam is aligned so as not to ablate the polymer poppet that opens and closes the pulsed nozzle. A second laser (a Nd/YAG-dye coupled system) is used to probe the radicals about 2 cm downstream from the tube end. Fluorescence is collected perpendicularly to the laser/molecular jet plane using a C31034 RCA photomultiplier tube. A U330 filter is employed to attenuate scattered light from the excimer laser.

There are mainly two ways to deal with the difficult problem of cluster structure calculations: atom-atom empirical potential energy surfaces and *ab initio* quantum calculations. The second approach is perhaps the more difficult one. In principle, *ab initio* calculations are powerful techniques, capable of describing any type of interaction between molecules, no matter what its character. But the flatness of the potential energy surface and the weakness of the interaction in van der Waals clusters can be overwhelmed by basis set superposition corrections and errors [19]. Large basis sets, including diffuse functions, and high theory levels have to be employed in order to account for the long-range forces that form the cluster, increasing dramatically the cpu time needed to carry out the calculations. Quartic CI (double excitations on each molecule) is the minimum order required to generate those dispersion interactions. When more than one isomer is possible for a given cluster, the situation becomes even more complicated, due to the difficulties of finding a structure that is not an absolute minimum using *ab initio* calculations.

The other approach is based on the use of a semiempirical potential energy surface that parametrizes and describes the interaction between atoms in both molecules. One of the most popular forms is given by [20],

$$E=\sum_{i=1}^{n}\sum_{j=1}^{m}\left\{\left(\frac{A_{ij}}{r_{ij}^{12}}-\frac{C_{ij}}{r_{ij}^{6}}\right)\left(1-\delta_{ij}^{hb}\right)+\frac{q_i q_j}{D r_{ij}}+\left(\frac{A_{ij}^{hb}}{r_{ij}^{12}}-\frac{C_{ij}^{hb}}{r_{ij}^{10}}\right)\delta_{ij}^{hb}\right\} \tag{2.1}$$

in which

$$A_{ij}=C_{ij}r_{min}^{6}/2 \quad \text{and} \quad C_{ij}=\frac{3/2e\left(\hbar/m^{1/2}\right)\alpha_i\alpha_j}{\left(\alpha_i/N_i\right)^{1/2}+\left(\alpha_j/N_j\right)^{1/2}} \quad . \tag{2.2}$$

In (2.1) and (2.2), m is the electron mass, δ_{ij} is 1 when the atoms can form hydrogen bonds and 0 in the rest of the cases, q_i, q_j are the atomic charges, D is the dielectric constant, r_{ij} is the sum of van der Waals radii for atoms i and j and is different for each pair of atoms, α_i are the atomic polarizabilities and are obtained experimentally, and N_i is the effective number of electrons for each atom type.

The use of experimental α_i and r_i gives a good description of the interaction between atoms. Furthermore, each different interaction (dispersion, coulomb, hydrogen bonding) is given by a different term in (2.1). One can thereby evaluate which one is the most important contribution to the cluster binding energy. The main disadvantage of these types of potentials is the lack of parameters specifically for radicals. In open shell systems, the unpaired electron generates electronic distributions very different from stable molecules. As α_i and r_i depend strongly on the charge distribution, these parameters will be different than for stable molecules. This difficulty can be partially overcome by calculating *ab initio* the atomic partial charges. The final method is then a combination of *ab initio* and atom-atom potential calculations.

To apply (2.1) to the cluster excited state is yet a more difficult task, due to the absence in the literature of α_i, r_i and q_i for excited state atoms. Again, the molecular excited state geometry and atomic partial charges (q_i) can be calculated using *ab initio* procedures. But polarizabilities and van der Waals radii have to be estimated. The procedure we have employed is as follows: a first calculation is performed in both cluster states, using the correct geometries and charges for ground and excited electronic states, but ground α_i and r_i values for both states. This leads to a correct geometry for the ground state cluster, and to an approximate geometry for the excited state cluster. The calculation is good enough to allow the assignment of the correct structure to the cluster. Then excited state α_i and r_i are modified until the calculated bare radical to cluster shift is achieved for a particular chosen cluster ((2.1), radical$(Ar)_1$). These radical potential parameters are used for the rest of the clusters without further modifications. The cluster used for this purpose is the one formed with one atom of Ar, due to its simplicity and the absence in the Ar atom of any charge (leading in (2.1) only to the dispersion interaction contribution).

In this Sec. we discuss CH_3O clusters with four nonpolar solvents: Ar, N_2, CH_4, and CF_4. The structure for each cluster is presented, and we analyze the nature of the forces responsible for cluster formation.

In Fig. 1, the CH_3O 3^1_0 spectrum, together with $CH_3O(X)_1$ $\overline{3}^1_0$ (X = Ar, N_2, CH_4, and CF_4) spectra, are shown. As can be seen, the cluster spectra are very simple: most of the solvents have only one peak for each cluster stoichiometry. All the cluster spectra present a red shift compared with the bare radical spectrum (see Table 1). This means that the chromophore excited state is better solvated, the cluster is more tightly bonded in this state than in the ground state. For one of the clusters ($CH_3O(N_2)_1$) a second peak is very evident and can be assigned as a cluster vibration, possibly the stretching mode.

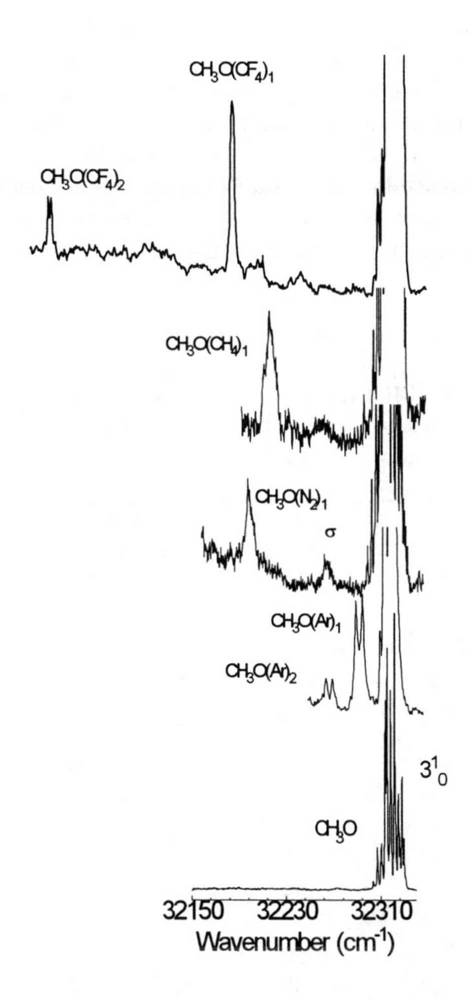

Fig. 1. LIF spectra of CH_3O 3^1_0 (lower trace) and $CH_3O(X)$, X = Ar, N_2, CH_4, and CF_4. Notice the low intensity of the cluster spectra compared with the bare molecule. The estimated rotational temperature for the bare molecule is ca. 5 K.

The results from the atom-atom potential calculation show two possible isomers for each cluster and more than thirty different isomers for the $CH_3O(CF_4)_2$ cluster (see Fig. 2). A pattern arrives for cluster structure of all these clusters. For each cluster, the solvent molecule can be placed either close to the hydrogens (Isomer 1 for all clusters) or attached to the C-O electronic cloud (Isomer 2).

As pointed out earlier, only a single origin feature is present for the cluster spectra, so only one of the isomers is formed (or at least detected). In principle we can assume that the spectrum can be assigned to the most stable Isomer for èach cluster (if the energy difference between isomers is large enough). In Table 1 we present the comparison between experimental and calculated shifts, together with the calculated binding energies for the different cluster isomers. For the most tightly bonded cluster ($CH_3O(CF_4)_1$) the large difference in ground state binding energies between isomers is enough to consider that only the most stable one is formed. Furthermore, the barrier for the isomerization process should prevent the less stable Isomer from interconverting to

Table 1. Calculated binding energies, shifts, and isomerization barriers for all the clusters studied. All quantities are given in cm^{-1}. The excited state parameters (polarizabilities and van der Waals radii) are adjusted for $CH_3O(Ar)_1$ Isomer 2 to match the experimental results, as described in the text. The corrected parameters are then used "as is" for the rest of the excited state calculations.

CH_3O/X	Isomer	E_B (cm^{-1})	E^*_B (cm^{-1})	Shift (cm^{-1}) Exp.	Shift (cm^{-1}) Calc.	Barrier (cm^{-1})
Ar	1	170	172		-2	20
	2	190	219	-29	-29	
$(Ar)_2$[a]	1	444	480		-36	88
	2	457	513	-55	-56	
N_2	1	270	278		-8	150
	2	265	354	-124	-89	
CH_4	1	245	250		-5	64
	2	261	298	-107	-27	
CF_4	1	371	380		-9	500
	2	560	807	-141	-247	

[a]E_B for Ar dimer = 99 cm^{-1}; E_B = ground state binding energy; E^*_B = electronic excited state binding energy; Barrier = calculated potential energy barrier for the isomerization process; Shift = Cluster spectral shift = $E_{mole} - E_{clust} = E_B - E^*_B$

the more stable one.

For the rest of the clusters, both isomers are close enough in energy that one can not *a priori* rule one of them out. We use two different tests to discern which isomer leads to the experimental spectrum. One is the comparison between experimental and calculated shifts and the other is a comparison between simulated and experimental rotational contours. The first test allows us to assign the $CH_3O(N_2)_1$ and $CH_3O(Ar)_2$ spectra to their respective Isomer 2 structures (see Table 1). For $CH_3O(CH_4)_1$ neither of the isomer calculated shifts is close enough to the experimental value to ensure a safe

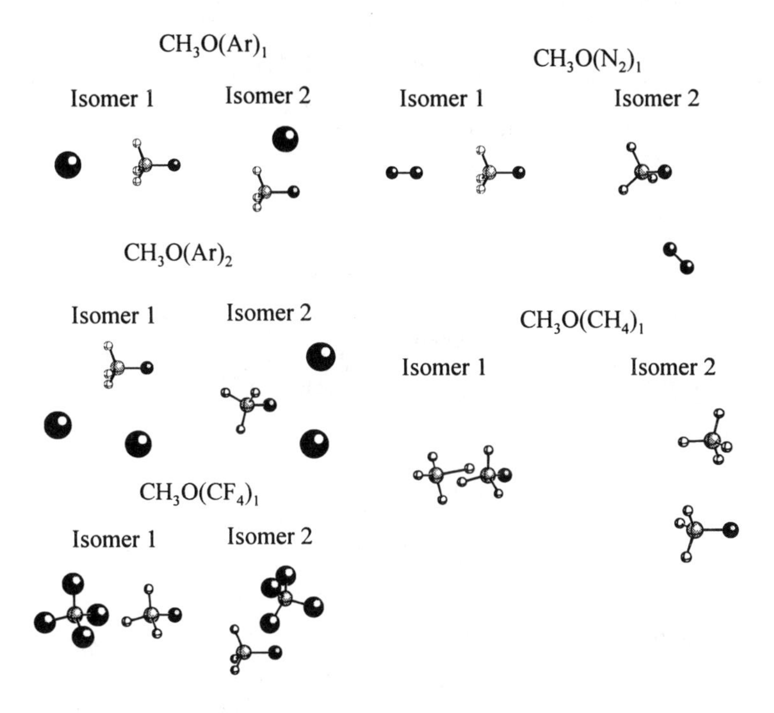

Fig. 2. Calculated structures for the CH_3O clusters studied in this work. See Table 1 for binding energies.

assignment.

To assign CH_4 and Ar cluster spectra, a rotational contour simulation has been carried out using the calculated cluster geometries given in Fig. 2. The comparison between experimental and simulated rotational contours is shown in Fig. 3 for $CH_3O(X)_1$, X = Ar, N_2, and CH_4. As can be seen, only the Isomer 2 simulation for all the clusters is in good agreement with the experimental results. The comparison reinforces the earlier assignment for $CH_3O(N_2)_1$ and $CH_3O(Ar)_1$ and allows us to identify $CH_3O(CH_4)_1$ Isomer 2 as the correct minimum energy structure.

The calculated $CH_3O(CH_4)_1$ spectral shift is surprisingly much smaller than the experimental value. This lack of agreement is not found in similar systems containing an open-shell species and CH_4 as a solvent. For example, the calculated shift of -35 cm^{-1} for benzyl radical$(CH_4)_1$ matches very well the experimental value of -29 to -39 cm^{-1} [16]. The calculated $NCO(CH_4)_1$ shift (see Table 2) is also in good agreement with the experimental value: -64.4 cm^{-1} compared with -58.5 cm^{-1}. A good explanation of why

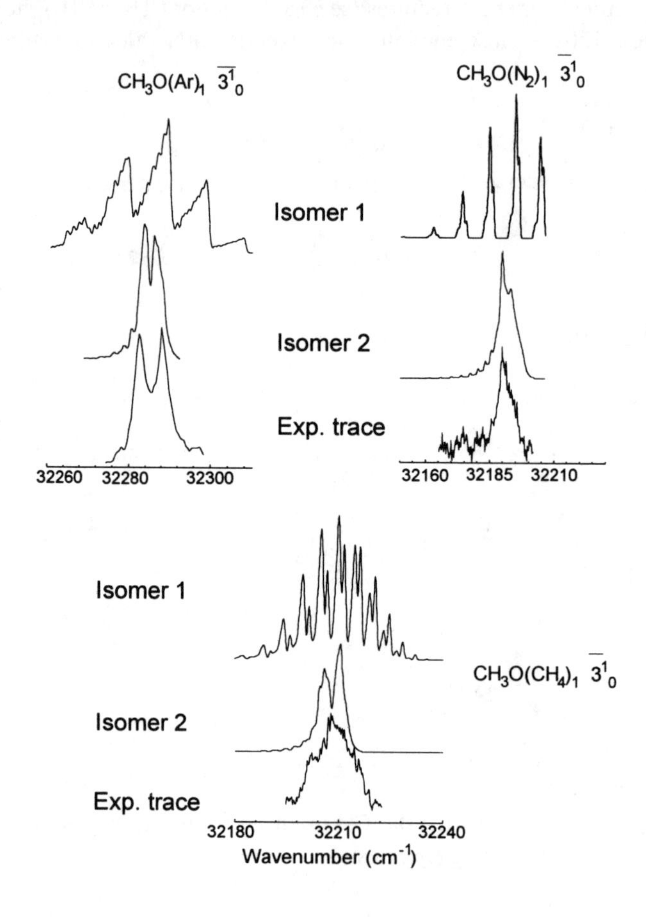

Fig. 3. Comparison between simulated and experimental rotational contour for CH_3O clusters with Ar, N_2, and CH_4. The comparison allows one to assign the experimental spectrum to one of the isomers.

the atom-atom potential calculation fails to predict the $CH_3O(CH_4)_1$ experimental shift is the existence of an interaction that is present in this system, but not in the benzyl radical$(CH_4)_1$ or $NCO(CH_4)_1$ systems, and is not taken into account in [18]. The most attractive possibility is the existence of an incipient chemical reaction between CH_3O

and CH_4 in the CH_3O excited electronic state. In fact, Wantuc et al. [21] found the existence of a reactive process between excited state CH_3O and CH_4, but no reaction for the CH_3O ground state. So probably there is an important chemical interaction between CH_3O (A) and CH_4 that stabilizes the cluster in the excited state, leading to a larger than expected shift. We can consider $CH_3O(CH_4)_1$ a pre-reactive cluster in the excited electronic state of CH_3O.

The main interaction can be found for the rest of the CH_3O 1:1 clusters by analyzing the contributions from the different terms in (2.1). For example, forces in the $CH_3O(Ar)_1$ system are dispersive, and there is no contribution from the coulomb term. This is the weakest bound cluster of the group studied and is also probably highly dynamic. The barrier between the two possible isomers for $CH_3O(Ar)_1$ is only ca. 20 cm^{-1}. If the cluster has only a small internal energy (perhaps even the zero point energy), the Ar atom can move around the radical almost freely.

As other interactions became important and the existing ones strengthen, the cluster becomes more rigid. For example, the N_2 molecule presents a quadrupolar moment that increases the binding energy significantly, and a coulomb interaction due to the partial charges in the nitrogen atoms must also be considered. The result is a stronger solute-solvent bond and a larger barrier between minima on the potential energy surface. The minima become more defined and the potential energy surface loses part of it flatness. The extreme of this development for these clusters is found for $CH_3O(CF_4)_1$. The large charge on the fluorine atoms increases the binding energy, leading to a calculated value ca. $\approx$3 times the one for $CH_3O(Ar)_1$. Here the different isomers are represented by deep local minima on the potential energy surface with such a large barrier between them that the cluster would need an extra energy close to the cluster binding energy to isomerize from one minimum to another.

3 A and B States of NCO and its Clusters

NCO is a linear triatomic molecule with an open shell structure containing 15 electrons. The ground and first two excited states of NCO have electronic configurations as $...(1\pi)^4(\sigma)^2(2\pi)^3$, $...(1\pi)^4(\sigma)^1(2\pi)^4$, and $...(1\pi)^3(\sigma)^2(2\pi)^4$. The rovibronic structure of NCO is notoriously complicated: it arises from spin-orbital interactions, the Renner-Teller effect, Herzberg-Teller coupling, Fermi resonance, and other perturbations [22-28].

LIF spectra, fluorescence lifetime, and wavelength range are recorded for the transition of NCO and its clusters to A and B excited state levels in order to understand more about the vibronic structure of NCO, to study the geometry and binding energy of NCO clusters, and to compare different fluorescence decay behavior of NCO and its clusters. The hole burning technique is used to verify vibronic transitions from the same ground state energy level. Potential energy surface calculations and other computational techniques are employed to determine cluster geometry and binding energy in the X, A, and B states of NCO.

3.1 Fluorescence Spectra of NCO

Supersonic jet fluorescence excitation (FE) spectra of NCO are recorded for both A-X and B-X transitions. The spectra are much simpler compared to those probed previously [22-40] due to intensive cooling of the NCO radical in the jet expansion.

The FE spectrum of the NCO A-X transition is shown in Fig. 4. The three strong features are assigned as the (000)←(000), (010)←(010), and (010)←(000) bands of the

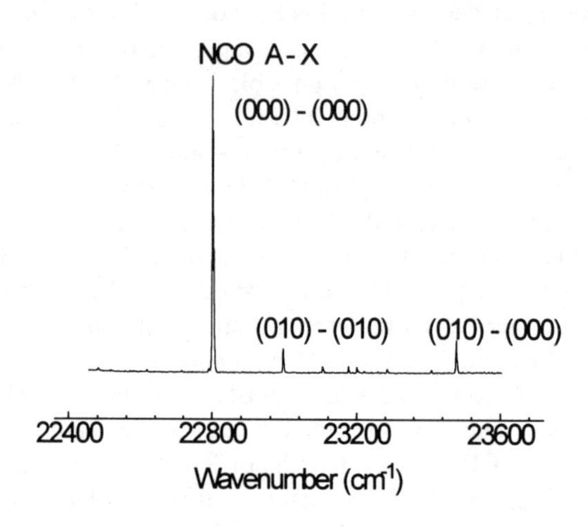

Fig. 4. FE spectrum of the NCO $A^2\Sigma^+ \leftarrow X^2\Pi_{3/2}$ transition.

$A^2\Sigma^+ \leftarrow X^2\Pi_{3/2}$ transition. No bands of the $A^2\Sigma^+ \leftarrow X^2\Pi_{1/2}$ transition are observed.

Spectra of the NCO B←X transition are more complicated due to the Renner-Teller effect, Fermi resonance, and the vibronic mixing between the A and B states [22, 23, 25, 26]. In Figs. 5 and 6 we present FE and hole burning spectra for the (000)←(000) and (100)←(000) band regions of the $B^2\Pi \leftarrow X^2\Pi$ transition. In the hole burning experiment, the probe laser is set at 22,800 cm^{-1}, the origin of the $A^2\Sigma^+ \leftarrow X^2\Pi_{3/2}$ transition. All the features appearing in hole burning spectra should arise from a common source, the $X^2\Pi_{3/2}(000)$.

Vibronic structure in the $B^2\Pi(100) \leftarrow X^2\Pi(000)$ transition region is reasonably well understood [22,26]. In Fig. 6, features b, c, and d at 32,761.4, 32,816.0, and 32,890.1 cm^{-1} can be assigned to the μ-$B^2\Pi_{3/2}(020) \leftarrow X^2\Pi_{3/2}(000)$, $B^2\Pi_{3/2}(100) \leftarrow X^2\Pi_{3/2}(000)$, and κ-$B^2\Pi_{3/2}(020) \leftarrow X^2\Pi_{3/2}(000)$ subbands, respectively, in agreement with Neumark et al. [26]. The lifetimes of these $B^2\Pi_{3/2}$ energy levels are all shorter than 10 ns, also in agreement with previous work [28a]. Features a and e at 32,690 and 32,901 cm^{-1}, respectively, may be A← X vibronic transitions arising from the $X^2\Pi_{3/2}(000)$ level.

The vibronic structure of the $B^2\Pi_{3/2}(000) \leftarrow X^2\Pi_{3/2}(000)$ region has been studied by Dixon et al[23] and Dagdigian et al[25] at high temperature. Dixon et al[23] have extrapolated a $B^2\Pi \leftarrow X^2\Pi$ origin transition energy of 31,751 cm^{-1}. The vibronic bands of the same region obtained in our supersonic jet expansion appear somewhat different from those obtained at higher temperature.[23,25] As is shown in Figure 3a, hole burning indicates that the two strong features at 31746.0 and 31768.5 cm^{-1} share the same ground state level, $X^2\Pi_{3/2}(000)$. These two strong bands can not be simply A vibronic bands since, within 650 cm^{-1} to the red of these two features, no other A vibronic feature is found which has an intensity comparable to these. Most likely they arise from two transitions originating at $X^2\Pi_{3/2}(000)$ and terminating at two vibronic levels that are generated by the mixing of $B^2\Pi_{3/2}(000)$ with nearby A vibronic levels. These features gain intensity mainly from the $B^2\Pi_{3/2}(000)$ level.

The feature at 31,768.5 cm^{-1} has apparently more B state character, which is indicated by the following facts: 1. the fluorescence lifetime of this transition is only 90 ns comparing to 169 ns for that of the feature at 31,746 cm^{-1} and 2. the fluorescence intensity of this transition is severely attenuated by the application of an A state band pass filter in front of the phototube (see Fig. 7).

Another band at 31,878.0 cm^{-1} has relatively high intensity with a lifetime comparable to that of the 31,746 cm^{-1} band. This feature may also be an A state

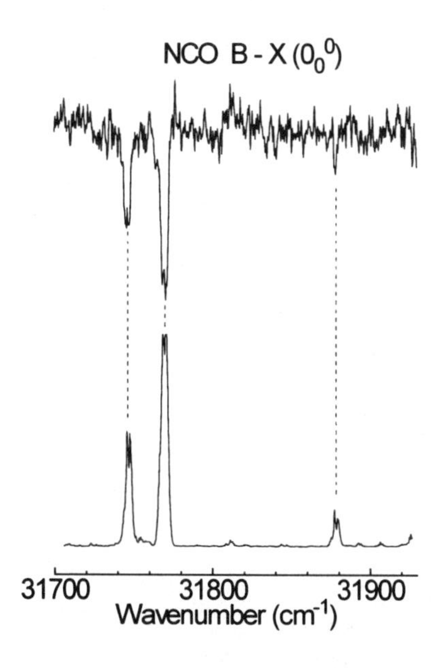

Fig. 6. FE and hole burning spectra of NCO for the $B^2\Pi_{3/2}$ (000)$\leftarrow X^2\Pi_{3/2}$ (000) transition region. The probe laser is set for $A^2\Sigma^+ \leftarrow X^2\Pi_{3/2}$ at 22,800 cm^{-1}.

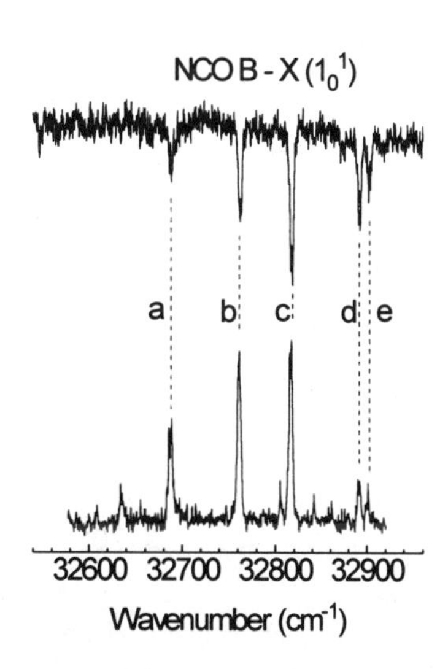

Fig. 5. FE and hole burning spectra of NCO in the $B^2\Pi_{3/2}$ (100)$\leftarrow X^2\Pi_{3/2}$ (000) transition region. The probe laser is set for the $A^2\Sigma^+ \leftarrow X^2\Pi_{3/2}$ transition at 22,800 cm^{-1}.

dominant vibronic band gaining intensity by mixing with the B state 0^0 through vibronic coupling.

3.2 NCO Clusters: Spectra and Calculation

1:1 cluster spectra. FE spectra are recorded for NCO clusters with one Ar, N_2, CH_4 or CF_4 molecule in both the NCO A← X and B← X transition regions. The A←X transition bands are observed only for NCO$(N_2)_1$ and NCO$(CF_4)_1$, as shown in Fig. 8. In the NCO B← X transition origin region, single features are observed for all the four clusters, as shown in Fig. 9. We assign this single in each cluster as arising from the 31,768.5 cm^{-1} bare radical feature for two reasons: 1. clustering with a nonpolar solvent like Ar, N_2, CH_4, and CF_4 should not remove the vibronic coupling between the A and B electronic states; and 2. cluster formation will certainly change the system dynamics and greatly shorten the lifetimes of the A dominant vibronically coupled levels and in general not perturb the B dominant level lifetime as much. Thus, the B dominant feature should survive in the cluster spectra, while the A dominant feature should not.

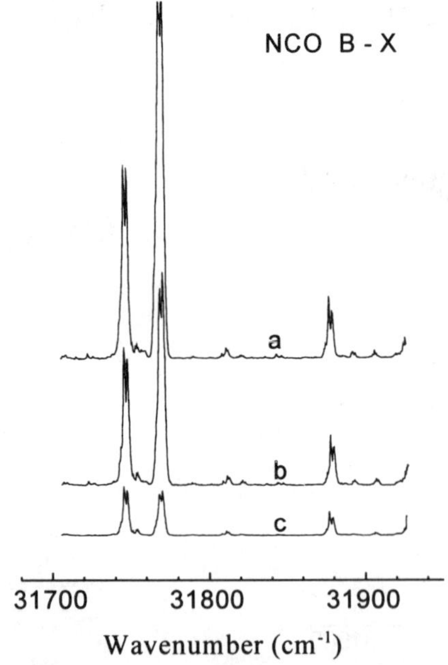

Fig. 7. Fluorescence signal collected for the NCO transition in the $B^2\Pi_{3/2}$ (000)←$X^2\Pi_{3/2}$ (000) region using a) UV30, b) L37, and c) L42 colored filters in front of the PMT. The transmission ranges for these filters are to the red of 280, 350, and 410 nm, respectively.

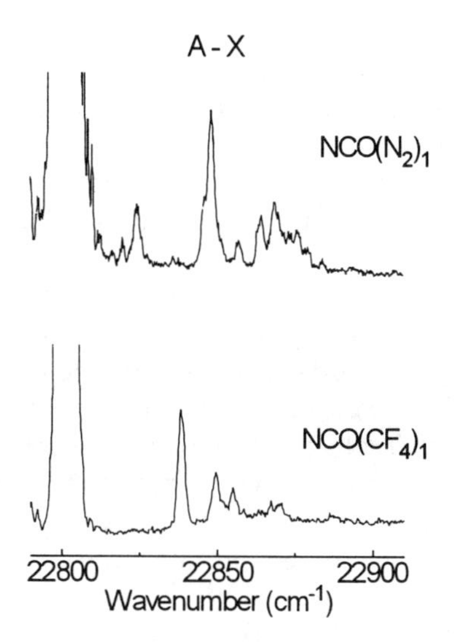

Fig. 8. FE spectra of the A←X transition of NCO clusters with N_2 and CF_4.

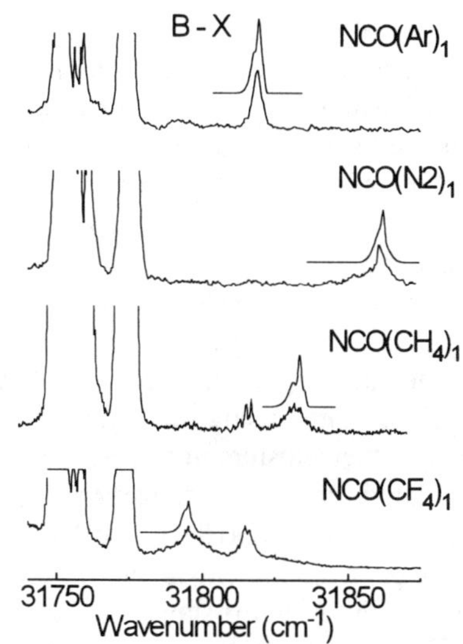

Fig. 9. FE spectra of the B←X transition of 1:1 NCO clusters. Simulated rotational envelopes are also shown for all cluster origin spectra.

An interesting phenomenon observed for the B←X transition of these NCO clusters is that their fluorescence can only be detected to the red of 390 nm, and all possess a decay lifetime within the range of 250 to 320 ns, which is much longer than that of the B state bare NCO radical. This will be discussed separately below.

1:1 cluster calculations. To extract information about the geometries and binding

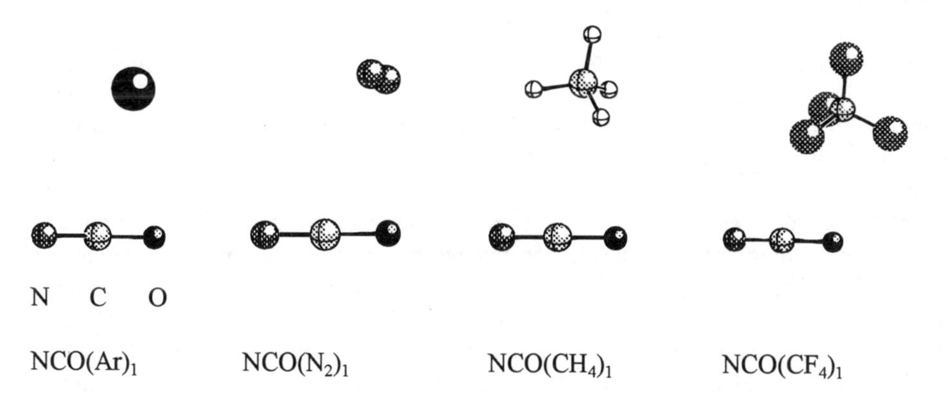

Fig. 10. Ground state geometry of NCO clusters obtained from potential energy calculations. Excited A and B state geometries would look identical.

energies of these clusters in different electronic states, potential energy calculations are performed. The atomic charges of ground and excited state NCO required for these calculations are obtained from *ab initio* calculations using a Gaussian 94 package [41]. The calculated binding energies are listed in Table 2, and the ground state geometries of the clusters are shown in Fig. 10. All ground state clusters have only one minimum energy geometry with a solvent molecule attached to the side of C=O bond in NCO.

For the B state NCO cluster potential energy surface calculations, we adjust the α_i and r_{ii} values of the atoms in NCO to calculate the $NCO(Ar)_1$ binding energy, so that the calculated spectroscopic shift matches exactly the experimental shift for the B←X transition. Then using the adjusted α_i and r_{ii} values, we calculate the binding energy for the other clusters in the B state. As is listed in Table 2, spectroscopic shifts thus calculated for the B-X transitions agree with the experimental values reasonably well in the case of $NCO(CH_4)_1$ and $NCO(N_2)_1$, but poorly in the case of $NCO(CF_4)_1$. The rather poor agreement in the case of $NCO(CF_4)_1$ may be due to vibronic coupling between the A and B states and its changes for different clusters.

The calculated geometries for these 1:1 clusters in the B electronic state are similar to those for the ground state.

Similar potential energy calculations are performed for $NCO(N_2)_1$ and $NCO(CF_4)_1$ in the A electronic state. Since the A-X transitions of $NCO(Ar)_1$ and $NCO(CH_4)_1$ are not observed, we sacrifice the $NCO(N_2)_1$ data for fitting the A state potential energy surface. α_i and r_{ii} of N, C, and O atoms in the NCO A state are adjusted so that the calculated spectroscopic shift of the A-X transition of $NCO(N_2)_1$ matches the experimental value of 22.6 cm^{-1}. With these adjusted parameters, the binding energies and geometries of the other clusters in the A state are calculated. As is listed in Table 2, the calculated spectroscopic shift for $NCO(CF_4)_1$ is quite close to the experimental value. For $NCO(Ar)_1$ and $NCO(CH_4)_1$, the calculations indicate that the binding energies for the A state are very close to those for the ground state. These small calculated shifts suggest that the cluster transition for A←X of $NCO(Ar)_1$ and $NCO(CH_4)_1$ may lie under the NCO origin feature.

The geometries of A state clusters obtained by these calculations are again very close to those of the ground state clusters. For $NCO(CF_4)_1$ another minimum energy position is found that places the CF_4 molecule near the N=C bond side of NCO and is roughly 20 cm^{-1} higher in energy. The barrier between these two A state minima is only a few wavenumbers. Probably only one structure is present for our experimental conditions.

Band contours are simulated for the B←X transition bands of the NCO clusters using rotational constants derived from cluster geometry obtained in the above calculations. Spectroscopic resolution used for the simulations is 0.8 cm^{-1}. Rotational bands for all clusters are simulated with a mixed band type since the electronic transition dipole moments are not parallel to any of the rotational axes. As can be seen from Fig. 9, all simulated band shapes match reasonably well with experimental ones. This again shows that the potential energy calculations performed in this work are qualitatively correct in predicting cluster geometry.

To estimate the contribution from the coulomb interaction, potential energy calculations are performed for all the clusters without adjustment in α_i and r_{ii} values. From the data in parentheses in Table 2, we can see that the excited state binding energies thus calculated are all within a few wavenumbers of the ground state ones. This clearly demonstrates that the van der Waals interaction contributes most to these changes in binding energies of the clusters at different electronic states.

1:n Clusters. With relatively higher backing pressures and higher solvent molecule concentration in the backing gas, we have observed cluster fluorescence signals for all the four solvents. Such spectra are observed for both A and B electronic states. The

Table 2. Binding energy (E_B) and spectroscopic shift (δ) of NCO clusters obtained from potential energy calculations (cm^{-1}).*

State		Solvent			
		Ar	CH_4	N_2	CF_4
X	E_B	187.7	248.5	230.5	418.4
A	E_B	176.2(188.6)	230.3(250.9)	207.9(227.3)	363.7(419.4)
	δ_{calc}	11.5	18.2	22.6	54.7
	δ_{exp}	----	----	22.6	37.4
B	E_B	142.0(185.9)	184.1(247.0)	168.3(229.1)	308.1(427.4)
	δ_{calc}	45.7	64.4	62.2	110.3
	δ_{exp}	45.7	58.5	88.0	21.4

* For B state, the α_i and r_{ii} parameters of NCO are adjusted for $NCO(Ar)_1$ to match experimental results.

For A state, the α_i and r_{ii} parameters of NCO are adjusted for $NCO(N_2)_1$ to match experimental results.

E_B values in parentheses are calculated without adjusting α_i and r_{ii} values.

spectroscopic features observed for NCO clusters containing multiple solvent molecules are all broad, and have fluorescence decay lifetimes and wavelengths similar to the 1:1 NCO clusters with the four solvents. The spectroscopic shifts of these larger clusters are somewhat surprising. As is shown in Figs. 11, 12 and 13, the transitions for these clusters can have either red or blue shift with réspect to the NCO transitions. The A-X transition bands of $CO(N_2)_n$ and $NCO(CF_4)_n$ are slightly to the blue of the 1:1 cluster bands. At high pressure and with high solvent concentration, the 1:1 cluster bands of NCO with CF_4 are submerged in their 1:n cluster bands, as is shown in Fig. 12.

Although we have not observed the A←X transition bands of $NCO(Ar)_1$ and $NCO(CH_4)_1$ to the blue of the NCO A←X origin, with 500 psi pressure and with high concentration of solvent we do observe weak, broad features to the red of the NCO A←X transition origin under these conditions, as shown in Fig. 11. These weak, broad bands are probably the 1:n cluster bands of NCO with Ar and CH_4.

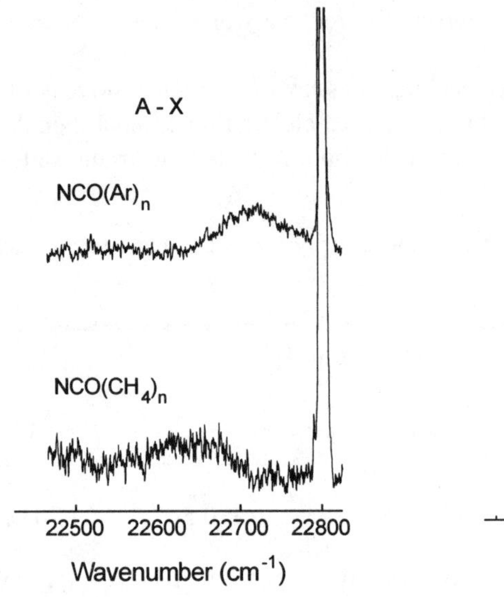

Fig. 11. FE spectra of A←X transition of $NCO(Ar)_n$ and $NCO(CH_4)_n$. The broad features to the red of the NCO $A^2\Sigma^+ \leftarrow X^2\Pi_{3/2}$ at 22,800 cm^{-1} belong to these clusters. Pure argon and 10% CH_4 in helium are used as backing gases for $NCO(Ar)_n$ and $NCO(CH_4)_n$, respectively.

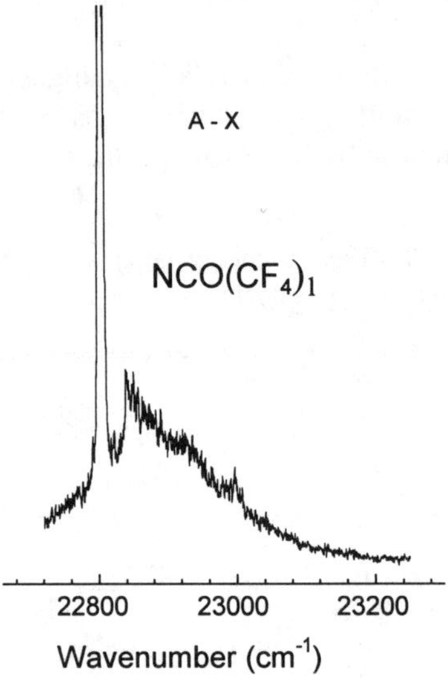

Fig. 12 FE spectra of the A←X transition of $NCO(CF_4)_n$. The broad feature to the red of the NCO $A^2\Sigma^+ \leftarrow X^2\Pi_{3/2}$ at 22,800 cm^{-1} belongs to these clusters. 0.5% CF_4 in helium is used as backing gas.

All 1:n clusters in the NCO B ← X transition region show a red shift in their spectra. Fig. 13 shows features of NCO clusters with Ar and CF_4 in this region. In the B(000)←X(000) transition region of NCO, in addition to the strong $NCO(Ar)_1$ and $NCO(CF_4)_1$ peaks observed at 31,814.2 and 31,789.9 cm^{-1}, respectively (marked with an * in Fig. 13), broad features appear to the red of the NCO origin peaks. These red shifted broad peaks belong apparently to $NCO(Ar)_n$ and $NCO(CF4)_n$. Note that in Fig. 13, the sharp features of the NCO radical show low intensities due to the insertion of an L42 filter (transmission of $\lambda > 410$ nm) in front of the phototube detecting fluorescence. The very low intensity ratio of the 31,768.5 cm^{-1} feature versus that at 31,746.0 cm^{-1} is due to both the insertion of the L42 filter and a 50 ns delay in fluorescence signal acquisition for the purpose of avoiding scattered light interference.

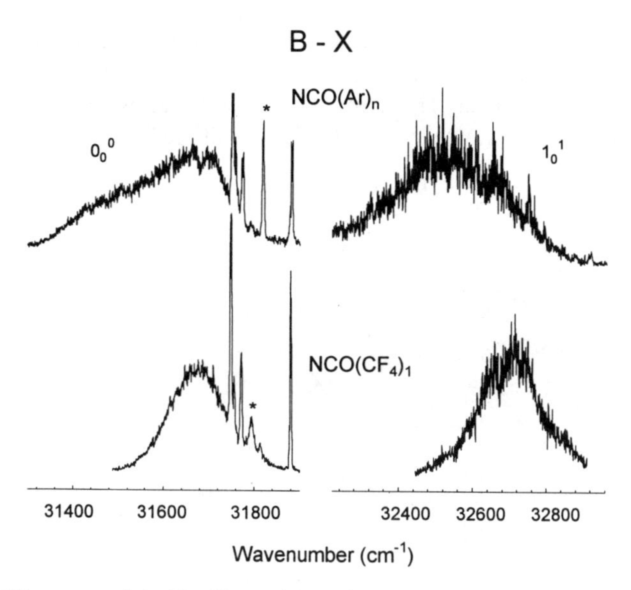

Fig. 13. FE spectra of the B←X transition of 1:n NCO clusters. The broad features belong to these clusters. Pure argon and 1% CF_4 in helium are used as backing gases for $NCO(Ar)_n$ and $NCO(CF_4)_n$ detection, respectively. Features of 1:1 clusters are labeled with asterisks.

Note that even in the absence of a B(100)←X(000) transition observed for the 1:1 clusters, features to the red of B(100) ←X(000) for NCO are observed with similar intensity and broadness as those observed to the red of the NCO B(000) ←X(000) transition (see Fig. 13). This will be further discussed in the next section.

We do not have a good explanation for the diverse behavior of the spectroscopic shifts for $NCO(X)_n$ (n > 1) clusters. Previous calculations for $CH_3O(X)_n$ (n > 1) clusters gave semiquantitative prediction of the spectroscopic shifts. The calculated results for $NCO(X)_n$ (n > 1) clusters do not match the experimental results: contrary to the experimental results, calculations predict blue shifts for all the cluster transitions of $NCO(X)_n$ (n > 1).

Decay Pathways. Previous studies [25, 28a], as well as our observations, have demonstrated that the lifetime of the NCO radical in the B state vibronic levels shortens dramatically as energy increases. Dissociation products are also detected by Neumark et al. [26]. Alexander and Werner [42] propose that a crossing between the B and A states of NCO exists, and that another crossing between $A^2\Sigma^+$ and the repulsive $^4\Sigma^-$ states also can be suggested. Both crossings are within a few hundred wavenumbers of the $B^2\Pi(000)$ level. The eventual crossing of the NCO radical from the A or B state potential surface to the dissociative potential surface $^4\Sigma^-$ is the cause of the unusual lifetime shortening and the formation of dissociation products in the B vibronic levels of the NCO radical near and above the crossing intersections.

All B state clusters fluoresce only to the red of 390 nm (25,641 cm^{-1}) and have a decay lifetime in the range of 250 to 320 ns compared to a lifetime of 90 ns for the B state origin of the bare radical. This strongly suggests that the B state NCO clusters have a different decay pathway than does the B state of the bare NCO radical itself.

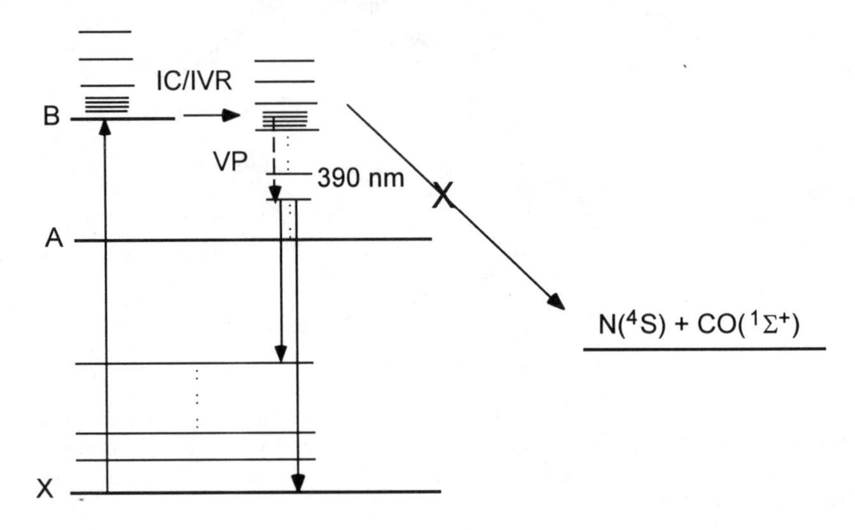

Fig. 14. A schematic diagram of the decay pathway of B state NCO clusters. Clusters at B vibronic levels first internally convert (IC) to nearby A vibronic levels, and then undergo IVR/VP (intracluster vibrational redistribution/vibrational predissociation). NCO fragments are formed at A state vibronic levels lower than ca. 25,640 cm^{-1} in energy, and subsequently fluoresce.

In the case of 1:1 NCO clusters at the $B^2\Pi_{3/2}(000)$ dominant level, we actually observe a fluorescence wavelength range and decay lifetimes that match those radiative decay of A state vibronic levels lower in energy than 25,640 cm^{-1}. A possible decay pathway for these B state NCO clusters is depicted in Fig. 14. After an NCO cluster is excited to a B vibronic level, it first internally converts (IC) to an A vibronic level, then undergoes further IVR to populate van der Waals modes and lower A state vibronic levels, and finally undergoes vibrational predissociation (VP), generating a translationally hot solvent molecule and a bare NCO A state radical. The NCO radical is in an A vibronic state with an energy lower than 25,640 cm^{-1} (the highest possible level is A(040) level). The A vibronic state NCO fragment then fluoresces with a decay lifetime longer than 250 ns. This fluorescence is actually what we detected for the $B \leftarrow X$ transition of NCO clusters. The IVR/VP rate is apparently fast and prevents NCO dissociative decay through crossing to $^4\Sigma^-$ state.

The energy difference between the B(000) level cluster and the 25,640 cm^{-1} level is more than 6,100 cm^{-1}. In the decay process, this energy is distributed into translational and rotational energy for both the NCO fragment and solvent molecule.

As is shown in Fig. 5, the NCO feature at 31,746.0 cm^{-1} has an intensity not much lower than that of the $B^2\Pi_{3/2}(000)$ dominant band at 31,768.5 cm^{-1} however, only one

1:1 cluster band is observed for each NCO cluster. A possible explanation is that the lifetime of the cluster in the A state dominant vibronic level corresponding to the 31,746 cm^{-1} feature is so short that the cluster at this level decays through IVR/VP on a sub-picosecond time scale. The A vibronic level cluster states at this energy may thus be lifetime broadened and may not be observed. For the cluster in the $B^2\Pi(000)$ dominant level corresponding to the 31,768.5 cm^{-1} feature, a long lifetime arises due to the slower IC/IVR process. Once the cluster crosses to an A vibronic level, it experiences the same IVR/VP and subsequent fragment fluorescence. Thus the increased lifetime of the B dominant level cluster enables us to detect a reasonably well defined spectroscopic feature for the cluster at ca. 32,000 cm^{-1}.

Clusters of $NCO(X)_n$ ($n > 1$) have decay lifetimes and wavelength ranges similar to that of the 1:1 clusters. These larger clusters at the B(000) dominant level probably experience a similar decay scheme by first crossing to nearly degenerate A vibronic levels, and then decaying through IVR/VP to generate A state bare NCO radicals.

For a 1:1 cluster at the B(100) level (B(000) + ca. 1000 cm^{-1}), the cluster can decay to NCO B(000) and a ground state solvent molecule. Since the absorption is broad and weak, no fluorescence excitation spectra are observed, as is also the case for the A state cluster level at 31,746.0 cm^{-1}. Larger clusters at the B(100) level do not totally fragment at this energy due to slower VP and thus are weakly observed through the B(000) $\rightarrow$A(lmn)$\rightarrow$IVR/VP pathway.

3.3 Summary

Using LIF techniques, we have investigated NCO and its clusters with the nonpolar solvents Ar, N_2, CH_4, and CF_4 for transitions from the $X^2\Pi_{3/2}(000)$ level to vibronic levels of both the A and B electronic states. Transitions observed in this work for the $B^2\Pi_{3/2}(100) \leftarrow X^2\Pi_{3/2}(000)$ and $A^2\Sigma^+(000) \leftarrow X^2\Pi_{3/2}(000)$ regions agree with previous work. We believe that the three bands in the region of $B^2\Pi_{3/2}(000) \leftarrow X^2\Pi_{3/2}(000)$ transition arise from the mixing of $B^2\Pi_{3/2}(000)$ and A state vibronic levels. The two bands with lower intensity and longer decay lifetime are more A dominant, while the feature at 31,768.5 cm^{-1} with the highest intensity and the shortest lifetime is more likely $B^2\Pi_{3/2}(000)$ dominant.

All observed transitions of 1:1 NCO clusters show blue shifts in their spectra; this shift indicates a smaller binding energy in the cluster excited states. Potential energy calculations are performed to estimate binding energies as well as to find cluster geometry in different electronic states. All four solvents tend to cluster with NCO at the side of the C=O bond. The change of cluster binding energy for different electronic states is mainly caused by changes in the α_i and r_{ii} potential parameters of the atoms in NCO; in another words, by changes in the van der Waals dispersion interaction.

While NCO radicals at low B state vibronic levels decay through radiative and dissociative (via crossing to a $^4\Sigma^-$ surface) pathways, the observed B state NCO clusters seem to decay through first crossing to A vibronic levels, followed by very fast IVR/VP processes. NCO fragments are formed at A vibronic levels lower than 25,640 cm^{-1} and fluoresce from there.

4 Solvation of Picolyl and Lutidyl Radicals

Isolation and detection of radical species is oftentimes difficult in all phases of matter. Using supersonic expansions we are able to isolate reactive species in the gas phase, and combined with spectroscopic techniques, we are able to study the electronic properties and reactivities of these isolated species. Two problems can be identified for assigning reactive intermediates that have not previously been detected. The first problem is to determine whether the product is actually formed from the original precursor species and is the product that is anticipated. A well–known example of this difficulty is the C_6H_5N: (phenylnitrene) *vs.* $C_5H_4CN^{\cdot}$ (cyanocyclopentadienyl) gas phase chemistry reported by a number of groups [43]. The second problem encountered arises for the determination of the electronic/vibronic energy levels of the new radicals. The combination of these two problems has limited our knowledge of the types of reactive species.

In an attempt to increase our understanding of organic radicals and their reactivity, we have isolated and identified two new radicals by employing photolysis as the technique to generate the radicals, which are then detected by either mass resolved excitation spectroscopy or fluorescence. The 3-picolyl and the 2,5-lutidyl radicals are

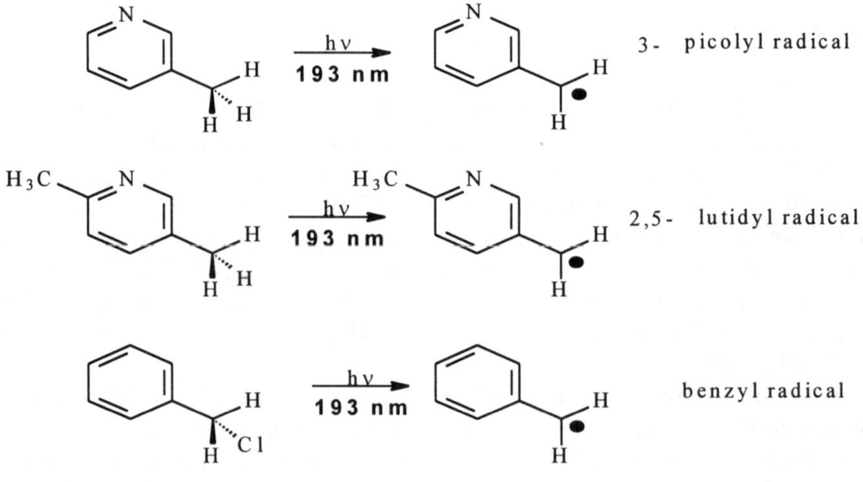

Fig. 15. Photolysis mechanism and structure for the generation and identification of the picolyl, lutidyl, and benzyl radicals.

created using the 3-methyl pyridine (3-picoline) and the 2,5-dimethyl pyridine (2,5-lutidine) as the precursor materials, respectively (see Fig. 15). The use of pyridine chromophores as the precursor to radical generation has afforded us the opportunity to examine how the nitrogen heteroatom effects the electronic properties and reactivity of benzyl and substituted benzyl radical analogs. The long history of benzyl radical studies both experimentally and theoretically has given us a great deal of information regarding this system and will therefore be used as our benchmark for comparison [44].

4.1 Picolyl Radical

Understanding the vibronic structure of the 3-picolyl radical spectrum shown in Fig. 16 has created a bit of confusion. Several features of this spectrum are worthy of note. The first is the intensity of the origin relative to all other features in the spectra. The origin peak has a relative intensity that is four to five times larger than any other spectral feature. A second observation of this spectrum that is of considerable interest is the lack of any spectral features beyond 550 cm^{-1} to the blue of the origin. Energies of $\approx$1000 cm^{-1} both to the blue and the red of the presented spectrum were scanned but neither direction showed further evidence of vibronic features for the 3-picolyl radical. This behavior is very different from that observed for benzyl the radical, which has an extensive vibronic structure. The ν_6 totally symmetric ring vibration in substituted benzene aromatic radical systems is observed with an intensity that is comparable to that demonstrated in the spectrum of the radical precursor. For example, fluorotoluenes [45] and the associated radicals [46, 47] all have weak ν_6 bands, while the methyltoluenes [46, 48] and their corresponding radicals are all observed with intense ν_6 vibrations. The picoline molecule is shown to have an intense ν_6 totally symmetric vibration, which is absent in the picolyl radical. Based upon the experimental observations for this radical we conclude that the Franck-Condon shift between the ground and first electronic states is very minimal.

Ab initio calculations for the 3-picolyl radical in both the ground and excited electronic states verify that little change in the radical structure occurs upon electronic excitation. The calculated geometries for both states are optimized using a 5×5 CASSCF algorithm with a Dunning/Huzinnaga (9s5p)/(3s2p) plus polarization basis set using the *ab initio* program HONDO 8.5 [49]. Utilizing Møller Plesset second order perturbation theory (MP2) in conjunction with the CAS calculation, we obtain an electronic transition for the 3-picolyl radical of 23,280 cm^{-1}, which is in good agreement with the experimental result of 21,965 cm^{-1}. Comparable results are obtained for the benzyl radical employing the same method. Calculations for these aromatic radicals thus become an effective tool for spectral assignments and species identification.

Using methyl-substituted pyridines as the precursor, we have the possibility of forming three distinct picolyl radicals for which we can observe the electron interactions between the nitrogen lone pair and the radical electron. We are unable to detect any pyridyl radical generated using a precursor with a methyl group in either the two or four position. Calculations, described previously, were performed on these potential radicals, 2- and 4-picolyl, to determine possible electronic effects that could substantially increase/decrease the transition energies for these unobserved radicals relative to the 3-picolyl. The calculated results of both the transition and ionization energies for these radicals are quite similar to those of the 3-picolyl calculated results.

Previous static cell gas phase experiments [50], confirmed using jet techniques and photolyzing with 193 nm light, have shown that 2-methyl pyridine rearranges to form both the 3- and 4-picolines when irradiated by photons of energy higher than 248 nm.

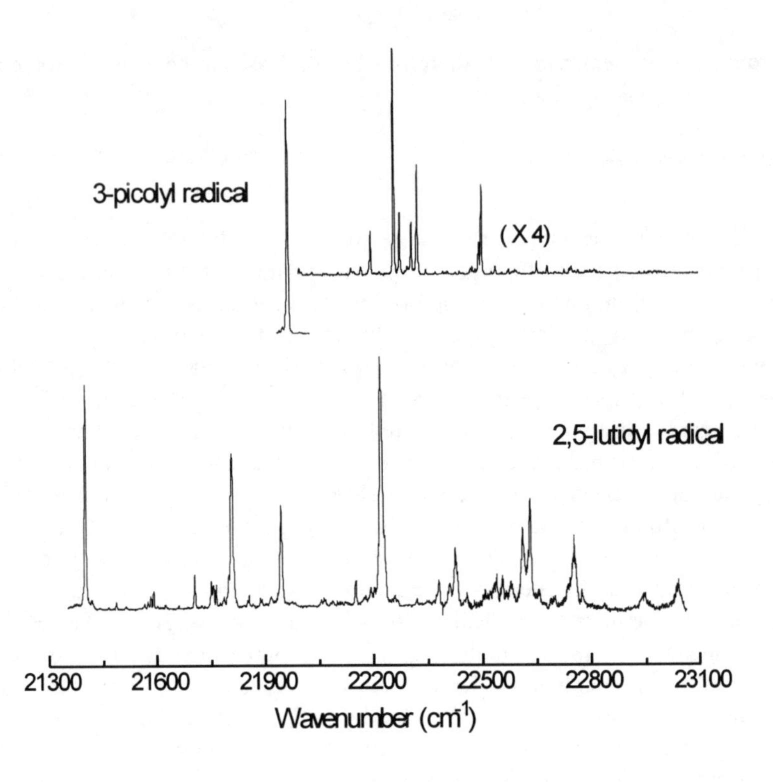

Fig. 16. Two color mass resolved excitation spectra for the 3-picolyl and 2,5-lutidyl radicals. The ionization energy is well in excess of the thresholds for both radicals. The upper trace has been enhanced so that the features to the blue of the origin are more pronounced.

Using any of the three picoline precursors we find a substantial increase in the weak background aniline signal upon photolysis with a 193 nm ArF laser. In addition to the aniline signal, cyanocyclopentadienyl is also detected with all three picoline precursors upon photolysis. The formation of these products can only result through substantial ring rearrangement during the supersonic expansion process. Because the photolysis is done with such high energy photons, and many photons may be absorbed, competing processes for product formation may exist that prevent the 2- and 4-picolyl radicals from being formed.

4.2 Lutidyl Radical

Understanding the spectrum of the 2,5-lutidyl radical is somewhat less difficult than for the picolyl described above. The spectrum, also shown in Fig. 16, for lutidyl radical exhibits many features that are sharp and well resolved. High energy vibronic transitions for the lutidyl radical start to broaden as is indicative of nonradiative decay processes at higher vibrational energy, such as internal conversion (IC), intramolecular vibrational redistribution (IVR), and intersystem crossing (ISC). Calculations on the 3-picolyl radical show that the lowest quartet state is substantially higher in energy than the first excited doublet; therefore, IC and IVR are much more likely relaxation candidates based on inferences from the picolyl work.

Table 3. Experimental values for the cluster transitions and energy shifts relative to the lutidyl radical 0_0^0. All values are given in cm^{-1}. Several values for the benzyl radical clusters are also presented as comparative shifts.

Species	origin transition (cm^{-1})	cluster shift relative to lutidyl origin (cm^{-1})	comparable shifts for benzyl radical (cm^{-1})
lutidyl radical (LR)	21,401.38		
$LR(Ar)_1$	21,378.6	-22.8	-23.9
$LR(N_2)_1$	21,354.9	-46.5	-32.4
$LR(CH_4)_1$	21,338.8	-62.5	-35.1
$LR(C_2H_6)_1$	21,302.6	-98.8	-39.5
$LR(C_3H_8)_1$	21,283.5	-117.9	-46.3
$LR(C_4H_{10})_1$	2,1272	-129.3	
$LR(C_2H_4)_1$	21,316.7	-84.6	
$LR(CF_4)_1$	21,360.7	-40.7	

Analysis of the vibronic spectrum for the lutidyl radical shows two intense peaks between 400-600 cm^{-1}, which are indicative of the symmetric stretch, $\nu_{6a,b}$, for a conjugated ring system. The intense peak near 820 cm^{-1} has to be examined more closely to determine if it is a second electronic origin or the 1_0^1 vibration. Using the benzyl radical as a comparison, one can expect to find two electronic excited states in this general region. Calculations for the excited states of this radical are not available.

Nonetheless, comparison of the 2,5-lutidine precursor spectrum with that of the 2,5-lutidyl radical suggests that the peak at 820 cm^{-1} is the 1_0^1 vibration and not a second low lying excited electronic state. The assignment of this peak as a vibration and not a second electronic state is further verified by the study of 2,5-lutidyl radical clusters, as discussed below. The three methyl benzyl radical isomers also show similar vibronic structure in which the 1^1 vibration is prominent in their spectra [49].

The formation and detection of the lutidyl clusters are accomplished using methods very similar to those described for the generation of the bare radical. The spectra shown in Fig. 17 are indicative of the clusters intensities for most clusters. Table 3 lists the relative origin shifts of the 1 to 1 radical/solvent clusters for Ar, N_2, CF_4, CH_4, C_2H_6, C_3H_8, and C_4H_{10}. These cluster spectra are all red shifted, but some of the clusters are observed with energy shifts substantially larger than those demonstrated in the benzyl radical studies [18].

2,5-Lutidyl cluster spectra at the radical 0_0^0 transition are presented in Fig. 17. The intense feature at 0_0^0 + 820 cm^{-1} (see Fig. 16) does not show any evidence of starting a second cluster transition as would be expected with a second origin for a new electronic state. Instead, what is observed are sharp cluster features built on the lutidyl radical intense transitions. The spectroscopic observation of these cluster transitions is limited by their binding energies. The N_2 cluster transitions are observed up to energies that correspond to the lutidyl symmetric stretch, ≈21,800 cm^{-1} (0_0^0 + 407 cm^{-1}) and ≈21,950 cm^{-1} (0_0^0 + 544 cm^{-1}). Further cluster transitions are not observed at the 1^1 vibration of the 2,5-lutidyl radical. This is contrary to what is observed for the lutidyl$(C_2H_4)_1$ cluster, for which the binding energy is much higher and cluster transitions are seen near all intense bare radical peaks. Given these observations, one can conclude that all the observed vibronic features in the 2,5-lutidyl spectrum are associated with a single electronic excited state. Thus, unlike the case for the benzyl radical with the D_1 and D_2 electronic states within ca. 400 cm^{-1} of each other, the lutidyl radical has only one low lying electronic excited state.

The most notable difference between the lutidyl and benzyl radical cluster spectra is observed for the ethylene cluster. The benzyl radical has been demonstrated to undergo an excited state addition reaction with C_2H_4 [16]. The 2,5-lutidyl$(C_2H_4)_1$ spectrum is very similar to the other cluster spectra, as shown in Fig. 17. Observation of a sharp cluster spectrum with vdW modes indicates that no reaction has taken place in this system in either the ground or excited state of the radical.

We have attempted to cluster the 3-picolyl radical as well, but no clusters have been observed; we believe this absence of spectra for picolyl clusters is related to relaxation dynamics in this system.

Substitution of a phenyl ring with a pyridine ring has been demonstrated to have significant effects on the radicals generated. These nitrogen containing radicals have shown both a difference in electronic structure as well as reactivity with solvent molecules, compared to comparable hydrocarbon systems.

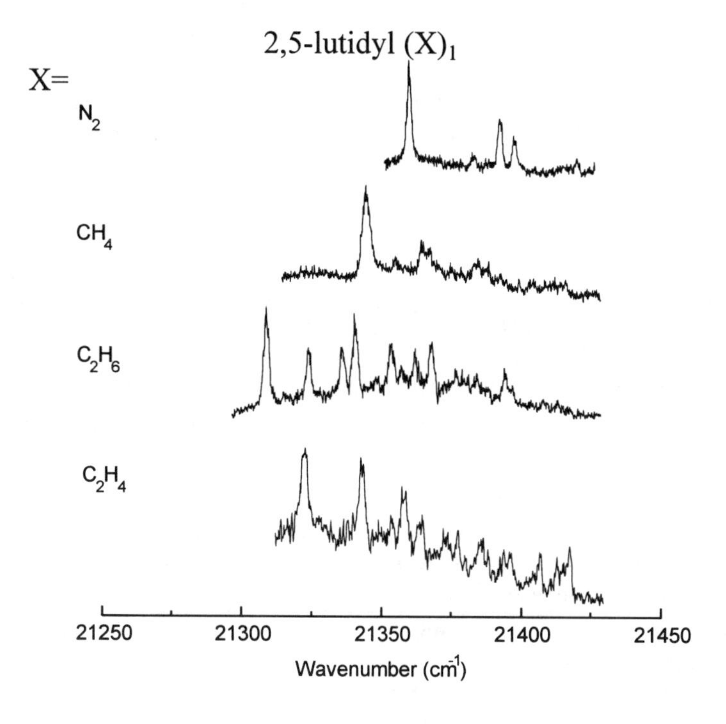

Fig. 17. Mass resolved spectra of 2,5-lutidyl radical with nonpolar solvents: $C_7H_8N\ (X)_1$, $X{=}N_2$, CH_4, C_2H_6, and C_2H_4. The 0^0_0 transition for the lutidyl radical is at 21401 cm^{-1}.

5 Solvation of Cyclopentadienyl and Substituted Cyclopentadienyl Radicals

Cyclopentadienyl radical (cpd) is an aromatic five-membered ring radical. Its high symmetry (belonging to the D_{5h} point group) leads to a very evident geometry for its possible clusters with small molecules. By comparison with studies on other aromatic molecules (benzene [51], benzyl radical [16], toluene [52], aniline [6], etc.) one expects that the solvent molecule will be located above the plane of the ring, approximately above the radical center of mass. Previous studies of $cpd(N_2)_1$ show [53] 1. a large cluster spectral red shift and 2. a complicated spectrum, derived from the chromophore high symmetry. The introduction of a functional group in the ring reduces the molecular symmetry, affects radical reactivity, and alters cluster structures. For example, the introduction of a CH_3 group should not greatly change the reactivity of the radical but will add a new "active" site at which to solvate the radical. If we introduce a CN group, the unpaired electron can delocalize through the C-N triple bond, changing the radical reactivity and its charge distribution for the atoms of the ring.

Will the CN group be more “attractive” for the solvent molecules than the aromatic ring itself? Can vdW cluster studies be employed to understand the changes in

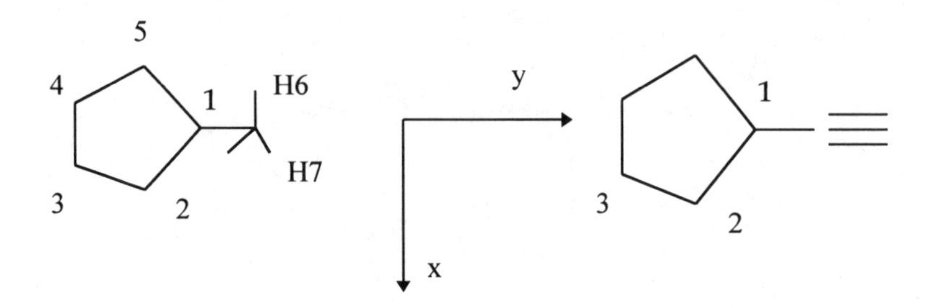

Table 4. Calculated structural parameters for xcpd. The atom labeling and axis orientation are shown in the diagram above. All the distances in Å, angles in degrees, and transition energies in cm^{-1}.

Radical	Bond	Ground State		Excited State		Transition Energy	
		This work	Other work	This work	Other work	Exp.	Calc.[g]
cpd	R_{CC}	1.423	1.421[a]	1.460	1.4473[f]	29,576.9	28,468
	R_{CH}	1.067	1.0774[b]	1.068	1.080[f]		
mcpd	R_{12}	1.450	1.437[b]	1.461	1.445[e]	29,769.5	30,453
	R_{23}	1.368	1.378[b]	1.459	1.445[e]		
	$R_{C\text{-}Me}$	1.501	1.501[b]	1.495	1.504[b]		
	$R_{H2C\text{-}H}$	1.085	1.093[e]	1.086	1.093[e]		
	R_{CH}	1.068	1.083[e]	1.068	1.069[e]		
	CCC	108.5		108.5			
	CCH	126.3		126.3			
cncpd	R_{12}	1.407	1.442[b]	1.461		27,144.6	29,527
	R_{23}	1.470	1.365[b]	1.454			
	$R_{C\text{-}CN}$	1.424	1.411[b] 1.4509[c]	1.409	1.3773[b]		
	$R_{C\text{-}N}$	1.148	1.158[b]	1.150			
	R_{CH}	1.069	1.094[d]	1.067			
	CCC	107.8		107.6			
	CCH	126.0		125.4			

[a]Ref. [55]; [b]Ref. [56]; [c]Ref. [57]; [d]Ref. [58]; [e]Ref. [59]; [f]Ref. [60]. [g]This work

reactivity produced by the introduction of a functional group? The present work tries to address these questions. We present here preliminary results of a more extensive work, which includes the study of clusters with polar solvents and molecules like CH_3OH and H_2O that can form strong hydrogen bonds with x-cpd radicals. In this Sec. we present

Table 5. Calculated atomic charges, polarizabilities and vdW radii for cpd, mcpd and cncpd. Only one value is shown for each group of equivalent atoms. r_{ii} in (2.2) is $2r_i$

Radical	Atom	Ground State			Excited State		
		q_i	α_i (Å^3)	r_i (Å)	q_i	α_i (Å^3)	r_i (Å)
cpd	C1	-0.1611	1.150	1.85	-0.1989	1.341	1.715
	C2	-0.2336			-0.1989		
	C3	-0.1876			-0.1989		
	H1	0.2086	0.420	1.465	0.1989	0.420	1.465
	H2	0.1938			0.1989		
	H3	0.2037			0.1989		
mcpd	C_{Me}	-0.4830	0.930	2.06	-0.4773	0.930	2.06
	C1	-0.278	1.150	1.85	-0.0298	1.375	1.700
	C2	-0.1936			-0.1862		
	C3	-0.2032			-0.2040		
	C4	-0.1923			-0.1890		
	C5	-0.2044			-0.2028		
	H2	0.1957	0.420	1.465	0.1972	0.420	1.465
	H3	0.1987			0.1976		
	H4	0.1978			0.1973		
	H5	0.1963			0.1972		
	H6	0.1681	0.420	1.46	0.1643	0.420	1.460
	H7	0.1738			0.1677		
cncpd	N	-0.2729	0.930	1.755	-0.2706	0.930	1.755
	C_{CN}	0.0403	1.150	1.85	0.0068	1.222	1.800
	C1	-0.0592	1.150	1.85	-0.0111	1.294	1.750
	C2	-0.1092			-0.1173		
	C3	-0.2049			-0.1997		
	H1	0.2405	0.420		0.2353	0.420	1.465
	H2	0.2195			0.2191		

results obtained for cyclopentadienyl (cpd), methylcyclopentadienyl (mcpd), and cyanocyclopentadienyl (cncpd) with Ar, N_2, CH_4, and CF_4. We elucidate the cluster structures and binding energies, through an analysis of cluster spectra. The final cluster geometries and shifts will serve as an indication of the chemistry of these radicals.

The radicals are prepared using phenyl azide ($C_6H_5N_3$) and phenyl isocyanate (C_6H_5NCO) as precursors for cncpd, and methylcyclopentadiene dimer for both mcpd and cpd. The phenyl azide is synthesized using the procedure described in [54].

Ground and excited state charges and structures for cpd, mcpd, and cncpd are calculated at the CASSCF/6-31G level, using the *ab initio* package Gaussian 94 [41] (see Tables 4 and 5). All the ring π orbitals and electrons (5 electrons and 5 orbitals) are included in the active space, leading to correct quantitative transition energies for cpd and mcpd (28,468 and 30,453 cm^{-1} compared with the experimental values of 29,576.9 and 29,769 cm^{-1} for cpd and mcpd, respectively) and a reasonably good value for cncpd (29,527 compared with 27,144.6 cm^{-1}). For this radical one must probably include in the calculations the CN π orbitals and electrons (four electrons and four orbitals).

The agreement between calculated and experimental structures for the radicals is also very good for cpd and mcpd and acceptable for cncpd (see Table 4). As stated above, these are preliminary results. More accurate calculations are in progress on these systems, including a better basis set that accounts for the electron correlation (cc-pVDZ) and a more completed active space for cncpd. The results will be presented in forthcoming papers.

As we will show, xcpd cluster spectra show very clear progressions in vdW modes. In order to assign these progressions, a normal mode analysis was performed for each cluster. The procedure has been described in previous work [61]. The results are not always quantitatively accurate, but are qualitatively correct, allowing a vdW vibrational assignment of the cluster spectra.

5.1 Experimental Results: Structure Determination

Parent molecules. In Fig. 18 we present the fluorescence spectra for the parent molecules a) cpd and mcpd and b) cncpd. As can be seen, the introduction of a methyl group in the ring has little influence in the electronic transition (that is, in the stability of the ring π orbitals) and shifts the transition energy by only $\approx$ 200 cm^{-1} to the blue.

The introduction of a CN group, however, has a larger effect on the π system of the ring, as shown by a comparison of Figs. 18a and b. The cncpd origin is shifted $\approx$

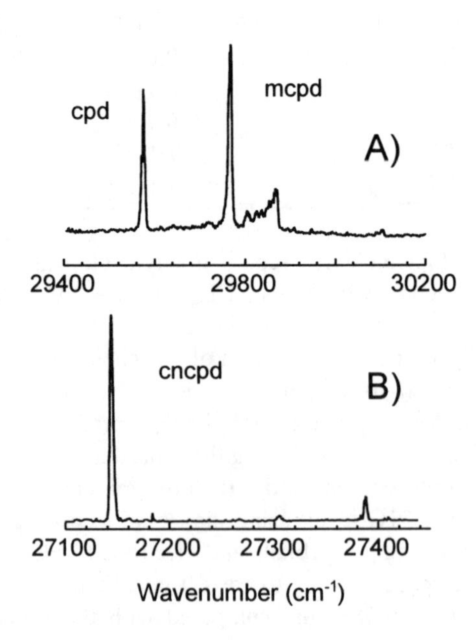

Fig 18. (18) Excitation spectra for A) cpd and mcpd and B) cncpd.

-2500cm^{-1} with respect to the cpd origin. The CN group also affects the charge distribution on the carbon atoms: due to the resonance between the triple bond and the ring, the unpaired electron can delocalize through the CN group, and the π system of the CN group can delocalize through the ring π system. The CH_3 and CN groups seem to have opposing effects on the π–π* transition observed. This will be reflected in the cluster binding energies and shifts. The radical identifications are confirmed using mass-resolved spectroscopy, but the spectra (similar to those shown in Fig. 18) are not shown here.

xcpd(Ar)$_1$. Fig. 19 depicts a) cpd(Ar)$_1$ and mcpd(Ar)$_1$ and b) cncpd(Ar)$_1$ cluster spectra. All three clusters present extremely large red shifts (for a 1:1 radical/Ar solvation [6,62]): -92, -124.6 and -139.3 cm^{-1} for cncpd(Ar)$_1$, cpd(Ar)$_1$ and mcpd(Ar)$_1$, respectively. They also show progressions in vdW vibrations. The calculated ground state cluster structures are shown in Fig. 19.

The cpd(Ar)$_1$ spectrum is a progression of three peaks, with a spacing of 46.8 and 44 cm^{-1} between them. The spectrum assignment is carried out taking into account 1. the calculated frequencies for the vdW modes for the excited state [61] and 2. that the strongest peaks in this kind of systems are due usually to the totally symmetric vibrations (out-of-plane stretch σ and bend b_p, where p= x, y or z and refers to the axis

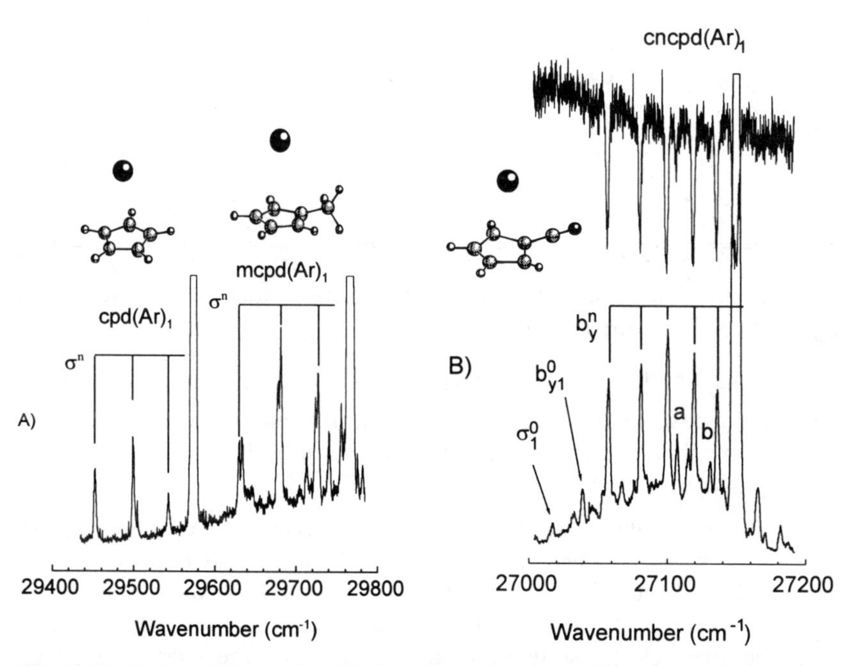

Fig. 19. A) Excitation spectra of cpd(Ar)1 and mcpd(Ar)1 and B) excitation spectrum of cncpd(Ar)$_1$ (lower trace) and hole burning (HB) spectrum (upper trace). The features labeled a and b in Fig. B, can be assigned as σ^1 and $b^n_y\sigma^1$ respectively. The calculated cluster structures are also depicted.

along which the displacement takes place). Keeping the above considerations in mind, one can assign the progression in cpd(Ar)$_1$ spectrum as due to the stretching vibration (calculated value of 56 cm^{-1}, compared with the experimental value of 46.8 cm^{-1}).

In the mcpd(Ar)$_1$ spectrum there is also a progression of three peaks, but they exhibit a non-totally resolved structure, probably due to the CH_3 torsional motion. Peaks labeled as a and b in this spectrum, at 82.6 and 109.8 cm^{-1} from the cluster origin, do not form part of the main progression. Based on the comparison of peak spacing (51 and 46 cm^{-1}) and the calculated vibronic frequencies, one can assign the progression as due to the stretching mode vibration (calculated value of 56 cm^{-1}).

cncpd(Ar)$_1$ shows a number of peaks (see Fig. 19b). In order to distinguish between bare molecule and cluster features, a hole burning (HB) spectrum is recorded (upper trace in Fig. 19b). A detailed analysis of peak spacing shows one progression, as depicted in Fig. 19. A comparison with the calculated energies allows one to assign the

Table 6. Calculated binding energies and shifts for xcpd clusters. All quantities are shown in cm^{-1}.

Radical	Solvent	Binding Energies				Shift	
		Ground State	Excited State p1	p2	p3	Calc.	Exp.
cpd	Ar	358.4	357.7	483.1		a	-124.6
	N_2	506.1	500.0	661.8		-156.7	-168.2
	CF_4	874.3	862.9	1117.9		-243.6	-182.2
mcpd	Ar	402.7	400.7	527.8	542.0	a	-139.3
	N_2	556.8	552.4	715.2	734.5	-177.7	-160.5
	CF_4	972.5	959.8	1213.5	1243.4	-270.9	-146.6
cncpd	Ar	397.9		525.2	489.9	a	-92.0
	N_2	536.5		692.5	647.6	-111.1	-132.7
	CF_4	913.1		1152.3	1088.2	-175.1	-167.3
	CH_4	513.9		679.2	634.9	-121.0	-184.3

a: Used to fit excited state potential parameters; p1: results using ground state parameters; p2: results obtained using cpd excited state parameters; p3: results obtained using the excited state modified parameters

progression as b_y^n (b_y = 23.5 cm^{-1} compared with the calculated value of 17 cm^{-1}). There is also a peak at 49.9 cm^{-1} from the origin (labeled a), which can be assigned as the stretching vibration (calculated value of 53 cm^{-1}). The peak labeled b is then assigned as $b_y^n \sigma_0^1$.

As expected, for the cpd(Ar)$_1$ cluster the Ar is centered above the middle of the ring, attracted by the π electron density. The addition of a CH_3 or a CN shifts the solvent to the substituted carbon position. As the Ar atom has no net charge, the coulombic term in (2.1) is 0, and the binding energy is due entirely to dispersive forces. The large cluster shift must then be due to large changes in the atomic parameters α_i and r_i. We will return to this point later.

The calculated shifts and binding energies for all the clusters are shown in Table 6. Ar clusters are used to estimate excited state potential energy parameters (see Sec. 2).

The same parameters are used for all the aromatic carbons in the radicals for the ground state, while for the excited state, different parameters are employed in order to fit the experimental results (see Tables 5 and 6). As the excited state is formed through a π-π^* transition, one can consider that the carbon atom parameters will experience the largest changes. To a first approximation, we consider that the change in hydrogen atoms, and CH_3 group atoms' parameters is very small or zero. A confirmation of this hypothesis can be generated by using cpd excited state modified carbon parameters for the mcpd excited state calculation. This approximation leads to cluster shifts in relatively good agreement with the experimental values (see Table 6, the column labeled p2). Of course, the use of mcpd/Ar modified parameters (column labeled as p3) improves the agreement. A relatively poor agreement obtains if cpd excited state parameters are used for the cncpd calculations, because the CN group has a stronger interaction with the ring than does the CH_3 group.

Excited state parameters for cncpd are difficult to evaluate, because a balance must be struck between the ring and the CN group orbitals. In a first approximation N parameters can be considered unaffected by the electronic excitation, while the C parameters change.

To match experimental shifts, an increase in the ring carbon excited state atomic polarizabilities is required for mcpd and a decrease for cncpd (compared with cpd, see Table 5). This is consistent with the fact that the CH_3 group gives electronic density to the ring, while the CN group withdraws it.

xcpd$(N_2)_1$. The cpd$(N_2)_1$ spectrum is very similar to that of cpd$(Ar)_1$ (see Figs. 19a and 20a): a progression of three peaks with a spacing of 55 and 53.5 cm^{-1} between

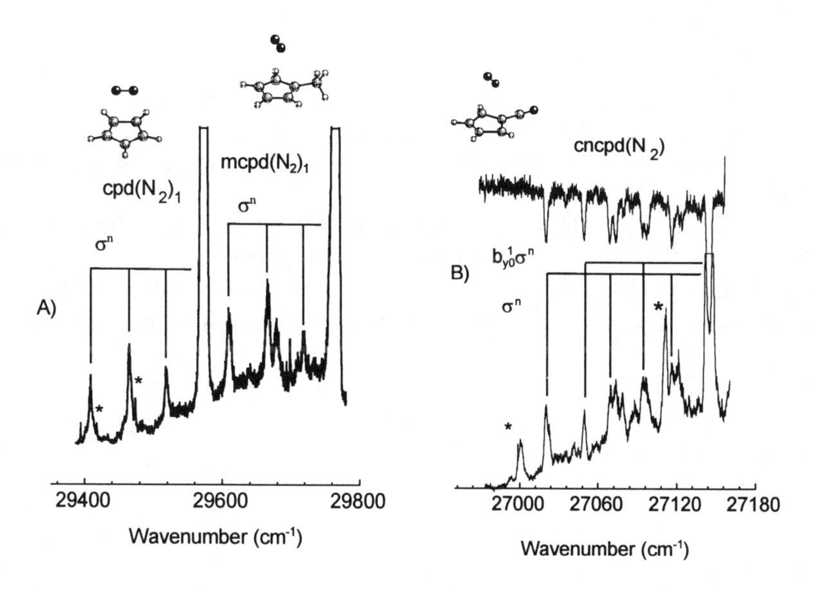

Fig. 20. A) Fluorescence excitation spectra of cpd$(N_2)_1$ and mcpd$(N_2)_1$. The asterisks denote peaks due to the rotation of the N_2 over the ring plane. B) Excitation (lower trace) and HB (upper trace) spectra of cncpd$(N_2)_1$. The asterisks denote bare molecule peaks. The calculated cluster structures are also depicted.

them. The small peaks labeled with asterisks are due to the rotation of the N_2 molecule in a plane parallel to the ring. The assignment of the main progression is not as straightforward as for cpd$(Ar)_1$, due to the fact that the experimental frequency (55 cm^{-1}) lies halfway between the calculated values for σ (77 cm^{-1}) and b_x (39 cm^{-1}). By comparison with cpd$(Ar)_1$, we tentatively assign this progression as σ^n. This assignment is also coincident with the one in [5].

The same analysis can be performed for mcpd$(N_2)_1$. The peak spacing (56 cm^{-1}) lies between σ and b_x calculated values (77 and 35 cm^{-1}, respectively). Using the same arguments as for cpd$(N_2)_1$, we assign the progression as σ^n. As can be seen, the calculation overestimates the vdW frequencies.

The cncpd vdW vibrational spectrum is more complicated; most of the peaks appear as doublets and triplets. One can distinguish two different progressions in σ: one built on the origin and one built on the b_y vibration (σ= 49 cm^{-1}, calculated value of 70 cm^{-1} and b_y = 29 cm^{-1}, calculated value of 18 cm^{-1}). As can be seen the agreement between experimental and calculated values is not very good. This is mainly due to two reasons: 1. the potential wells for the clusters are very shallow, so the harmonic approximation does not hold any more and 2. the excited state calculation accumulates the errors in both ground and excited calculations. Keeping this in mind, the only information that one can extract from the calculated frequencies is that $E(\sigma)>E(b_x)>E(b_y)$.

Cluster geometries are also shown in Figs. 20a and b. As for the clusters with Ar, the solvent molecule is attracted by the ring π electron density. Both CH_3 and CN groups shift the solvent position toward the substituted carbon, but the ring is still the preferred solvation site. The peaks labeled with asterisks in Fig. 20b are due to the bare molecule, as the HB shows.

xcpd$(CF_4)_1$. The spectra depicted in Fig. 21 show a large profusion of features. cpd$(CF_4)_1$ shows a progression of at least four peaks, and a number of small peaks probably due to combination bands. The main progression can be assigned as σ^n by comparison with previous clusters and calculated values (74 cm^{-1} calculated value, compared with 44.4 cm^{-1}).

Features for mcpd$(CF_4)_1$ exhibit a partially resolved structure that can be attributed to CH_3 rotor motion. We can only safely assign the main progression as σ^n (46.2 cm^{-1}, compared with the calculated value of 70 cm^{-1}). The calculated value for this vdW mode is overestimated, probably as a result of the overestimation of the excited state binding energy.

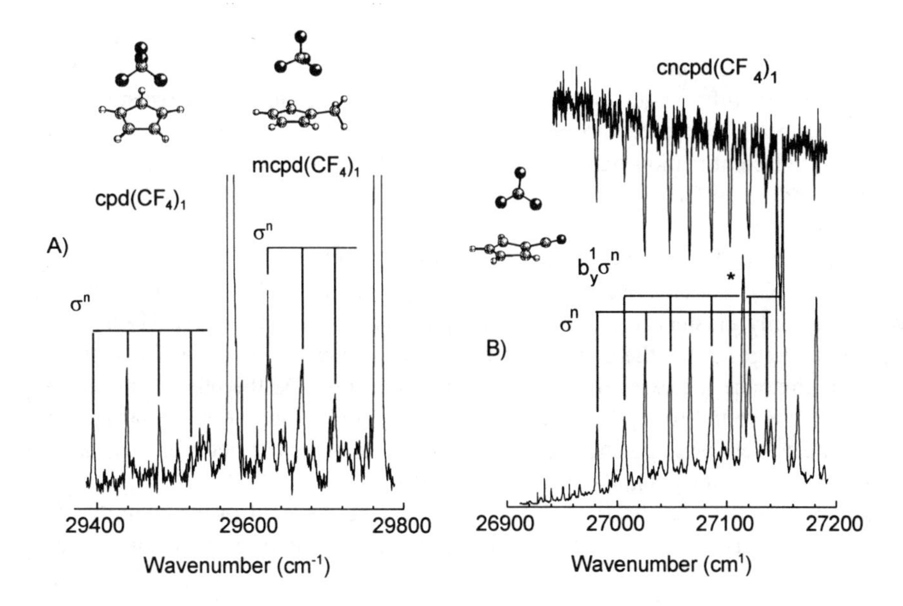

Fig. 21. A) Excitation spectra of cpd(CF_4)1 and mcpd(CF_4)1 and B) excitation (lower trace) and HB (upper trace) spectra of cncpd(CF_4)1. The peak labeled with an asterisk is assigned to the bare molecule. Calculated cluster structures are also depicted.

The extensive spectrum in Fig. 21b shows two progressions that could be assigned to two different conformers for the cncpd$(CF_4)_1$ cluster. The HB spectrum however shows that both sets of features have the same origin (that is, both originate with a single ground state energy level). An HB spectrum can also discriminate between peaks belonging to the cluster and to the bare molecule. For example, the peak at 27,110.2 cm^{-1}, labeled in Fig. 21b with an asterisk, is due to the bare molecule and not to the cluster, as it is not present in the HB, and only one isomer for the cluster is expected according to the atom-atom potential energy calculations. Through comparison with previous assignments, one notes that totally symmetric vdW vibrations tend to be the most intense ones in the cluster spectra. Thus the progressions can be assigned as σ^n and $b_y^1\sigma^n$; however, the experimental values of 25 and 44 cm^{-1} do not match the calculated values of $b_y = 15$, $b_x = 34$, and $\sigma = 66$ cm^{-1} very well.

xcpd$(CH_4)_1$. Up to this point, all spectra shown for a given radical seem to follow similar patterns: progressions in σ for cpd and mcpd and in one of the bending modes and σ for cncpd. Also the observed shifts are in good agreement with the theoretical predictions (except for mcpd$(CF_4)_1$). xcpd$(CH_4)_1$ clusters do not match any of these characteristics (see Fig. 22): 1. the cncpd spectrum is very crowded; 2. the peaks exhibit only partially resolved structure; 3. the first three peaks seem to form a progression which abruptly ends; 4. the observed cluster shift (-184.3 cm^{-1}) is 50% larger than the predicted value of -121 cm^{-1}; and 5. no cluster spectra are found for cpd$(CH_4)_1$ and mcpd$(CH_4)_1$.

All these observations can be explained if we assume a prereactive interaction between excited state cncpd and CH_4. The interaction stabilizes the cluster excited state, leading to shifts larger than expected and to large changes in cluster structure. Then the Franck-Condon factors no longer favor transition to levels with low vdW excitation, and combination bands and overtones appear in the spectrum. Absence of cpd$(CH_4)_1$ and mcpd$(CH_4)_1$ spectra is then due to the larger reactivity of these radicals toward CH_4.

A map of cncpd reactivity can be attempted by studying its clusters with CH_3Cl, CH_2F_2, CHF_3, C_2H_6, CH_3OH, and CH_3CH_2OH. The results will be presented in future papers. The only clusters not found are cncpd$(C_2H_6)_1$ and cncpd$(CH_3CH_2OH)_1$. This observation suggests that the cncpd cluster reaction is electrophilic (the inclusion of a halogen in methane increases the barrier for the reaction). On the other hand, none of these clusters is found for cpd or mcpd, indicating a higher and very similar reactivity between the solvent and cpd and mcpd. Most likely, the reaction between CH_4 and cpd and mcpd occurs in the excited electronic state of the radical.

Additionally, the cncpd$(CH_4)_1$ cluster could not be detected in the vicinity of the cncpd vibration at 27,674 cm^{-1}. This observation can be interpreted to imply that the

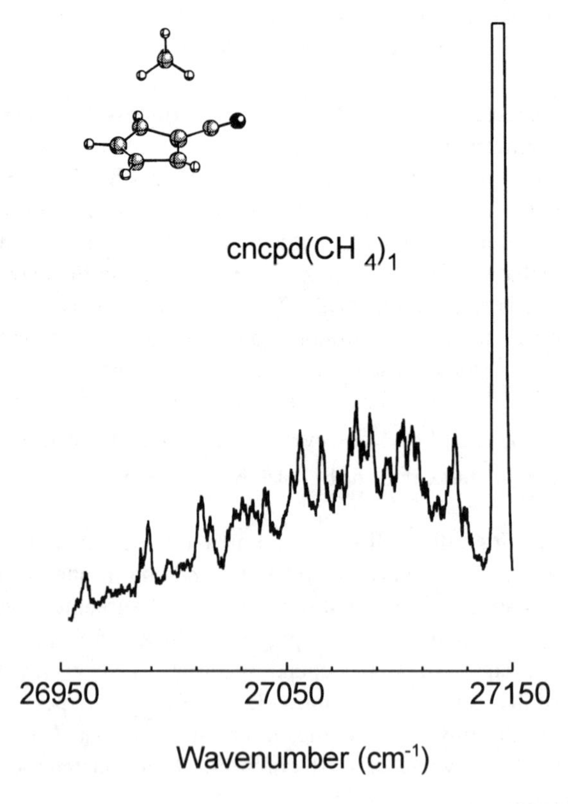

Fig. 22. cncpd(CH4)1 spectrum, together with the calculated cluster geometry.

chromophore extra vibrational energy of ~ 500 cm^{-1} is sufficient activation energy to start the reaction.

A vibrational assignment of the spectrum in Fig. 22 is very difficult with the present degree of approximation in our calculations and the present spectral resolution, so it will not be attempted.

5.2 Discussion

As pointed out earlier, the shifts found for all three xcpd radical clusters are quite large compared with other aromatic ring clusters. For example, the shift for aniline$(Ar)_1$ [6] is -40 cm^{-1}, for $C_6H_5Cl(Ar)_1$ [62] -27 cm^{-1}, for $C_6H_5F(Ar)_1$ [62] -24 cm^{-1} and for $C_6H_5OH(Ar)_1$ [62] -36 cm^{-1}. The shifts are larger even compared with what was found for other radicals; for example, for benzyl radical $(Ar)_1$ = -23.8 cm^{-1} [18]. The shifts found are of the same order as those for some Rydberg states [63]. For example, for dioxane$(CF_4)_1$ the shift is 188.3 cm^{-1}, but in this case the electronic excitation energy is 50,663 cm^{-1}, compared with ≈ 29,000 cm^{-1} for xcpd. The calculated ground state binding energies are, however comparable to those found for other clusters of aromatic molecules. For example, the ground state binding energy for benzyl radical$(Ar)_1$ cluster is 399.5 cm^{-1} (see Table 6). This means that the shift derives from a change in the atomic parameters comparable with what we observe for Rydberg states. The atomic parameters α_i, and r_{ii} are closely related to atomic charges and atomic volumes. As the change in q_i is not large enough to justify the large shifts, this suggest that the increase in binding energy upon electronic excitation is due to a large increase in atomic volume.

The structures found for all the clusters show that even with the addition of a CH_3 or CN group, the preferred position for the solvent molecule is above the aromatic ring. The calculated binding energies suggest that the CH_3 group increases the cluster binding energy more than the CN group, probably because the CH_3 group gives electronic density to the ring, while the CN group withdraws it.

Finally, the addition of the CH_3 group does not change the radical reactivity very much, but the CN group has a dramatic effect on the radical reactivity.

5.3 Summary

In this Sec. we have reported the clusters of cpd, mcpd, and cncpd with Ar, N_2, CH_4, and CF_4 solvent molecules. All the clusters exhibit large spectral shifts to the red. This observation is consistent with our theoretical model, implying a large increase in atomic volume upon electronic excitation. The inclusion of a CH_3 or a CN group affects (increases) slightly the cluster binding energy. The CH_3 group has a larger effect in the cluster binding energy that the CN group; however, the CN group has a strong influence in the radical reactivity.

6 Conclusions

We demonstrate in these radical cluster studies that a simple theoretical and experimental approach to the study of radical solvation can be useful in understanding how radicals interact with their environments and how they begin to react. The combined empirical potential energy/*ab initio* calculation, taken with experimental values of radical/Ar cluster spectral shifts, seems to generate a self- consistent picture of radical solvation behavior.

We are now exploring radical reactions in these cluster systems and improving our understanding of their potential surfaces and van der Waals modes.

Acknowledgments. We want to thank Dr. Ana B. Bueno for her help in the synthesis of phenyl azide. This work is supported in part by grants from the USNSF and USARO. One of us (JAF) thanks the Basque Government for a postdoctoral fellowship.

References

1 (a) R. Ahamad-Bitar, W. P. lapatovich, D. E. Pritchard, and I. Renhom, Phys. Rev. Lett. **39**, 1657 (1997); (b) R. E. Smalley, D. A. Auerbach, P. S. H. Fitch, D. H. Levi, and L. Wharton, J. Chem. Phys. **66**, 3778 (1977); (c) R. R. Freeman, E. M. Mattison, D. E. Pritchard, and D. Kleppner, J. Chem. Phys. **64**, 1194 (1976); (d) M.–C. Duval, O. B. D'Azy, W. H. Breckenridge, C. Jouvet, and B. Soep, J. Chem. Phys. **85**, 6324 (1986); (e) J. Tellinguishen, Adv. Chem. Phys. **60**, 299 (1985).

2 I. W. M. Smith, *Kinetics, and Dynamics Elementary Gas Reactions*, Butterworths, London (1980).

3 (a) J. A. Fernández, R. Martínez, M. N. Sánchez, and F. Castaño, J. Phys. Chem. **100**, 12305 (1996); (b) R. A. Moss, Acc. Chem. Res. **22**, 15 (1988); (c) *ibid.* **13**, 58 (1980); (d) J. A. Fernández, P. Puyuelo, D. Husain, M. N. Sánchez, and F. Castaño, J. Chem. Phys. **106**, 7090 (1997).

4 (a) E. F. van Dishoeck, and A. Dalgarno, J. Chem. Phys. **79**, 873 (1983); (b) H. -J. Werner, P. Rosum, and E.-A. Reinsch, J. Chem. Phys. **79**, 905 (1983).

5 (a) R. N. Schwartz, and K. F. Herzfeld, J. Chem. Phys. **22**, 767 (1954); (b) R. N. Schwartz, Z. I. Slawsky, and K. F. Herzfeld, J. Chem. Phys. **20**, 1591 (1952); (c) D. C. Clary, Chem. Phys. **65**, 247 (1982); (d) *ibid.* J. Phys. Chem. **91**, 1718 (1987).

6 (a) M. F. Hineman, S. K. Kim, E. R. Bernstein, and D. F. Kelley, J. Chem. Phys. **96**, 4904 (1992); (b) B. Coutant, and P. Brechignac, J. Chem. Phys. **91**, 1978 (1989); (c) M. R. Nimlos, M. A. Young, E. R. Bernstein, and D. F. Kelley, J. Chem. Phys. **91**, 5268 (1989).

7 (a) Y. Oshima, H. Koguchi, Y. Endo, Chem. Phys. Lett. **184**, 21 (1991); (b) D. J. Nesbit, Chem. Rev. **88**, 843 (1988); *ibid.* Faraday Discuss. **97**, 1 (1994); (c) T. G. Fraser, Int. Rev. Phys. Chem. **10**, 189 (1991).

8 See for example (a) M. I. Lester, Adv. Chem. Phys. **46**, 51 (1996), and references therein; (b) R. Nowak, J. A. Menapace, E. R. Bernstein, J. Chem. Phys. **89**, 1309 (1988); (c) S. Sun, and E. R. Bernstein, J. Chem. Phys. **103**, 4447 (195); (d) Th. Weber, A. M. Smith E. Riedle, J. H. Neuser, and E. W. Schlag, Chem. Phys. Lett.

175, 79 (1990); (e) P. Hobza, Q. Bludsky, H. L. Selzle,, and E. W. Schlag, J. Chem. Phys. **98**, 6223 (1993).
9 See for example (a) E. R. Bernstein, K. Law, and M. Schauer, J. Chem. Phys. **80**, 207 (1984); *ibid.* 634 (1984); (b) J. L. Knee, and P. M. Johnson, J. Chem. Phys. **80**, 13 (1984); (c) A. W. Garret, and T. S. Zwier, J. Chem. Phys. **96**, 3402 (1992).
10 D. Fulle, H. F. Hamann, H. Hippler, and J. Troe, J. Chem. Phys. **105**, 983 (1996).
11 C. Chakravarty, D. C. Clary, A. D. Esposti, and H. -J. Werner, J. Chem. Phys. **95**, 8149 (1991); *ibid.* **93**, 3367 (1990).
12 L. C. Giancarlo, R. W. Randall, S. E. Choi, and M. I. Lester, J. Chem. Phys. **101**, 2914 (1994).
13 See for example (a) S. Fei, X. Zheng, and M. C. Heaven, J. Chem. Phys., **97**, 1655 (1992); (b) L. C. Giancarlo, and M. I. Lester, Chem. Phys. Lett. **240**, 1(1995); (c) R. T. Carter, I. M. Povey, H. Bitto, and J. R. Hubert, J. Chem. Phys. **104**, 5365 (1996); (d) B. -C. Chang, J. M. Williamson, D. W. Cullin, J. R. Dunlop, and T. A. Miller, J. Chem. Phys. **97**, 7999 (1992); (e) T.-San Ho, H. Rabit, S. E. Choi, and M. I. Lester, J. Chem. Phys. **104**, 1187 (1996).
14 (a) G. W. Lemire, M. J. McQauid, A. J. Kotlar, and R. C. Sausa, J. Chem. Phys. **99**, 91 (1993); (b) M. J. McQauid, G. W. Lemire, and R. C. Sausa, Chem. Phys. Lett. **210**, 350 (1993).
15 H. Hettema, and P. E. S. Wormer, J. Chem. Phys. **93**, 3389 (1990).
16 D. Disselkamp, and E. R. Bernstein, J. Phys. Chem. **98**, 7260 (1994).
17 (a) D. R. Woodward, D. A. Fletcher, and J. M. Brown, Mol. Phys. **62**, 453 (1987); (b) *ibid.* **68**, 261 (1989).
18 R. Disselkamp, and E. R. Bernstein, J. Chem. Phys. **98**, 4339 (1993).
19 P. Hobza, H. L. Selzle, and E. W. Schlag, Chem. Rev. **94**, 1767 (1994).
20 (a) R. A. Scott, and H. A. Scheraga, J. Chem. Phys. **45**, 2091 (1966); (b) F. A. Monamy, L. M. Carruthers, R. F. McGuire, and H. A. Scheraga, J. Chem. Phys. **78**, 1595 (1974); (c) F. A. Momany, R. F. McGuire, A. W. Burguess, and H. A. Scheraga, J. Chem. Phys. **79**, 2361 (1975); (d) G. Némethy, M. S. Pottle, and H. A. Scheraga, J. Chem. Phys. **87**, 1883 (1998).
21 P. J. Wantuc, R. C. Oldenborg, S. L. Baughcom, and K. R. Winn, J. Phys. Chem. **91**, 3253 (1987).
22 (a) R. N. Dixon, Phil. Trans. Roy. Soc. (London) **A 252**, 165 (1960); (b) R. N. Dixon, Can. J. Phys. **38**, 10 (1960).
23 R. N. Dixon, M. J. Trenouth, and C. M. Western, Mol. Phys. **60**, 779 (1987).
24 (a) D. Patel-Misra, D. G. Sauder, and P. J. Dagdigain, J. Chem. Phys. **93**, 5448 (1990), and reference therein; (b) D. Patel-Misra, D. G. Sauder, P. J. Dagdigain, and D. R. Crosley, J. Chem. Phys. **95**, 2222 (1991);
25 S. A. Wright, and P. J. Dagdigian, J. Chem. Phys. **104**, 8279 (1996).
26 D. R. Cyr, R. E. Continetti, R. B. Metz, D. L. Osborn, and D. M. Neumark, J. Chem. Phys. **97**, 4937 (1992).
27 (a) F. J. Northrup, M. Wu, and T. J. Sears, J. Chem. Phys. **96**, 7218 (1992); (b) M. Wu, F. J. Northrup, and T. J. Sears, J. Chem. Phys. **97**, 4583 (1992).
28 (a) B. J. Sullivan, G. P. Smith, and D. R. Crosley, Chem. Phys. Lett. **96**, 307 (1983); (b) R. A. Copeland, and D. R. Crosley, Can. J. Phys. **62**, 1448 (1984).

29 (a) J. A. Fernández, Ph. D. Thesis (Universidad del País Vasco, 1995); (b) J. A. Fernández, I. Merelas, D. Husian, M. N. Sánchez Rayo, and F. Castaño, Chem. Phys. Lett. **267**, 301 (1997); (c) J. A. Fernández, P. Puyuelo, D. Husain, M. N. Sánchez Rayo, and F. Castaño, J. Chem. Phys. **106**, 7090 (1997).
30 D. E. Milligan, and M. E. Jacox, J. Chem. Phys. **47**, 5157 (1967).
31 V. E. Bondybey, and J. H. English, J. Chem. Phys. **67**, 2868 (1977).
32 H. Reisler, M. Mangir, and C. Wittig, Chem. Phys. **47**, 49 (1980).
33 G. Hancock, and G. W. Ketley, J. Chem. Soc. Faraday Trans. **278**, 1283 (1982).
34 T. R. Charlton, T. Okamura, and B. A. Thrush, Chem. Phys. Lett. **89**, 98 (1982).
35 (a) P. S. H. Bolman, J. M. Brown, A. Carrington, I. Kop, and D. A. Ramsay, Proc. Roy. Soc. (London) **A 343**, 17 (1975); (b) P. Misra, C. W. Mathews, and D. A. Ramsay, J. Mol. Spectrosc. **130**, 419 (1988).
36 D. R. Woodward, D. A. Flether, and J. M. Brown, Mol. Phys. **62**, 517 (1987);
37 R. Bruggemann, M. Petri, H. Fischer, D. Mauer, D. Reinert, and W. Urban, Appl. Phys. **B 48**, 105 (1990);
38 (a) C. E. Barnes, J. M. Brown, A. D. Fackerell, and T. J. Sears, J. Mol. Spectrosc. **92**, 485 (1982); (b) J. Werner, W. Seebass, M. Koch, R. F. Curl, W. Urban, and J. M. Brown, Mol. Phys. **56**, 453 (1985);
39 P. B. Davies, and I. H. Davis, Mol. Phys. **69**, 175 (1990).
40 (a) K. Kawaguchi, S. Saito, and E. Hirota, Mol. Phys. **49**, 663 (1983); (b) **55**, 341 (1985).
41 Gaussian 94, M. J. Frisch, G. W. Trucks, H. B. Schlegel, P. M. W. Gill, B. G. Johnson, M. A. Robb, J. R. Cheeseman, T. A. Keith, G. A. Petersson, J. A. Montgomery, K. Raghavachari, M. A. Al-Laham, V. G. Zakrzewski, J. V. Ortiz, J. B. Foresman, J. Cioslowski, B. B. Stefanov, A. Nanayakkara, M. Challancombe, C. Y. Peng, P. Y. Ayala, W. Chen, M. W. Wong, J. L. Andrés, E. S. Replogle, R. Gomperts, R. L. Martin, D. J. Fox, J. S. Binkley, D. J. Defrees, J. Baker, J. P. Stewart, M. Head-Gordon, C. González, and J. A. Pople, Gaussian Inc., Pittsburgh, PA, 1995.
42 M. H. Alexander, and H. J. Werner, (private communication).
43 H.S. Im, and E. R. Bernstein, J. Chem. Phys. **95**, 6326 (1991).
44 (a) M. A. Hoffbauer, and Hudgens, J. Phys. Chem. **89**, 5152 (1985); (b) G.C. Eiden, and J.C. Weisshaar, J. Chem. Phys. **104**, 8896 (1996); (c) C. Cossart-Magos, and S. Leach, J. Chem. Phys. **64**, 4006 (1976); (d) F. Negri, G. Orlandi, F. Zerbetto, and M.Z. Zgierski, J. Chem. Phys. **90**, 600 (1990); (e) G. Orlandi, G. Poggi, and F. Zerbetto, Chem. Phys. Lett. **115**, 253 (1985).
45 K. Okuyama, N. Mikami, and M. Ito, J. Phys. Chem. **89**, 5617 (1985).
46 M. Fukushima, and K. Obi, J. Chem. Phys. **93**, 8488 (1990).
47 A. Goumri, J.-F. Pauwels, J.-P. Sawerysyn, and P. Devolder, Chem. Phys. Lett. **171**, 303 (1990).
48 T.-Y. D. Lin, and T. A. Miller, J. Phys. Chem. **94**, 3554 (1990).
49 M. Dupuis, F. Johnston, and A. Márquez, "Hondo 8.5 from Chemstation", (1994) IBM Corporation, Neighborhood Road, Kingston, NY 12401.
50 W. Roebke, J. Phys. Chem. **74**, 4198 (1970).
51 (a) Th. Brupbacher, J. Makarowich, and A. Bauder, J. Chem. Phys., **101**, 9736 (1994); (b) *ibid.* Chem. Phys. Lett. **173**, 435 (1990); (c) E. Aruna, T. Emilsson, and

H. S. Gutowsky, J. Chem. Phys. **101**, 861 (1994); (d) T. D. Klots, T. Emilsson, and H. S. Gutowsky, J. Chem. Phys. **97**, 5335 (1992).

52 (a) M. Schawer, K. Law, and E. R. Bernstein, J. Chem. Phys. **81**, 49 (1984); (b) *ibid.* **82**, 736 (1985); (c) S. Li, and E. R. Bernstein, **97**, 804 (1992).

53 S. Sun, and E. R. Bernstein, J. Chem. Phys. **103**, 4447 (1995).

54 R. O. Lindsay, and C. F. H. Allen in "Organic Synthesis Collective", **3**, 710. Edited by E. C. Horning. John Wiley, and sons. London 1955.

55 L. Yu, D. W. Cullin, H. M. Williamson, and T. A,. Miller, J. Chem. Phys. **98**, 2682 (1993).

56 D. W. Cullin, L. Yu, H. M. Wiliamson, and T. A,. Miller, J. Phys. Chem. **96**, 89(1992).

57 J. Casado, L. Nygaard, and G. O. Sorensen, J. Mol. Struct. **8**, 211 (1971).

58 D. W. Cullin, L. Yu, H. M. Williamson, M. S. Platz, and T. A. Miller, J. Phys. Chem., **94**, 3387 (1990).

59 L. Yu, D. W. Cullin, H. M. Williamson,, and T. A. Miller, J. Chem. Phys. **95**, 804 (1991).

60 L. Yu, H. M. Williamson, and T. A. Miller, Chem. Phys. Lett. **162**, 431 (1989).

61 J. A. Fernández , and E. R. Bernstein, J. Chem. Phys. **106**, 3029 (1997).

62 E. J. Bieske, M. W. Rainbird, I. M. Atkins, and A. E. W. Knight, J. Chem. Phys. **91**, 752 (1989).

63 P. O. Moreno, Q. Y. Shang, and E. R. Bernstein, J. Chem. Phys. **97**, 2869 (1992).

The Electronic Spectroscopy of Molecules Undergoing Large-Amplitude Motions: Acetaldehyde in the First Excited Singlet and First Triplet States

David C. Moule

Department of Chemistry, Brock University, St. Catharines, ON
L2S 3A1 Canada E-mail: dmoule@abacus.ac.brocku.ca, and

Edward C. Lim

Department of Chemistry, The University of Akron, Akron, OH
44325-3601 USA E-mail: elim@uakron.edu

Abstract. As a result of the coupling between the rotation of the methyl group and the other vibrational modes, and the implications to IVR (intramolecular vibrational redistribution), there has been great interest in the S_0 ground state spectroscopy of acetaldehyde. This paper focuses on the dynamics of methyl torsion and acetyl wagging in the first S_1 and T_1 excited electronic states. Acetaldehyde has sufficiently few atoms that it is possible to determine the torsion-wagging potential surfaces with relatively high level *ab initio* calculations. When combined with maps of the electronic transition moments, these surfaces yield vibronic torsion-wagging band progressions that simulate the red end of the ultraviolet spectrum. Along with these simulations there is also the prediction that the individual bands contained a complex rotational structure of hybrid bands. To test these predictions, a jet-cooled LIF (laser induced fluorescence) excitation spectrum was recorded of the low-energy bands in the $S_1 \leftarrow S_0$ system under conditions of very high resolution with an amplified frequency doubled CW laser system. For the 0_0^0 band, the *ab initio* calculations predicted that there should be five components in the band cluster. A rotational analysis was able to identify these transitions in the LIF spectrum. Moreover, the strengths of the observed transitions were in good agreement with the calculated strengths. The very weak $T_1 \leftarrow S_0$ system of acetaldehyde was observed as a LIP (laser induced phosphorescence) excitation spectrum with a rotating slit jet apparatus. An analysis of the spectrum showed good agreement to high level MP/UHF calculations. The critical data to emerge from these analyses are the barriers to methyl torsion and hydrogen wagging. For the S_1/T_1 states the barriers (in cm^{-1}) were respectively found to be 712.5/638.6 and 609.7/869.0.

1 Introduction

Methyl torsions and molecular inversions have long been of interest and a challenge to spectroscopists. Because of their large vibrational amplitudes, their motions cannot be described satisfactorily within the harmonic oscillator approximation. Moreover, the violation of the conditions for separability of vibrations into normal modes, namely that the vibrational amplitudes be small, leads to coupling between the large-amplitude coordinates and the small-amplitude harmonic oscillator coordinates. This intermode coupling between low-frequency motion and other high-frequency vibrations complicates the analyses of the spectra (electronic, vibrational, and rotational), and it also greatly influences the efficiency of intramolecular vibrational redistribution. Thus, the presence of large-amplitude vibrations in a molecule has important consequences for spectroscopy as well as intramolecular dynamics.

Acetaldehyde CH_3CHO is an excellent prototype for a detailed study of large-amplitude motions and their effects on rovibronic spectra for several reasons. Firstly, it is the smallest carbonyl compound that exhibits both the methyl torsion and acetyl inversion upon electronic excitation to an $n\pi^*$ state. Secondly, it has near ultraviolet $S_1(n\pi^*) \leftarrow S_0$ and $T_1(n\pi^*) \leftarrow S_0$ spectra that can be conveniently probed with high-resolution tunable dye lasers. Thirdly, the molecule is small enough that direct simulations of the spectra are possible by *ab initio* calculations. Despite these desirable features, the analyses of the gas-phase acetaldehyde spectra recorded under room temperature conditions have proved to be intractable due to the severe thermal (hot band and sequence) congestions involving the low-frequency large-amplitude vibrations. The band structure can, however, be greatly simplified by cooling the molecules to very low temperatures in a supersonic expansion, to the extent that individual bands can be identified. Combined with *ab initio* quantum-chemical calculations, the supersonic-jet laser spectroscopy is therefore providing detailed information concerning the large-amplitude potentials for a number of molecules, including acetaldehyde.

In this chapter, we present a summary of our recent work on the electronic structure and spectra of jet-cooled acetaldehyde. Specifically, we discuss the results of spectroscopic and *ab initio* quantum-chemical studies of torsion-inversion structure in the $S_1(n\pi^*) \leftarrow S_0$ and $T_1(n\pi^*) \leftarrow S_0$ electronic spectra.

2 The First Excited Singlet State, S_1

In the S_0 ground electronic state of acetaldehyde, the torsion of the methyl group about the C–C bond is the single large-amplitude coordinate. The internal rotation of the methyl group about the CHO frame generates a sinusoidal potential that contains three minima for each full rotation. The shape of the threefold potential and the height of the barrier have become the object of intense scrutiny [1].

On $n \rightarrow \pi^*$ electronic excitation of an electron from the n nonbonding orbital on the oxygen to the π^* orbital, antibonding density is introduced into the carbonyl bond with the result that the molecular CHC=O frame distorts into a pyramidal conformation similar to the well-known formaldehyde system [2]. Thus, the excited S_1 and T_1 states are nonplanar in the frame, and the potential function describing the wagging motion of the hydrogen atoms contains two minima separated by a central barrier. To further complicate the picture, the methyl group undergoes a conformational change on excitation and rotates from its eclipsed form in the lower S_0 state to a staggered form in the $n\pi^*$ state. Thus, the first excited S_1 and T_1 states are highly flexible along the internal coordinates of torsion and wagging [3].

To describe the torsion-wagging problem it is necessary to express the potential energy as a function in two dimensions, θ (torsion) and α (wagging), as shown in Fig. 1. The two-dimensional potential may be described as the V(θ, α) surface

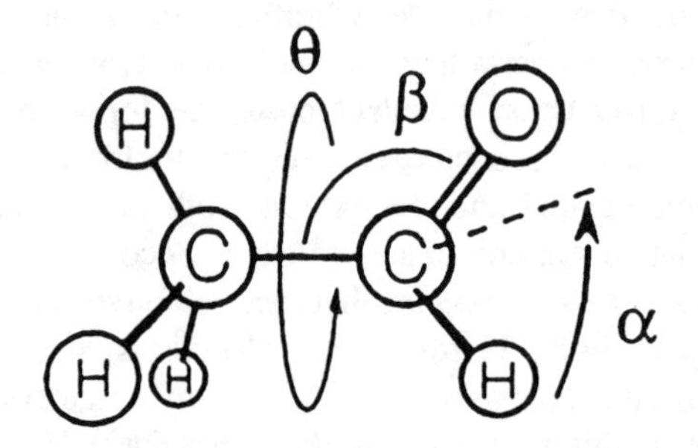

Fig. 1. Structural parameters used in the three-dimensional study of acetaldehyde. The β angle measures the CCO bending, θ describes the methyl torsion, and α represents the out of plane wagging of the aldehydic hydrogen with respect to the CCO plane

defined by θ and α. Plots of the S_0 and S_1 surfaces derived from *ab initio* models are given in Fig. 2. The form of these plots follows without difficulty. The S_0 ground state surface contains the three minima corresponding to the wells created by the

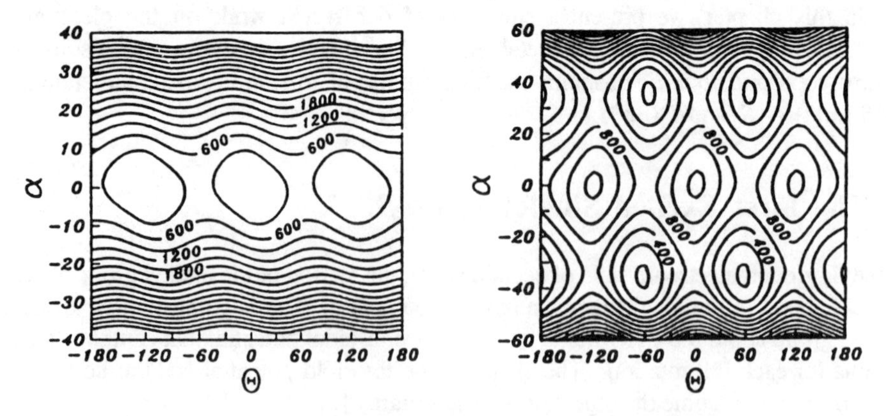

Fig. 2. The calculated potential energy surfaces of acetaldehyde in the ground (*left*) and excited (*right*) singlet states [16]. The intervals between the isopotential lines for the S_0 state is 300 cm ^{1}; for S_1, 200 cm^{-1}

threefold rotational symmetry of the methyl group. In the α direction, the wagging potential has a single minimum and is harmonic (quadratic in α). For the excited state, the positions of minima now become maxima, and it is clear that the torsion-wagging potential surface has three central humps with three deep minima to each side. Thus, for one full revolution of the methyl group, the $V(\theta, \alpha)$ surface in the upper S_1 state contains six minima.

The spectral consequences of having the wells in the T_1 and S_1 potential surfaces displaced in the θ and α directions from their S_0 positions is that long progressions in the ν_{15} (torsion) and ν_{14} (wagging) modes become active in the spectrum. As every level in the torsional manifold is a suborigin for a wagging progression, the density of vibronic bands increases rapidly from the electronic origin. Thus, under bulb conditions (room temperature), the spectrum is so highly congested that assignments of the spectra recorded with white light sources and long path length cells are impossible. The low temperatures of the supersonic jet are able to overcome the congestion problems to the extent that Lee and coworkers [4] and Baba et al. [5] have been able to make vibronic assignments for the first 600 cm^{-1} of the $S_1 \leftarrow S_0$ singlet-singlet system.

It is possible to simulate the vibronic structure in the $S_1 \leftarrow S_0$ system from ground and excited state potential surfaces and the molecular structures. In essence, ROHF and RHF calculations are used to establish the potential surfaces. In this method, the total energy is calculated for a set of molecular conformations defined by selected values of θ and α that form grid points on the $V(\theta, \alpha)$ surface. These *ab initio* data points are reduced to analytical form by fitting to a Fourier + Taylor series [6],

$$V = \sum_{k}^{N_v} V_k^0 \prod_{j}^{2} f_{kj}\,, \tag{1}$$

where V_k^0 are the coefficients in the expansion for the potential energy obtained from the fitting procedure and the f's represent trigonometric (torsion) or polynomic (wagging) terms. The kinetic terms, B_{ij}, are treated in a somewhat similar way. The kinetic energies evaluated at grid points defined by θ and α were obtained from the molecular geometry as elements of the rovibrational G matrix [7] using the KICO program [8]. The B_{ij} are the coefficients obtained from the fit of (2) to the points on the grid. N_B and N_A are the number of terms in the expansions for the kinetic and potential energies:

$$B_{ij} = \sum_{k}^{N_B} B_{ijk}^0 \prod_{f}^{2} f_{ijkl}\,. \tag{2}$$

For the treatment of the multidimensional problem the general Hamiltonian,

$$\hat{H} = \sum_{i}^{n}\sum_{j}^{n}\left(-B_{ij}\frac{\partial^2}{\partial q_i \partial q_j} - \frac{\partial B_{ij}}{\partial q_i}\frac{\partial}{\partial q_j}\right) + \hat{V}, \tag{3}$$

is used, where the number of vibrations, n, is two.

The two-dimensional Hamiltonian, (3), can be solved variationally for eigenvalues and eigenvectors using a hybrid free rotor + harmonic oscillator basis for methyl torsion and aldehydic hydrogen wagging coordinates, respectively.

Nonrigid group theory was used for the energy levels, factorization of the Hamiltonian matrix, and the generation of the selection rules. The existence of symmetry planes in the methyl rotors and in the molecular frame allows the nonrigid symmetry of the S_0 and S_1 states of acetaldehyde to be classified under the Altmann's $\hat{C}_3$ torsion and $\hat{V}$ switch operations of the G_6 nonrigid group [9]. This is a group that is isomorphic with the C_{3v} point group. The character table is shown in Table 1.

Table 1. The character table for the G_6 nonrigid group of acetaldehyde. The symmetry operations are in Altmann's notation [9]

G_6	E	$2\hat{C}_3$	$3\hat{V}$
A_1	1	1	1
A_2	1	1	-1
E	2	-1	0

The present method of defining the nonrigid symmetry by the Altmann formalism assumes that a supergroup can be obtained from the semidirect product of two subgroups. The first of these groups is the "isodynamic" group, I. It contains the operations that correspond to the internal nonrigid motions of the molecule. The second group, G, specifies the rigid symmetry operations for a specific conformation of the molecule. In the Altmann method, the feasibility of molecular flexing is taken into account through the "motions" collected in the isodynamic group. For the planar C_s eclipsed conformation of acetaldehyde, the nonrigid supergroup $G_6 = \mathrm{I} \wedge G = C_3\left\{E, C_3, C_3^2\right\} \wedge C_s\{E, \sigma\}$. The other formalism, proposed by Longuet-Higgins [10], defines the feasible internal rotation of the methyl group as a cyclic (1, 2, 3) permutation of the nuclei. For systems with several rotating tops attached to a molecular frame, Woodman [11] has shown that the construction of the semidirect product is a valid operation and for these types of problems the two methods are equivalent.

Before attempting the computer simulation of the red end of the fluorescence excitation spectrum, we felt that it was necessary to calibrate the *ab initio* method. For this purpose, we set out to calculate the six a, e components of the lowest three torsional levels in the S_0 state. Acetaldehyde is a molecule of intermediate size, and over the years it has become a test species for experimental as well as theoretical studies. Recently, very accurate energy data for the rotationless *a* and *e* levels has become available from high-resolution far infrared experiments [1(a)]. A long-standing problem that is currently being addressed is the extent of the coupling between the ν_{15} torsional mode and the higher-frequency vibrational modes. A useful way to consider the coupling is through the interactions between the CH_3 top and the CHO frame. As would be expected, the strengths of this coupling change as the methyl group rotates against the frame, and one consequence is that the C_3 symmetry of the top changes with the various conformations of the methyl group. Thus, the CH bond lengths and the HCH bond angles oscillate with advancing torsional angle θ. One way of compensating for this interaction is to use the technique of full relaxation of the molecular structure, whereby all of the structural parameters are optimized at each data point (θ, α) that defines the grid [12]. Of course, the correct method for accounting for the interaction is through the direct coupling of the kinetic and potential terms through the multidimensional Hamiltonian of (3). In the study discussed here we combine these two techniques.

The region of interest in the excitation spectrum (+ 600 cm^{-1} from the 0^0_0 origin) shows activity of three modes: ν_{15} (CH_3 torsion), ν_{14} (CHO wagging), and ν_{10} (C–C=O bending). For this reason our analyses were carried out in one, two, or three dimensions with θ, α, and β coordinates as defined in Fig. 1. A series of calculations was made at different levels of the *ab initio* approximation with the GAMESS [13] or G94 [14] suite of programs. In all of these calculations, the coordinates in question were fixed at their grid point values while the remaining structural parameters were optimized with the standard routines of the *ab initio* package. Table 2 shows the results [15]. What is remarkable about these calculations is that even for this

Table 2. Torsional energy levels, in cm^{-1}, for the S_0 state of acetaldehyde

ν_{10}	ν_{14}	ν_{15}	*Sym.*	*Calc.*[a]	*Calc.*[b]	*Calc.*[c]	*Obs.*[d]
0	0	0	a_1	0.00	0.00	0.00	0.00
			e	0.07	0.08	0.07	0.0690
0	0	1	e	142.06	141.43	140.81	141.9935
			a_2	143.78	143.37	143.43	143.7434
0	0	2	a_1	257.56	252.96	251.40	255.2243
			e	271.20	267.77	266.24	269.1121
0	0	3	e	353.66	346.45	344.38	
			a_2	412.32	407.80	407.31	
0	0	4	a_1	430.97	423.45	421.22	
			e	514.57	508.86	504.07	
1	0	0	a_1			506.74	
			e			508.76	

[a]One-dimensional model
[b]Two-dimensional model
[c]Three-dimensional model with the bending fundamental frequency fitted to the experimental value
[d]Observed frequencies

modest level of approximation, MP2(full)/6-311G(d,p), the fit to the observed levels with the two- and three-dimensional treatments is excellent. The agreement between the three-dimensional model of torsion + wagging + bending and the two-dimensional model of torsion + wagging is a consequence of the small coupling between the bending and the torsion modes. When the kinetic term for the bending mode was adjusted to reproduce the observed ν_{10} bending frequency, a Fermi interaction was found between the e components of the ($\nu_{10} = 0, \nu_{15} = 4$) and ($\nu_{10} = 1$, $\nu_{15} = 0$) levels. This interaction increases the $a_1 - e$ torsional separation of the first quantum of bending to 2.04 cm^{-1}.

The intensities of the various transitions were determined from the transition dipole moment between an n and m pair of torsion-wagging vibronic states belonging to electronic states e' and e'',

$$\mu_{nm} = \left\langle \varphi_n \varphi_{e'} \middle| \hat{\mu} \middle| \varphi_{e''} \varphi_m \right\rangle, \tag{4}$$

where μ represents the dipole moment operator. The electronic transition moments, $\mu_{e'e''}(\theta, \alpha)$ were obtained directly from the GAMESS package from the CISDT/ROHF/4-31G(d,p) length approximation,

$$\mu_{e'e''} = \left\langle \varphi_{e'} \middle| \hat{\mu} \middle| \varphi_{e''} \right\rangle, \tag{5}$$

which can be expanded as a series expression in terms of the same θ and α grid points selected for the potential surfaces [16],

$$\mu_{e'e''} = \sum_j^{N_\mu} \mu_j^0 \prod_k^n f_{kj} \,. \tag{6}$$

The relative intensities for the transitions within the manifold of levels were obtained from

$$I \propto \left(g_n - g_m\right) \mu_{nm}^2 . \tag{7}$$

Transition dipole moments were calculated from the above procedure by using the geometry of the ground state. The results are shown in Fig. 3 as contour maps that depend on the methyl torsion and wagging angles for each projection on a principal inertial axis. The μ_c component is perpendicular to the C–C=O frame and belongs to the A_1 representation, while μ_a and μ_b lie in the molecular plane and are of A_2 species. The threefold periodicity and the inversion of the sign of the component can be clearly seen from Fig. 3.

Selection rules were calculated from the symmetry of the torsion-wagging wave functions and the components of the transition dipole moment. The nonvanishing cases are given in Table 3.

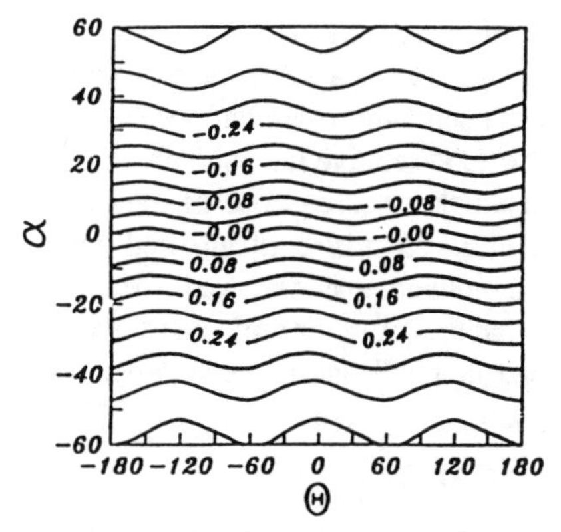

Fig. 3. A contour map for the μ_a component of the transition moment. The interval between the lines is 0.04 debye [16]

Table 3. Vibronic selection rules for the $S_1 \leftarrow S_0$ electronic transition of acetaldehyde [16]

ψ_n^a	μ^b	ψ_m^a
A_1	*c*	A_1
A_2	*c*	A_2
E	*c*	*E*
A_1	*a*	A_2
A_1	*b*	A_2
A_2	*a*	A_1
A_2	*b*	A_1
E	*a*	*E*
E	*b*	*E*

[a] ψ_n and ψ_m give the symmetries of the initial and final levels

[b] The corresponding transitions moment component along the *a, b, c* principal axes

The allowed electronic transition between the S_0 and S_1 states of acetaldehyde, $\tilde{X}\,^1A'' \leftarrow \tilde{A}\,^1A'$, is directed out of the C–C=O molecular plane and results in *c*-type polarization for the $a_1 - a_1$ components. The transitions to the higher inversion level 0^-, 14_0^1, are electronically forbidden, but they become allowed through a Herzberg-Teller vibronic coupling. This band is composed of *a* and *b* components and is in-plane polarized. As the 0_0^1 and 14_0^1 bands are of about the same size in the excitation spectrum, it follows that the allowed and forbidden components to the transition have about the same strength. This result can be traced back to the excitation process. An examination of the molecular orbitals of acetaldehyde shows that the *n* and π^* orbitals project at right angles to each other from the sides of the oxygen atom. As a result, the promotion of the electron involves more of a rotation of charge than a translation.

The consequence of a lack of charge translation is that the system has low electric dipole strength. Acetaldehyde, thus, is a textbook example of an electronic transition that is allowed by the overall selection rules but forbidden by the local symmetry of the C=O chromophore. Thus, the relative intensities of the vibronically induced bands in the spectrum are not so much a consequence of their absolute strength, but rather are related to the weakness of the electronically allowed transition.

Figure 4 shows the simulated and observed spectra under conditions of low resolution [16]. It is clear that the gross vibronic band features are accounted for. The interval between the 14_0^1 and 0_0^0 bands measures the inversion doubling separation,

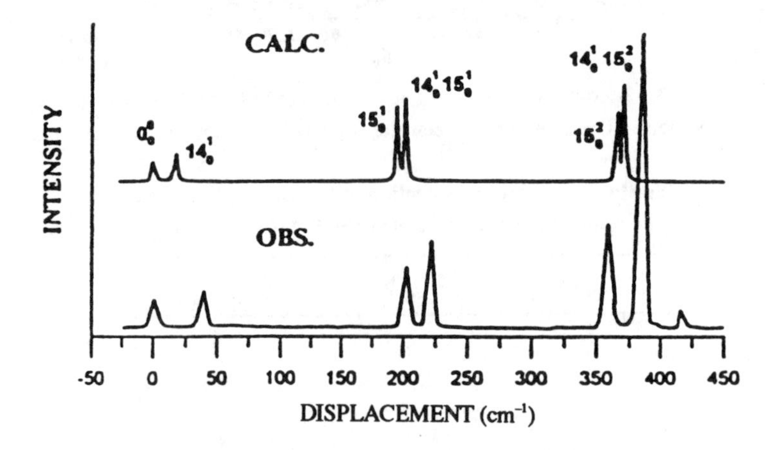

Fig. 4. Experimental [4a] and simulated [16] $S_1 \leftarrow S_0$ electronic spectra of acetaldehyde. The ν_{14} is the wagging-inversion mode, ν_5 is methyl torsion

$0^- - 0^+$, in the manifold of ν_{14} wagging levels. This interval attaches to the quanta of torsional levels to form sets of doublets in the spectrum that are not constant but change in separation with the torsional quantum number. This variation in the interval separation is a direct indication of the strength of the intermode coupling between the torsion and wagging in the upper state.

While it is satisfying to know that the overall band positions and intensities in the spectrum can be successfully reproduced by direct *ab initio* methods, the interesting aspect of the calculation is the prediction that all of the vibronic bands possess an underlying structure that is not evident in the low-resolution simulation. The tripling of the minima of the potential that defines the methyl group rotation creates a pair of a_1 and e levels for the $v'' = 0$, S_0 state that are separated by only 0.069 cm^{-1}. It would be anticipated that these levels would be populated at the temperature of the supersonic molecular beam and as a result the electronic transitions would arise from both levels. The selection rules given in Table 3 show for the $v' \leftarrow v''$, 0_0^0 band, the $a_1 - a_1$ and $e - e$ torsional transitions combine through the moments $\langle a_1|c|a_1\rangle$, $\langle e|a|e\rangle$, $\langle e|b|e\rangle$, and $\langle e|c|e\rangle$ to form a cluster of four components. The strongest of these hybrid components was found to come from transitions between the singly degenerate $a_1 - a_1$ levels and was directed along the c principal axis (out of the

C–C=O plane). The transitions between the doubly degenerate $e - e$ levels were calculated to have components along all three principal axes [16].

The prediction that the vibronic bands contained components with different polarizations was perhaps the most interesting aspect of the foregoing *ab initio* analysis. Lee and coworkers [4] and Baba et al. [5] have demonstrated that the $S_1 \leftarrow S_0$ system of acetaldehyde can be probed without too much difficulty in a supersonic jet with a tunable dye laser. The coupling of a high-resolution laser light source to the low temperature molecular beam device allows the individual lines in the band complexes to be clearly resolved. At the lowest temperature of the jet, 0.75 K, the populations of the rotational levels are so altered that each vibronic band is reduced to just five or six rotational lines. In such jet-cooled experiments, the problems of the line assignments are reduced, and the rotational analyses are able to clearly establish the *a*/*b*/*c* band type. This encouraged us to make a concerted search to see if these forbidden components could be observed directly in the spectrum.

For this purpose we re-recorded a group of vibronic bands at the red end of the singlet-singlet system of acetaldehyde [17] with a pulse-amplified CW dye laser [18]. A block diagram of the experimental setup is given in Fig. 5. The excitation source

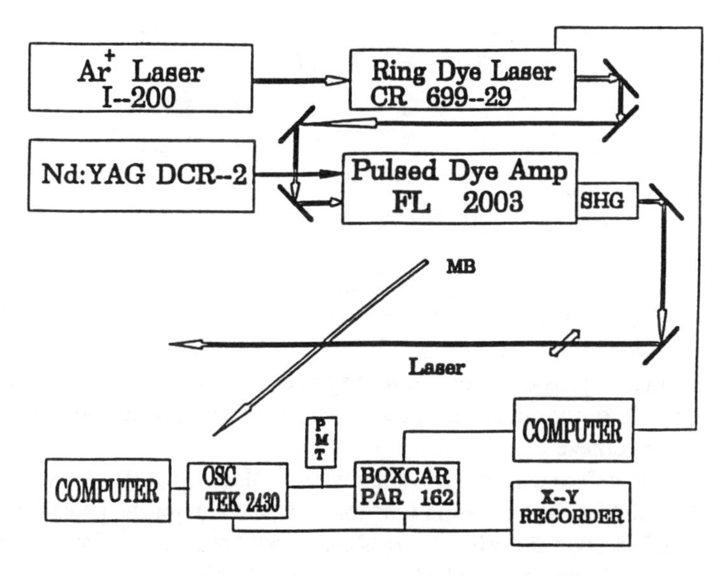

Fig. 5. A schematic representation of the experimental apparatus used to record high-resolution $S_1 \leftarrow S_0$ LIF excitation spectra of acetaldehyde

for the experiment was a CW ring dye laser (Coherent 699-29), pumped with an argon ion laser (Coherent Innova 200-15) and operated with DCM dye. The pulse amplifier (Lambda Physik FL 2003) was pumped with the second harmonic of an Nd:YAG laser (Quanta Ray DCR2). The output of the pulsed single-mode laser was frequency doubled into the near UV by a KDP crystal. Under molecular beam conditions, the experimental widths of the individual rotational lines were reduced to 160 MHz, allowing for the complete resolution of the vibronic bands.

Table 5. Observed and fitted energy levels for the S_1 state of acetaldehyde[a]

v_{14}	v_{15}	*Sym.* (Γ_{tw})	*Obs.*	*Calc.*
0	0	a_1	0.000	0.000
0	0	e	0.022	0.023
1	0	a_2	34.398	34.398
1	0	e	34.393	34.393
0	1	a_2	193.510	193.510
0	1	e	193.204	193.204
1	1	a_1	210.625	210.623
1	1	e	210.616	210.618
0	2	a_1	346.696	346.696
0	2	e	349.496	349.496
1	2	a_2	372.244	372.244
1	2	e	374.731	362.050[b]
2	0	a_1	482.387	482.387
2	0	e	479.494	479.494

[a]In cm^{-1}

[b]Not fitted; perturbed by 10^1_0

The model Hamiltonian of (3) that was used to treat the S_0 state was used to analyze the S_1 level data. As a starting point for the refinement, we used a kinetic expansion derived from the UHF/MP2(full)6-311G(*d,p*) basis, whereas the coefficients for the potential expansion were computed at the level ROHF/4-31G(*d,p*). Table 5 shows the results of the fitting procedure where both the kinetic and potential energy expansion were refined.

The excellent fit to the torsion-wagging term values is partly a consequence of the extreme flexibility of the expansion functions, (1) and (2), for the potential and kinetic energies. Table 6 gives the refined values that were obtained by simultaneously least squares adjusting 11 potential and 24 kinetic parameters.

The effects of the fitting procedure can be seen by comparing the experimentally derived $V(\theta, \alpha)$ surface of Fig. 9 with Fig. 2, obtained from the *ab initio* calculations. The general shapes of the two surfaces are similar in that they have three central maxima surrounded on the sides by six minima. The more complex contours for the experimentally derived surface are a consequence of the complicated intramolecular couplings.

The minima in the potential function of Fig. 9 appear at $\theta_{min} = 59.9°$ and $\alpha_{min} = 33.5°$, for the S_1 state compared to their planar $\theta_{min} = 0.0°$ and $\alpha_{min} = 0.0°$ values for the S_0 state. The barriers to torsion and inversion were extracted as the differences to the corresponding values derived from the *ab initio* method. In this table, the relative strengths of both the observed and calculated $\langle a_1 | c | a_1 \rangle$ components were scaled to 1.00. The relative strengths of the $\langle e | a | e \rangle$ and $\langle e | b | e \rangle$ transitions that connect the e levels were calculated to be 0.33 and 0.39, respectively. The corresponding observed strengths are in good agreement, 0.25 and 0.40. The fourth member of this group, $\langle e | c | e \rangle$, comes from the *c*-type allowed component. Its relative strength of 0.5 is controlled by the nuclear statistical weights of the S_0 state.

The relative intensities of the $\langle e | a | e \rangle$ and $\langle e | b | e \rangle$ transitions are related to the strength of the Herzberg-Teller interaction and to the coupling between the θ and α torsion-wagging internal coordinates. Weersink et al. [19] have considered this problem, and they label the transitions that result from coupling of the θ and α coordinates that leads to transition moments $\langle e | a | e \rangle$ and $\langle e | b | e \rangle$ by the term *indirect*, whereas they refer to the $\langle a_1 | c | a_1 \rangle$ and $\langle e | c | e \rangle$ moments as *direct*.

The rotational constants for the S_1 levels are not in themselves useful since they were derived from a semirigid Hamiltonian that did not explicitly include the coupling between the torsion and the overall angular momentum. The band origins extracted from the analysis are for the rotationless molecule, $J = K = 0$, and consequently, they are free of the angular momentum coupling problems. A simultaneous least squares fit to all of the band components within the 0^0_0 band yielded the values 29,769.022 and 29,768.975 cm^{-1} for the $a_1 - a_1$ and $e - e$ transitions, respectively. The separation of –0.047 cm^{-1} between these components can be combined with the 0.069 cm^{-1} splitting between the $v'' = 0$, $a_1 - e$ levels of the S_0 state to predict a torsional splitting of 0.022 cm^{-1} for the upper S_1 state. The fact that the $a - e$ splitting for the zero point level in the upper state is about one-third that of the ground electronic state is a clear indication of the differences in barrier heights in the two states.

In general, the *direct* transitions (those that do not require torsion-wagging coupling) create the dominant lines in the spectrum. For the Franck-Condon allowed bands, 0^0_0, 15^2_0, etc., these would be the c-type components $\langle a_1 | a | a_1 \rangle$ and $\langle e | c | e \rangle$, whereas for the Herzberg-Teller bands 14^1_0, $14^1_0 15^2_0$, etc., it is the *a*/*b* type transitions that come from the $\langle a_1 | a | a_2 \rangle$ and $\langle a_1 | b | a_2 \rangle$ moments that form the *direct* transitions. Thus, the observation of the a/b/c band types are a powerful tool for the identification and assignment of the *a* or *e* components within a vibronic band.

The other equally powerful tool for testing the band assignments is the $e - a$ splitting between the components. With the excitation of successive quanta of torsion this separation increases dramatically. For example, from Table 2, the separations for the torsional quanta in the S_0 state are $v_{15} = 0$ (0.0689 cm^{-1}), $v_{15} = 1$ (1.7499 cm^{-1}), and $v_{15} = 2$ (13.8878 cm^{-1}). Thus, the magnitude of the observed torsional splitting directly provides the quantum numbering of the torsional excitation.

As a second phase of our studies, we recorded the bands in the $S_1 \leftarrow S_0$ system labeled #2 to #23 (Lee notation) under the same jet conditions that were used for the origin band, #1. The motivation for this study [20] was the realization that the band polarizations and the $a - e$ splittings along with the rotationless origin bands extracted from *ab initio* calculations should provide enough information to firmly establish the vibronic assignments of the higher bands in the spectrum.

Figure 6 shows the results for the 0_0^0 band [17], where the temperature for the experimental frame (at the bottom) is estimated to be 0.75 K from the relative

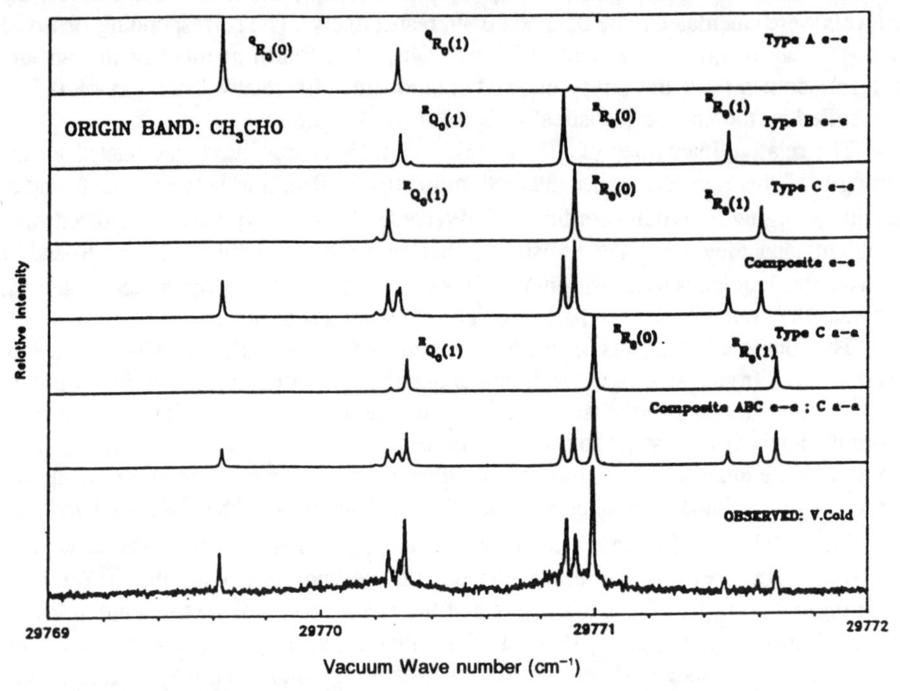

Fig. 6. A computer simulation of the very cold spectrum of the 0_0^0 origin band of the $S_1 \leftarrow S_0$ transition of acetaldehyde [17]. The computed spectra were calculated at a rotational temperature of 0.75 K

strengths of the rotational lines. The upper frames give the calculated contributions of the four components to the overall band profile. The weights given to each of the band types in the fitting procedure are tabulated in Table 4 and are compared

Table 4. Observed and calculated strengths and displacements of the hybrid components within the 0_0^0 origin cluster of acetaldehyde [17]

S_0	μ[a]	S_1	*Obs.*[b]	*Calc.*[b]	*Obs.*	*Calc.*
E	a	E	0.33	0.25	–0.047	–0.059
E	b	E	0.39	0.4	–0.047	–0.059
E	c	E	0.44	0.5	–0.047	–0.059
A_1	c	A_1	1.00	1.00	0.000	0.000

[a]Direction of the transition along the principal axis
[b]Intensity relative to the $\langle a_1 | c | a_1 \rangle$ component

Figure 7 compares the band #5, recorded under cold jet conditions of $T = 0.7$ K with the same band under warm jet conditions, $T = 5$ K. Thus, this band was subjected to a rotational analysis for both temperatures. Under the low-temperature conditions, this band complex consists of six lines that are grouped into two distinct

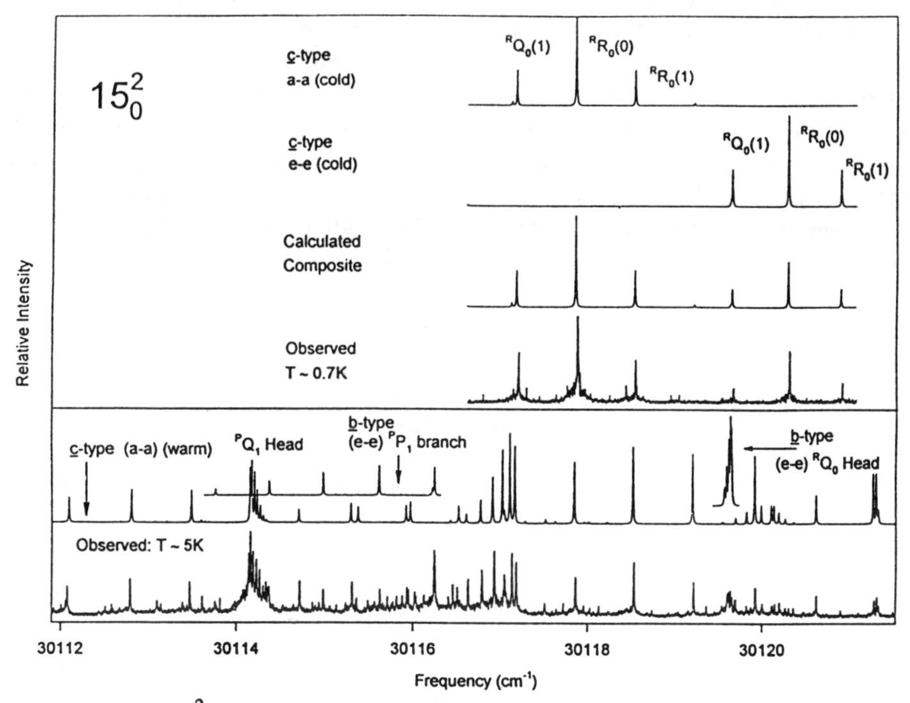

Fig. 7. The 15^2_0 band cluster of acetaldehyde [20]. The large $(a - e)$ splitting of the 2 ν_{15} level accounts for the simplicity of the cold spectrum

triplet structures that are each analyzable as c-type bands. The rotational analysis places the origin of the lower-energy transition at 30,115.718 cm^{-1} and the higher member at 30,118.449 cm^{-1}. The two positions, when combined with the $(a - e)$ values for the electronic origin band 0^0_0, give term values of 346.696 and 349.496 cm^{-1}. The 2.800 cm^{-1} separation between the levels immediately provides the assignment, namely, that the two levels attach to the origin by the double quantum of torsion, 2 ν_{15}, with $15^2(e)$ the higher-energy component and $15^2(a_1)$ the lower component. The large $(a_1 - e)$ splitting for this level accounts for the simplicity of the spectra. The calculated spectra are a 2:1 intensity mix of c-type $(a_1 - a_1)$ and $(e - e)$ components and are the result of the nuclear weights of the a_1 and e levels. The warm spectra consist mainly of $a_1 - a_1$ c-type bands with a small contribution from the $^RQ_0$ band head and the $^PP_1$ branch of the $(e - e)$ b-type transition. What is notable is that the b-type band appears only in the warm jet spectra suggesting that the torsion-wagging coupling is somehow affected by the temperature of the jet. At the elevated temperature of 5 K, the e – e transitions appear to be weaker than the a_1 – a_1 transitions. This phenomenon is unexpected and does require further study.

As an illustration of the second class of transition, we turn to band #2. Figure 4 clearly shows that under low resolution, this band has about the same intensity in the excitation spectrum as does the origin band. It comes about from the activity of the wagging-inversion ν_{14} and is given the assignment 14^1_0. This band is forbidden by the electric dipole selection rules but is vibronically allowed as an *a*/*b* in-plane polarized transition with moments $\langle a_1|a|a_2\rangle$ and $\langle a_1|b|a_2\rangle$. These moments along with $\langle e|a|e\rangle$ and $\langle e|b|e\rangle$ constitute the *direct* transitions. The $\langle e|c|e\rangle$ moments come from torsion-wagging mixing and is the simple *indirect* moment in this cluster. Figure 8 shows how the overall band profile for the very low temperature can be synthesized from all three *a*/*b*/*c* band types of the e – e transition and the *a*/*b* contour of the a_1 – a_2 transition.

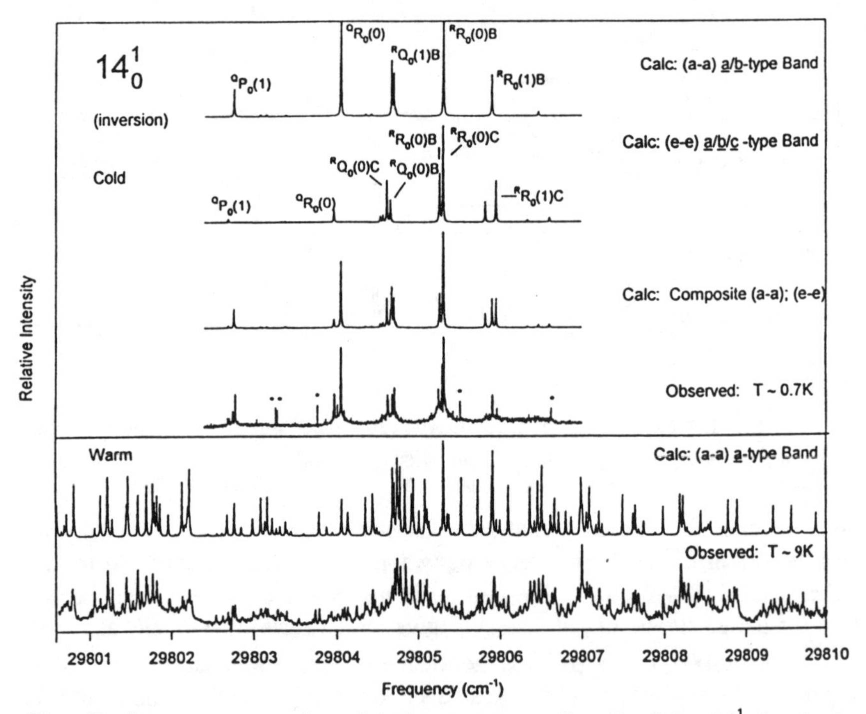

Fig. 8. Low temperature (*upper*) and warm spectra (*lower*) of the 14^1_0 band of acetaldehyde [20]

The results of the rotational analysis are collected together in Table 5 in the form of a set of term values attached to the $\nu = 0$ (a_1) zero point level of the S_1 state. These origin values were obtained by simultaneously least squares fitting the *a*/*b*/*c* band types to their common origins. The resulting term values then provided the input data for the refinement of the torsion-wagging potential surface for the S_1 state.

$V(0.0, \alpha_{min}) - V(180.0, \alpha_{min}) = 712.5\ cm^{-1}$ and $V(\theta_{min}, 0.0) - V(\theta_{min}, \alpha_{min}) = 638.6\ cm^{-1}$.

Table 6. Fitted kinetic and potential expansion coefficients (summed from k = 0 to 10) for torsion and wagging in the S_1 state of acetaldehyde (θ and α refer to the torsion and wagging angles, respectively)[a]

Coeff.	$B^0_{11}(\theta,\alpha)$	$B^0_{12}(q,a)$	$B^0_{22}(\theta,\alpha)$	$V^0(\theta,\alpha)$
constant	7.0535	-3.6605	22.1676	791.51
cos (3θ)	1.6516	0.3732	1.1541	226.70
cos (6θ)	----	----	----	19.92
α^2	0.0032	0.18×10^{-3}	-0.53×10^{-3}	-0.99
α^4	-0.19×10^{-5}	-0.21×10^{-6}	-0.35×10^{-6}	0.55×10^{-3}
α^2 cos (3θ)	-0.0015	-0.0011	0.51×10^{-3}	0.27
α^4 cos (3θ)	0.14×10^{-6}	0.40×10^{-6}	-0.18×10^{-5}	-0.13×10^{-3}
α^2 cos (6θ)	----	----	----	0.37
α^4 cos (6θ)	----	----	----	-0.34×10^{-3}
α sin (3θ)	-0.0444	-0.0390	-0.0173	10.17
α sin (3θ)	0.78×10^{-4}	0.41×10^{-6}	0.23×10^{-4}	-0.70×10^{-2}

[a]All values are in cm^{-1}

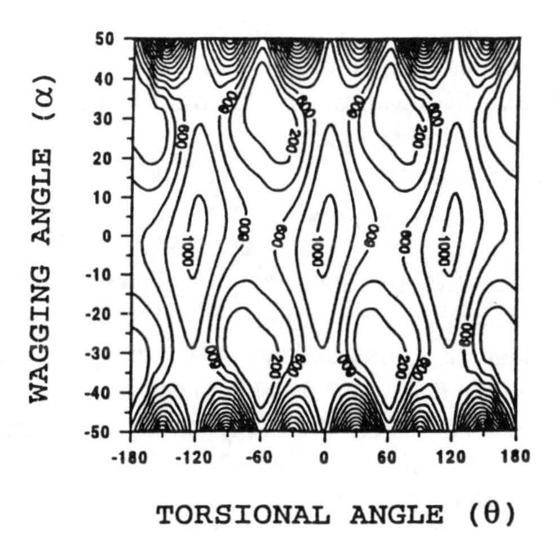

Fig. 9. The potential surface of S_1 acetaldehyde derived from fitting the observed levels to the calculated torsion-wagging energy levels [20]. The interval between isopotential lines is 200 cm^{-1}

Figure 10 shows schematically the Newman projections for the equilibrium positions for the electronic states along with the *ab initio* one-dimensional potential function $V(\theta)$ for torsion derived from *ab initio* calculations. What is clear is that the

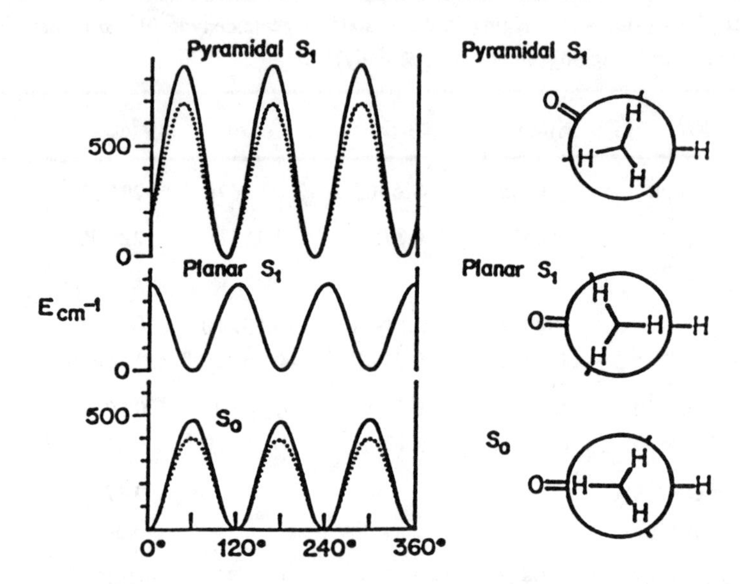

Fig. 10. The Newman projections for the conformers of S_0 and S_1 acetaldehyde and the one-dimensional potential functions for methyl torsion. The solid and dashed curves represent computational and experimental potential functions, respectively

shift from eclipsed to staggered configuration on $S_1 \leftarrow S_0$ excitation is not a consequence of the pyramidalization of the acetyl hydrogen. Rather, the nonplanar distortion of the aldehyde hydrogen actually has the effect of doubling the height of the barrier for the staggered conformation.

The shift in methyl conformation that exists between the two states could have its source in either the loss of an electron from the n nonbonding electron or from the addition of this same electron to the π^* antibonding orbital. The loss of the n electron gives rise to the first ion state that is eclipsed planar, as judged from the structures of the Rydberg states converging on the $^2A'$ ionization limit. For this reason, we assume that the conformational shift comes from the insertion of the electron into the π^* orbital [21]. It is possible to construct pseudo-π type orbitals on the two methyl hydrogen atoms that project out of plane that are able to hyperconjugate with the π orbitals of the C–C=O group. Figure 11 shows the arrangement for thioacetaldehyde.

In this picture, the σ(CH) MOs of the methyl group can be decomposed into the $2a' \oplus a''$ direct sum of the representations of the C_s rigid point group. The out of plane a'' AO is constructed from a pair of 1s orbitals and has a node at the molecular plane. It has the resemblance of a p orbital. Three other p orbitals on the carbon and oxygen atoms complete the description of the pseudo-π network. Thus, the

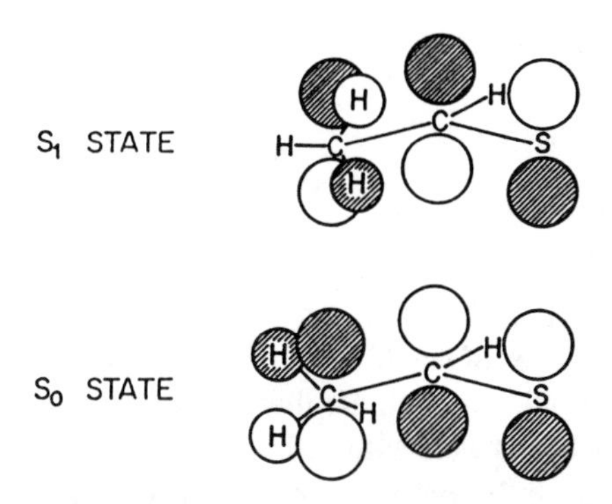

Fig. 11 Conjugative model for the pseudo-π orbitals in the S_0 and S_1 states of thioacetaldehyde

arrangement of the orbitals is not unlike that of the butadiene system. These orbitals can be constructed from the LCAO of the pseudo-π system, and in order of increasing energy they would be defined by four orbitals with nodal structure (a) ++++, (b) ++−−, (c) +−−+, (d) +−+−. In the S_0 state the (a) and (b) MOs would be fully occupied, $(a)^2(b)^2$, whereas the S_1 state would have configuration $a)^2(b)^2(c)^1$. If the relative stabilities of the two states are determined by π-type through-space intramolecular interactions between the end atoms, then (b) and (d) would be antibonding, whereas (a) and (c) would be bonding. This simple hyperconjugative model predicts that the eclipsed conformation is the preferred structure for the S_0 state, whereas it is the staggered conformation that would be stable for the S_1 state. The second major through-space interaction arises from the two electrons in the *n* orbital and the eclipsing in-plane hydrogen of the methyl group. On electronic excitation, the loss of one of these electrons reduces this interaction and thereby destabilizes the structure. Recently, we have explored these ideas with the *ab initio* ROHF/RHF method [22]. The general conclusion that was reached was that while both of these in-plane and out of plane mechanisms are responsible for the observed conformation shifts, the magnitude of the torsional barriers is the result of global changes in the electron density that arises largely from the CHO frame, rather than from the CH_3 group itself.

The Franck-Condon activity of ν_{14} acetyl wagging, the other large-amplitude mode, comes about from the nonplanar distortion of the aldehyde hydrogen. This pyramidalization of the S_1 state of carbonyl molecules was first predicted by Walsh [23] and then worked out in detail by Brand [24] for formaldehyde. A correlation of the one electron MO energy (Walsh's diagram) for the carbonyl chromophore shows that the π* orbital becomes more stable when the structure distorts from the planar 120° plane to a 90° right pyramid. (Walsh's rule works on the principle that s orbitals are more stable than *p* orbitals for the same principal quantum number). With the distortion of the frame from the plane, the form of the π* orbital changes from a pure *p* to an sp^3 hybrid, which results in a decrease in energy. Thus, it is the presence of the single electron in the π^* orbital that is responsible for the nonplanar acetyl structure in the $^1A''$ state.

3 The First Triplet State, T_1

With very long path lengths (168 m) and acetaldehyde vapor pressures close to saturation, we observed a series of bands at the red end of the UV absorption spectrum that have very different rotational contours than those of the higher energy singlet-singlet bands [25]. These bands were assigned to the $T_1 \leftarrow S_0$ electronic spectrum of acetaldehyde. At that time, little could be done in the way of vibronic analysis since at room temperature the spectrum quickly became congested at higher wave numbers.

As phosphorescence signals had been obtained earlier [26], we felt that a pulsed-slit jet might create an ideal molecular source for the detection of weak excitation signals. The setup for excitation experiments is shown as a schematic diagram in Fig. 12. The rotating slit nozzle [27] of dimension 0.22 x 46 mm was excited

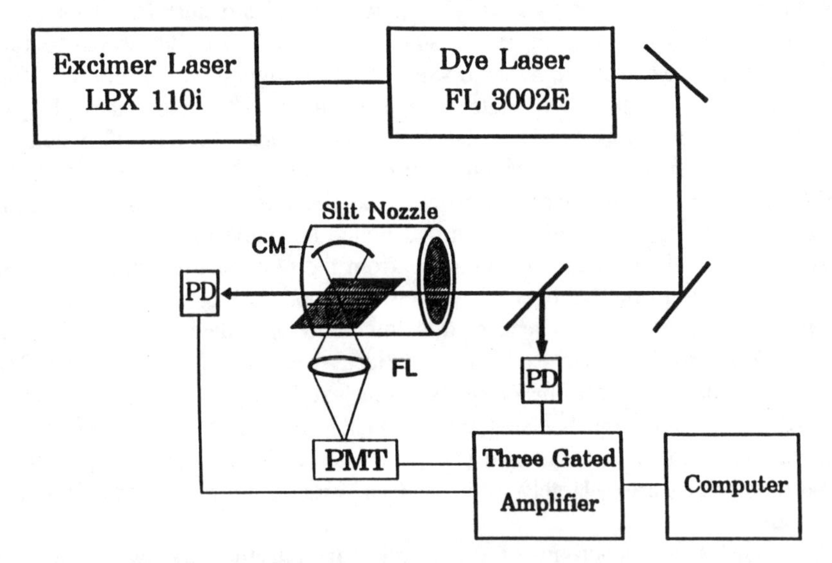

Fig. 12. A schematic diagram of the experimental apparatus used for $T_1 \leftarrow S_0$ LIP excitation spectroscopy [28]

crossways by the pulsed laser, and the phosphorescence molecules were detected 10 mm downstream from the slit nozzle. Figure 13 shows the spectra [28]. As a transition that involves the interchange of the electron spin direction during the electron promotion, it is forbidden by the $\Delta S = 0$ selection rule. The strength of the triplet-singlet transition comes from a spin-orbit coupling between the triplet and the higher singlet excited states. In the language of perturbation theory, the intensity is said to be borrowed or stolen from neighboring singlet-singlet transitions, provided that the combining electronic states are of the correct symmetry. Such a mechanism

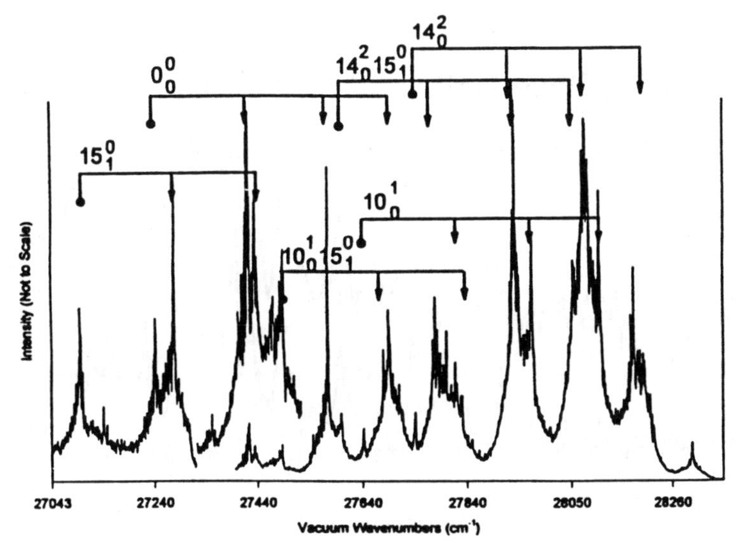

Fig. 13. Survey LIP excitation spectra of acetaldehyde [28]. Progressions in the Franck-Condon active mode ν'_{15} are indicated by horizontal leaders

operating in acetaldehyde would allow the in-plane *a*/*b* borrowing of intensity from the higher $\pi \rightarrow \pi^*$ valence transitions. Thus while even quanta of the ν_{14} mode (transitions such as 14^2_0) would be Franck-Condon active, odd quanta of ν_{14} (such as 14^1_0) would be absent in the spectrum. As a consequence of the more restrictive selection rules, the phosphorescence spectrum should be more open and less congested.

As a starting point, the LIP spectrum of acetaldehyde was simulated using anharmonic models generated from *ab initio* data at the UMP2/6-311G(*d,p*) level [29]. The results of the simulation predicted that the first member of acetyl torsion, 14^1, should be separated from the zero point level, 0^0, by only 4.58 cm^{-1}. The small value of the $0^- - 0^+$ inversion doubling indicates that the barrier to wagging inversion is much higher in the T_1 state than it is in the S_1 state. Moreover, as a consequence of the torsion-wagging coupling, the *indirect* transitions to the 14^1 level were calculated to have nearly 40% of the intensity of those to the 0^0 origin level. Figure 14 shows the 14^1_0 and 0^0_0 band clusters in the origin region of the LIP spectrum. The lower trace is a simulation of the band from a singlet–triplet band contour program. The observed and fitted levels are given in Table 7.

The experimentally derived potential energy surface for triplet acetaldehyde is given in Fig. 15. The height of the barriers to inversion and torsion are 869.0 and 609.7 cm^{-1} and the minima in the potential appear at $\theta_{min} = 61.7°$ and $\alpha_{min} = 42.2°$.

Table 8 compares the methyl torsion and acetyl inversion barriers, calculated at different levels of the *ab initio* approximation. The best overall agreement between the calculated and observed barrier parameters is found with the set created by the addition of five *d* functions UMP2(full)/6-311G(2*d,p*) or with additional *sp* diffuse functions, UMP2(full)/6-311+G(*d,p*).

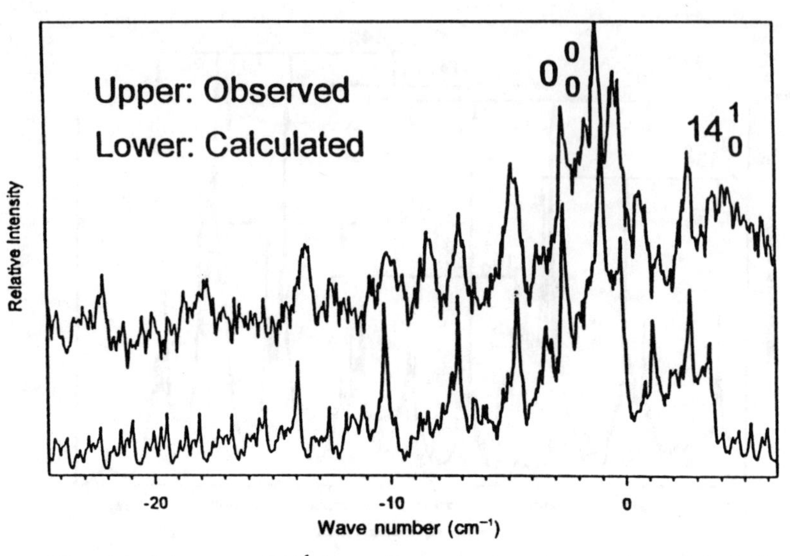

Fig. 14. The 0^0_0 and 14^1_0 band clusters in the origin region of the LIP excitation spectrum of acetaldehyde [28]

Table 7. Observed and calculated torsion-wagging energy levels in the lowest triple (T_1) state of acetaldehyde, in cm^{-1}

Q_{14}[a]	Q_{14}[b]	*Sym.*[c]	*Calc.*	*Obs.*
0	0	a_1	0.00	0
		e	0.01	
1	0	e	180.10	180
		a_2	180.70	
2	0	a_1	332.50	333
		e	334.20	
3	0	e	448.70	449
		a_2	475.25	
0	1	a_2	3.70	
		e	3.70	4
1	1	a_1	180.25	
		a_1	180.48	
2	1	a_2	331.33	
		e	335.94	
3	1	e	449.80	
		a_1	472.21	
0	2	e	501.96	
		a_1	502.20	502

[a]Torsional quantum number, high barrier notation
[b]Acetyl wagging quantum number, high barrier notation
[c]Irreducible representation of the G_6 nonrigid group

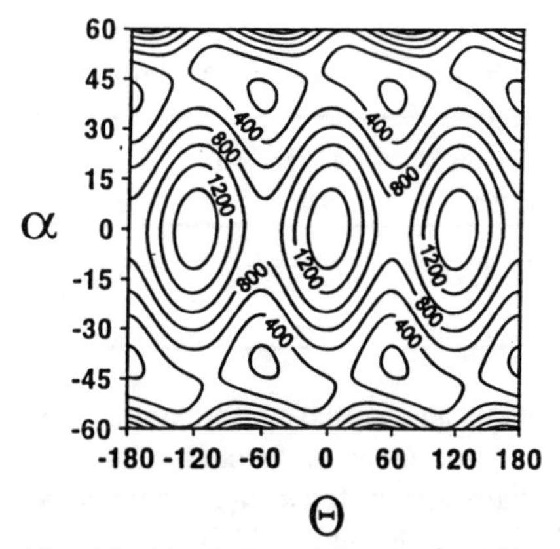

Fig. 15. The fitted potential surface of T_1 acetaldehyde [28]. The interval between isopotential lines is 200 cm^{-1}

Table 8. Calculated and observed methyl torsion and acetyl inversion barriers for the T_1 acetaldehyde, in cm^{-1}, [28]

Method	Torsion barrier	Inversion barrier
Ab initio		
4-31G//STO-3G	326.0	...
UHF/STO-3G//UHF-STO-3G	349.8	804.5
CISD//3-21G//RHF/3-21G	586.9	854.6
CISDQ/3-21G/RHF/3-21G	588.2	867.4
3-21G//3-21G	521.2	...
CIS/6-31+G(*d*)//6-31+G(*d*)	...	1112.7
UHF/6-31+G(*d*)//UHF/6-31+G(*d*)	...	948.1
UHF/6-311G(*d*,p)//UHF/6-311G(*d*,*p*)	612.1	930.8
UMP2/6-311G(*d*,*p*)//UMP2/6-311G(*d*,*p*)	647.4	968.1
UMP3/6-311G(*d*,*p*)//UMP2/6-311G(*d*,*p*)	608.8	854.8
UMP4/6-311G(*d*,*p*)//UMP2/6-311G(*d*,*p*)	633.8	965.9
UMP2/6-311G(2*d*,*p*)//UMP2/6-311G(2*d*,*p*)	603.6	873.3
UMP2/6-311+G(*d*,*p*)//UMP2/6-311+G(*d*,*p*)	644.2	887.8
Abs. (One-dimensional fit)	624.8	...
Abs. (One-dimensional fit)	590.0	1110.0
LIP (T_1 state two-dimensional fit)	609.7	869.0

The comparison of the T_1/S_1 barrier heights (in cm^{-1}) in Table 9 gives 869.0/ 638.6. for the acetyl inversion and 609.7/712.5 for methyl torsion. The insensitivity of the methyl torsion barrier and minimum of the methyl torsional potential (θ = 61.7°/59.9°) to spin interchange is interesting in view of the conformational changes that occur between the T_1 and S_1 states of acetophenone [30]. Analyses of the LIF and LIP spectra have shown that for this molecule, the methyl

Table 9. Comparison of the barriers to torsion and inversion and the minima in the potential hypersurface for $S_1(n\pi^*)$ and $T_1(n\pi^*)$ acetaldehydes

State	Torsion barrier	Inversion Barrier	θ	α
S_1	712.5 cm^{-1}	638.6 cm^{-1}	59.9°	33.5°
T_1	609.7 cm^{-1}	869.0 cm^{-1}	61.7°	42.2°

group in the $T_1(n\pi^*)$ state adopts an eclipsed conformation; whereas in the companion $S_1(n\pi^*)$ state, the staggered conformation is favored. Thus, the two states that have the same nπ^* configuration have very different equilibrium structures in acetophenone. Based on the present results for acetaldehyde, which has no $\pi\pi^*$ state near the lowest-energy $n\pi^*$ states, we propose that the $T_1(n, \pi^*)$ acetophenone has a significant $\pi\pi^*$ character due to its vibronic coupling with the close-lying $\pi\pi^*$ triplet state. The infusion of $\pi\pi^*$ character into the $T_1(n, \pi^*)$ state would lead to the conformational preference of the methyl group similar to that for S_0 acetophenone. In acetaldehyde, which has a very large $n\pi^*$ (S_1, T_1) - $\pi\pi^*$ (S_2, T_2) electronic energy gaps, the $n\pi^*$ - $\pi\pi^*$ vibronic coupling is unimportant, thus accounting for the insensitivity of the methyl torsional barrier to the difference in spin multiplicity of the $n\pi^*$ electronic state. It should be noted, however, that the direction of the barrier height change from S_1 to T_1 is towards the conformation of the S_0 state as in the case of acetophenone. The effect of spin interchange on the barrier height to inversion at the acetyl center occurs in the correct direction. For example, in the molecular prototype H_2CO, the T_1/S_1 barrier has been measured (in cm^{-1}) to be 667/354. In this case, the barrier height difference can be related to the Fermi correlation energy $K(n,\pi^*)$ that is responsible for the energy separation of the T_1 and S_1 states [31]. In the case of acetaldehyde, the effect is to reduce the correlation energy by extending the electron densities on the n and π^* orbitals in to the methyl group and thereby reducing the barrier height variation.

Summary

We have demonstrated in this chapter that the combination of supersonic-jet laser spectroscopy and *ab initio* calculations provides a powerful method for the precise determination of the structure of small molecules that undergo large amplitude torsion and inversion. Our analyses of the $S_1(n\pi^*) \leftarrow S_0$ and $T_1(n\pi^*) \leftarrow S_0$ spectra of acetaldehyde constitute the first detailed UV investigation of such species under rotational resolution. The origins of vibronic bands extracted from rotational analyses were fitted to a set of levels that were obtained from a Hamiltonian that employed flexible torsion-wagging large-amplitude coordinates. These analyses reveal that whereas the torsional barriers and the torsional minima are similar in the $S_1(n\pi^*)$ and $T_1(n\pi^*)$ states, the barrier height for the acetyl inversion is substantially greater in $T_1(n\pi^*)$ relative to $S_1(n\pi^*)$. The rotationally resolved $S_1(n\pi^*) \leftarrow S_0$ spectra are dominated by hybrid transitions that have components along the a, b, and c principal axes that originate as either $a-a$ or $e-e$ transitions. The appearance of the a- and b-type $e-e$ torsional bands is entirely consistent with the transitions arising from the torsion-wagging intermode coupling.

The study of methyl internal rotation in $S_1(n\pi^*)$ and $T_1(n\pi^*)$ acetaldehydes described here complements a significant body of information that is available about the dynamics of the methyl torsion for the S_0 ground electronic state. Presently, we are extending the experimental and theoretical investigations of the acetaldehyde $S_1(n\pi^*) \leftarrow S_0$ system to the bands lying higher than about 489 cm^{-1} above the 0^0_0 band origin.

Acknowledgments. We would like to express thanks to our colleagues who have made this work possible: Camelia Muñoz-Caro and Alfonso Niño, Universidad de Castilla-La Mancha, Ciudad Real, Spain; Richard Judge, University of Wisconsin-Parkside, Kenosha, WI; and Haisheng Liu, The University of Akron, Akron, OH. This work was supported by the National Sciences and Engineering Council of Canada, the U.S. Department of Energy, and in part by the U.S. National Science Foundation. Edward C. Lim is the holder of the Goodyear Chair in Chemistry at The University of Akron.

References

1 (a) S. Belov, M. Yu Tretyakov, I. Kleiner, and J. T. Hougen, J. Mol. Spectros. **160**, 1 (1993); (b) S. Belov, G. T. Fraser, J. Ortigoso, B. H. Pate, M. Tretyakov, and M. Yu, Mol. Phys. **81**, 359 (1994); (c) H. Gu, T. Kundu, and L. Goodman, J. Phys. Chem. **97**, 7194 (1993); (d) A. Bauder and Hs. H. Gunthard, J. Mol. Spectros. **60**, 290 (1976); (e) J. S. Crighton, and S. Bell, J. Mol. Spectros. **112**, 285 (1985); (f) L. Goodman, J. Leszczynski, and T. Kundu, J. Chem. Phys. **100**, 1274 (1994).

2 (a) D. C. Moule and A. D. Walsh, Chem. Rev. **75**, 67 (1975); (b) D. J. Clouthier and D. A. Ramsay, Ann. Rev. Phys. Chem. **34**, 31 (1983).

3 D. C. Moule, A. Niño, C. Muñoz-Caro, and Y. G Smeyers, *Structures and Conformations of Nonrigid Molecules*, edited by J. Laane et al. (Kluwer Academic, Amsterdam, 1993), pp. 591-602.

4 (a) M. Noble, E. C. Apel and E. C. K. Lee, J. Chem. Phys. **78**, 2219 (1983); (b) M. Noble and E. C. K. Lee, J. Chem. Phys. **81**, 1632 (1984); (c) M. Noble and E. C. K. Lee, J. Chem. Phys. **80**, 134 (1984).

5 (a) M. Baba, H. Hanazaki, and U. Nagashima, J. Chem. Phys. **82**, 3928 (1985); (b) M. Baba, U. Nagashima, and H. Hanazaki, J. Chem. Phys. **83**, 3514 (1985).

6 (a) A. Niño, C. Muñoz-Caro, and D. C. Moule, J. Mol. Structure, **318**, 237 (1994); (b), C. Muñoz-Caro, A. Niño, and D. C. Moule, J. Chem. Soc. Faraday Trans. **91**, 399 (1995).

7 C. Muñoz-Caro and A. Niño, Comput. Chem. **18**, 413 (1994).

8 C. Muñoz-Caro and A. Niño, QCPE Bull. **13**, 4 (1993).

9 S. L Altmann, Proc. R. Soc. A, **298**, 184 (1967).

10 H. C. Longuett Higgins, Mol. Phys. **6**, 445 (1963).

11 C. M. Woodman, Mol. Phys. **19**, 753 (1970).
12 C. Muñoz-Caro, A. Niño, and D. C. Moule, J. Mol. Structure, **350**, 83 (1995).
13 M. Dupuis, D. Splanger, and J. J. Wendolski, in: *National Resource for Computations in Chemistry*. Software Catalog (University of California, Berkeley, CA, 1980) Program QG01; M. W. Schmidt, J. A. Boatz, K. K. Bladridge, S. Koseki, M. S. Gordon, S. T. Elbert, and B. Lam, QCPE Bull. **7**, 115 (1987); M. W. Schmidt, K. K. Baldridge, J. A. Boatz, J. H. Jensen, S. Koseki, M. S. Gordon, K. A. Nuygen, T. L. Windus, and S. T. Elbert, QCPE Bull. **10**, 52 (1992).
14 M. J. Frisch, G. W. Trucks, M. Head-Gordon, P. M. W. Gill, M. W. Wong, J. B. Foresman, B. G. Johnson, H. B. Schegel, M. A. Robb, E. S. Replogle, R. Gomperts, J. L. Andres, K. Raghavachari, J. S. Binkley, J. J. P. Stewart, R. L. Martin, D. J. Fox, D. J. Defrees, J. Baker, J. J. P. Stewart, and J. A. Pople, *Gaussian 92*, Revision E2, Gaussian Inc., Pittsburgh, PA, 1992.
15 A. Niño, C. Muñoz-Caro, and D. C. Moule, J. Phys. Chem. **99**, 8511 (1995).
16 C. Muñoz-Caro, A. Niño, and D. C. Moule, Chem. Phys. **186**, 221 (1994).
17 H. Liu, E. C. Lim, R. H. Judge, and D. C. Moule, J. Chem. Phys. **102,** 4315 (1995).
18 P.-N. Wang, H. Liu, and E. C. Lim, J. Chem. Phys. **105**, 5697 (1996).
19 R. A. Weersink, D. T. Cramb, S. C. Wallace, and R. D. Gordon, J. Chem. Phys. **102**, 623 (1995).
20 H. Liu, E. C. Lim, C. Muñoz-Caro, A. Niño, R. H. Judge, and D. C. Moule, J. Mol. Spectros. **175**, 172 (1996).
21 D. J. Clouthier and D. C. Moule, Top. Current Chem. **150**, 167 (1989).
22 C. Muñoz-Caro, A. Niño, and D. C. Moule, Theor. Chim. Acta **88**, 299 (1994).
23 A. D. Walsh, J. Chem. Soc. 2306 (1953).
24 J. C. D. Brand, J. Chem. Soc. 858 (1956).
25 D. C. Moule and K. H. K. Ng, Can. J. Chem. **63**, 1378 (1985).
26 M. D. Schuh, S. Speiser, and G. H Akinson, J. Phys. Chem. **88**, 2224 (1984).
27 A. Amirav, C. Horwitz, and J. Jortner, J. Chem. Phys. **88**, 3092 (1988).
28 H. Liu, E. C. Lim, C. Muñoz-Caro, A. Niño, R. H. Judge, and D. C. Moule, J. Chem. Phys. **105**, 2547 (1996).
29 A. Niño, C. Muñoz-Caro, and D. C. Moule, J. Phys. Chem. **98**, 1519 (1994).
30 J. L Tomer, L. H. Spangler, and D. W. Pratt, J. Am. Chem. Soc. **110**, 1615 (1988); W. Siebrand, M. Z. Zgierski, A. Amirav, B. Fuchs, A. Penner, and E. C. Lim, J. Chem. Phys. **95** 3074 (1991).
31 S. P. McGlynn, T. Azumi, and M. Kinoshita, *Molecular Spectroscopy of the Triplet State* (Prentice-Hall, Englewood Cliffs, NJ, 1969).

Control

Excited State Dynamics and Chemical Control of Large Molecules

Valentin D. Vachev and John H. Frederick

Department of Chemistry and Program in Chemical Physics, University of Nevada, Reno, NV 89557, USA

Abstract. A detailed knowledge of the vibrational dynamics that occurs on excited states is important in understanding photochemical processes and in controlling their outcome. We examine ways in which spectroscopic observations can be used to construct models for electronic excited states and infer the relevant nuclear dynamics on them. Our methods are applied to the well-studied photoisomerization reactions of stilbene and used to explore two photon (IR + UV) schemes for manipulating the photoproduct branching ratio.

1 Introduction

Control over the outcome of molecular processes has been a long-standing goal of chemists. Among the oldest and most versatile tools at our disposal for exercising chemical control is radiation, and for this reason, photochemistry has long been a prominent area of study. However, many photochemical processes involve first an electronic excitation, followed by nuclear motion on the excited electronic state(s), radiationless crossing back to the ground electronic state, and finally relaxation into one or more potential minima representing different photoproducts. It is therefore difficult, using only a single photon, to manage this process to a single desired outcome (product).

Several imaginative schemes for chemical control using various forms and combinations of radiation have been proposed [1–6]. However, for these schemes to be successfully implemented in the laboratory, much must be known about the molecular systems to be manipulated. First, the (nonstationary) state prepared by the interaction of the molecule with radiation and the potential surface(s) on which it evolves in time should be known. If the radiation promotes the molecule to an electronic excited state, then one must also know the coupling between this state and other states, particularly the ground state, so that the relevant nonadiabatic dynamics can be included as part of the control scheme. Finally, the relaxation of the molecule due to intramolecular vibrational redistribution (IVR) or interaction with collision partners or external fields must also be taken into account. All of these details contrive to make chemical control a difficult undertaking in all but the most ideal systems.

Much work over the years has been devoted to understanding the interaction of molecules with radiation and the states that can be prepared by coherent, pulsed light sources [7, 8]. There has also been a significant effort to understand IVR, collisional relaxation processes, and relaxation in condensed phases of highly excited molecules [9–11]. As a result, the greatest area of uncertainty in applying control schemes to photochemical processes is the potential surface of the excited state on which the initial dynamics takes place and the nonadiabatic processes that convey the system back to the ground state for final disposition of the photoproducts. In this chapter, we describe methods for inferring the structure of the excited state from spectroscopic observations and using this structure to simulate the excited state dynamics.

It should be mentioned that many photochemical reactions are believed to be the statistical result of "hot" ground state dynamics. In other words, the effect of the photoexcitation is simply to place enough energy in the system so that barriers to isomerization or dissociation can be surmounted. When this is the case, the excited state dynamics are irrelevant to the final disposition of products. This does not apply, however, if the dynamics on the excited state remain coherent, and spectroscopic means can be used to induce the transition back to the ground state. In this event, knowledge of the excited state dynamics can be potentially useful for controlling the outcome of even statistical photochemical processes.

Our main source of detailed information about excited electronic states comes from electronic spectroscopy. High-resolution absorption spectroscopy can provide information about vibrations activated by the transition to the excited state, including their frequency and the displacement of their equilibrium position relative to the ground electronic state [12, 13]. This is extremely useful because these active vibrations are the same ones that contribute to the photochemical process. One can also derive useful information from ultrafast probes of the excited state dynamics [14]. These measurements provide time scales for various dynamical events and corresponding clues about the slope of the potential surface in the region where the dynamics takes place.

We begin in the next section by describing how our theoretical models for the excited state are constructed from experimental observations and applying these techniques to benzophenone and stilbene. In Sect. 3, we consider the dynamics of nonadiabatic transitions between electronic states, particularly the process of internal conversion, and apply quantum and semiclassical techniques to study simple models for internal conversion in π-conjugated systems. We then review results obtained from simulating the photochemical dynamics of *cis*-stilbene on its lowest excited state and discuss the connection between the excited state dynamics and the yields of various photoproducts in Sect. 4. Finally, in Sect. 5, we simulate the use of various two-photon (IR + UV) excitation schemes for *cis*-stilbene and explore the consequences this has on the final photoproduct yields.

2 Absorption Spectra and Potential Energy Surfaces

A prerequisite for any dynamical study is the Born–Oppenheimer potential energy surface(s) (PES) on which the reaction takes place. In the last decade there has been significant progress in using electronic structure calculations to predict the pathways a chemical reaction may follow by calculating cross-sections of the PES along certain coordinates or along a mimimum energy path [15–17]. Despite this progress, it remains impractical to construct a global potential surface for any but the smallest molecular systems using *ab initio* methods. It is therefore important to make use of qualitative trends predicted from electronic structure calculations, as well as spectroscopic information and explicit dynamical calculations in building more quantitatively precise surfaces.

In this section we describe a practical method for constructing empirical PESs in which the potential parameters are optimized to fit available electronic structure and spectroscopic data. The method is then illustrated with realistic applications to the $S_1 \leftarrow S_0$ transitions in benzophenone and stilbene.

2.1 Calculation of Absorption Spectra and Mode Displacements

If we consider a polyatomic molecule with two adiabatic electronic states $|\mathrm{g}\rangle$ and $|\mathrm{e}\rangle$ and N_v vibrational degrees of freedom, we can write the Hamiltonian as

$$\mathcal{H} = |\mathrm{g}\rangle\mathcal{H}_\mathrm{g}\langle\mathrm{g}| + |\mathrm{e}\rangle(\hbar\omega_\mathrm{ge} + \mathcal{H}_\mathrm{e})\langle\mathrm{e}| \quad . \tag{1}$$

Two approximations are often used to simplify the treatment of large molecular systems: *(i)* $|\mathrm{g}\rangle\mathcal{H}_\mathrm{g}\langle\mathrm{g}|$ and $|\mathrm{e}\rangle(\hbar\omega_\mathrm{ge} + \mathcal{H}_\mathrm{e})\langle\mathrm{e}|$ are both harmonic oscillator Hamiltonians, and *(ii)* the normal mode coordinates are the same in both states (i.e. there is no Dushinsky rotation). In this case, the harmonic oscillator Hamiltonians may be written

$$\mathcal{H}_\mathrm{g} = \frac{\hbar}{2}\sum_{j=1}^{N_\mathrm{v}} \omega_j^{''}(\tilde{p}_j^{''2} + \tilde{q}_j^{''2} - 1), \qquad \mathcal{H}_\mathrm{e} = \frac{\hbar}{2}\sum_{j=1}^{N_\mathrm{v}} \omega_j^{'}(\tilde{p}_j^{'2} + \tilde{q}_j^{'2} - 1) \quad . \tag{2}$$

In (1) and (2), ω_ge is the electronic transition frequency between the lowest vibronic states of $|\mathrm{g}\rangle$ and $|\mathrm{e}\rangle$ (0–0 transition). All ground-state quantities are labeled by a double prime, while excited-state quantities are denoted by a single prime. In each state, $\boldsymbol{p}$ and $\boldsymbol{q}$ represent the dimensionless normal mode momentum and position vectors, respectively. The vector of normal mode displacements between the electronic surfaces, $\boldsymbol{D}$, is defined as $\boldsymbol{q}^{'} - \boldsymbol{q}^{''}$ and is related to the relative intensities in the electronic absorption spectrum.

The absorption spectrum is calculated by Fourier-transforming the linear response function $\chi(\tau)$ [18–20],

$$P(\omega_\mathrm{L}) = A\ \mathrm{Re}\left\{\int \mathrm{d}\tau\ \exp[\mathrm{i}(\omega_\mathrm{L} - \omega_\mathrm{ge})\tau]\ \chi(\tau)\right\} \quad , \tag{3}$$

where ω_L is the radiation frequency and A is a constant proportional to the strength of the radiation field and the dipole moment. Using the approximations described above, the response function can be treated as a product of single mode response functions and is given by [21, 22]

$$\chi(\tau) = \frac{\det[\sinh(\hbar\boldsymbol{\omega}''/kT)/(\mathbf{s}_1\mathbf{s}_2)] \; \exp[-\frac{1}{\hbar}\boldsymbol{D}^T \cdot (\mathbf{c}_1/\boldsymbol{\omega}'' + \mathbf{c}_2/\boldsymbol{\omega}')^{-1} \cdot \boldsymbol{D}]}{\det^{1/2}(\mathbf{c}_1/\boldsymbol{\omega}'' + \mathbf{c}_2/\boldsymbol{\omega}')\det^{1/2}(\mathbf{c}_1\boldsymbol{\omega}'' + \mathbf{c}_2\boldsymbol{\omega}')} \quad , \tag{4}$$

where $\mathbf{c}_i$ and $\mathbf{s}_i$ are diagonal matrices defined by

$$\mathbf{s}_1 = \sinh[\boldsymbol{\omega}''(\hbar/kT - \mathrm{i}\tau)], \qquad \mathbf{s}_2 = \sinh(\mathrm{i}\boldsymbol{\omega}'\tau/2),$$

$$\mathbf{c}_1 = \coth[\boldsymbol{\omega}''(\hbar/kT - \mathrm{i}\tau)/2], \qquad \mathbf{c}_2 = \coth(\mathrm{i}\boldsymbol{\omega}'\tau/2) \quad .$$

Here, $\boldsymbol{\omega}''$ and $\boldsymbol{\omega}'$ are the diagonal frequency matrices of the ground and excited states, T is the temperature, and "det" denotes a matrix determinant.

Equations (3) and (4) provide the basis for a very efficient calculation of the absorption spectra of polyatomic molecules. One needs to know the frequency matrices of the ground and excited states and the displacements along each normal coordinate in order to calculate the absorption spectrum. In many cases the experimental data (fluorescence excitation and dispersed fluorescence spectra) allow for identification of the optically active modes and their frequencies in the ground and excited states. The displacements that give a best fit to the absorption spectrum are then determined by iterative calculations. The actual result, however, is not only a more quantitative interpretation of the available experimental data, but it also helps to support (or disprove) its initial interpretation. Here, we demonstrate this procedure with an application to benzophenone [23–26].

The vibrational structure of the $S_1 \leftarrow S_0(n, \pi^*)$ transition of benzophenone can be resolved only using supersonic expansion techniques. Due to the very weak fluorescence from the S_1 state, Ito and coworkers have used a sensitized phosphorescence excitation detection scheme to obtain the absorption spectrum indirectly [27, 28], while Holtzclaw and Pratt have succeeded in recording the fluorescence excitation spectrum [29]. The spectrum is dominated by several long progressions whose members are separated by approximately 60 cm^{-1}. To model the (n, π^*) spectrum, we use a model with four displaced normal modes, two of which correspond to low-frequency torsion and bending motions of the phenyl rings and two of which represent higher-frequency stretching vibrations [30].

The frequencies of the normal modes can be deduced in most cases from an analysis of the line positions in the experimental spectrum. To determine displacements, the relative intensities of the observed lines must be fit. These displacements are found by an iterative procedure that involves simulating the spectrum using a harmonic oscillator approximation. The best-fit frequency and displacement parameters for benzophenone are given in Table 1, and the final calculated harmonic spectrum is shown in Fig. 1. This procedure depends critically upon the availability of good quality high-resolution spectroscopic data.

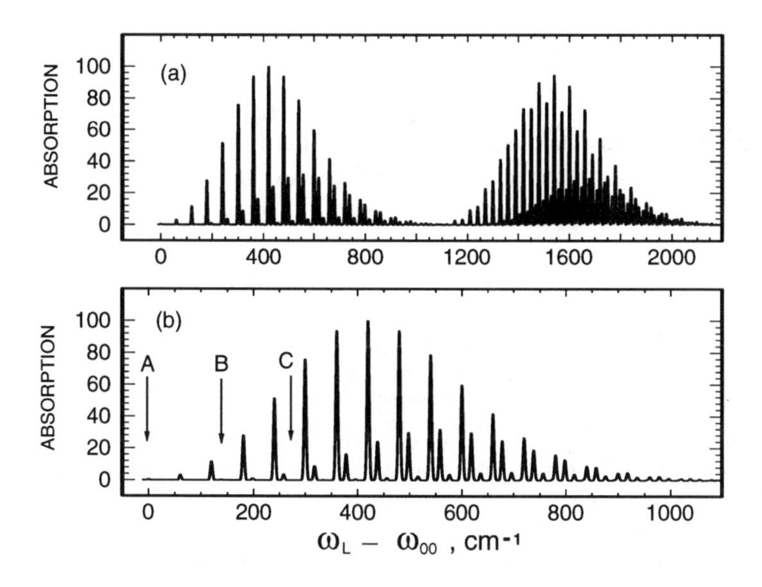

Fig. 1. (**a**) The $S_1 \leftarrow S_0$ absorption spectrum of benzophenone calculated with the parameters from Table 1; (**b**) Low-frequency part of the spectrum: **A** denotes the 0^0 origin, and **B** and **C** the origins of the progressions with one and two quanta in the bending mode, respectively.

Table 1. Parameters used for the calculation of the benzophenone $S_1 \leftarrow S_0$ absorption spectrum. Frequencies are given in cm^{-1}, and displacements are for dimensionless normal mode coordinates.

Mode	ω''	ω'	**D**
ν_{s1}	1120.0	1120.0	1.50
ν_{s2}	1090.0	1090.0	1.35
ν_b	165.0	140.0	0.80
ν_t	57.0	60.0	3.79

2.2 Model Molecular Mechanics Potential Energy Surfaces

The next step in using spectroscopic data to define a realistic PES is to link the dimensionless normal mode displacements to actual molecular changes and parameters used in defining the PES. We use a molecular mechanics approach in which the force field is formulated in terms of internal coordinates, that is, bond lengths, valence bond angles, dihedral torsional angles, and interatomic distances between nonbonded atoms [31, 32]. The resulting potential has the form

$$U(\boldsymbol{r}_1, \ldots, \boldsymbol{r}_N) = U_\mathrm{s} + U_\mathrm{b} + U_\mathrm{tor} + U_\mathrm{vw} \quad , \tag{5}$$

where $U_{\rm s}$, $U_{\rm b}$, $U_{\rm tor}$, $U_{\rm vw}$ are contributions due to bond stretches, valence angle bends, torsional (dihedral) angle twists, and nonbonded interatomic van der Waals interactions, respectively. In our view, the best strategy is to use the simplest possible functional forms for the individual terms in (6), yet they should be flexible enough to be able to represent both ground and excited state potentials.

For the valence bond stretch and bending forces, we assume harmonic potentials

$$U_{\rm s} = \frac{1}{2}\sum_b k_b(b - b_0)^2 \quad , \tag{6}$$

$$U_{\rm b} = \frac{1}{2}\sum_\varphi k_\varphi(\varphi - \varphi_0)^2 \quad , \tag{7}$$

in which b_0 and φ_0 are "equilibrium" values for the bond b and the angle φ, respectively, and k_b and k_φ are the force constants. The equilibrium position parameters above represent ideal values for the local interaction—the actual equilibrium values are determined from a balance between these local potentials and the van der Waals and torsional interactions. One could also use more complicated functions for these local vibrations if more data is available or if anharmonicity is important.

The potentials for the torsional angles θ are modeled by a truncated Fourier series,

$$U_{\rm tor} = 0.5\sum_\theta \{V_2[1 - \cos(2\theta)] + V_4[1 - \cos(4\theta)]\} \quad , \tag{8}$$

a form that has been widely applied to large molecules with extended π-systems. Here for each different type of torsional angle in the molecule, one must set the values of the force constants V_2 and V_4. For this functional form, V_2 determines the height of the barrier to rotation around a bond in the absence of steric interactions. In conjugated systems, it provides a measure of the disruption of the π-system as rotation occurs about the bond. V_4 affects the shape of the potential around the barriers and the minima. When $V_4 > \frac{1}{4}|V_2|$, the torsional potential develops additional extrema near $\theta = 0, \pi$.

Van der Waals (nonbonded) interactions are calculated for all pairs of atoms (i, j) that do not belong to the same chemical bond or the same valence angle using a Hill-type potential involving two parameters for each atom. One parameter is the van der Waals radius of the atom, and the other is an energy scale factor. The function representing this interaction is written

$$U_{\rm vw} = f\left[-2.25/r_n^6 + 8.28 \times 10^5 \exp(-r_n/0.0736)\right] \quad . \tag{9}$$

Here f is proportional to the depth of the potential and depends on the identity of the two atoms; and $r_n = r_{ij}/s$, where s is sum of the van der Waals radii for atoms i and j, and $r_{ij} = |\boldsymbol{r}_i - \boldsymbol{r}_j|$.

Using this potential surface, the object is to determine the relationship between changes in the potential parameters and the displacements between the two electronic surfaces. First, we express the positions of the atoms in Cartesian coordinates and, through minimization of the potential energy (5), find the equilibrium geometry of the molecule. After mass-weighting the Cartesian coordinates, the force constant matrix is diagonalized giving the squared normal mode frequencies (eigenvalues) and the unitary transformation matrix (mass-weighted Cartesian to normal coordinates) $\mathbf{V}$, whose columns are the eigenvectors corresponding to each normal mode. This allows the transformations from Cartesian positions $\mathbf{r}$ to normal mode coordinates, $\mathbf{Q}''$, to be performed via

$$\boldsymbol{Q}'' = \mathbf{V} \cdot \mathbf{M}^{1/2} \cdot \boldsymbol{r} \quad , \qquad \boldsymbol{P}'' = \mathbf{V} \cdot \mathbf{M}^{-1/2} \cdot \boldsymbol{p} \quad , \tag{10}$$

where $\mathbf{M}$ is the diagonal matrix of the atomic masses, and eigenvectors corresponding to "zero" frequencies (for translation and rotation) are omitted from $\mathbf{V}$. The displacement vector of the normal coordinates is simply $\bar{\boldsymbol{D}} = \boldsymbol{Q}' - \boldsymbol{Q}''$, while the components of the dimensionless normal mode displacement vector are calculated from

$$D_j = \left(\frac{\omega_j'}{\hbar} \right)^{1/2} \bar{D}_j \quad . \tag{11}$$

By adjusting the corresponding force constants and equilibrium parameters in (5–9) one can generally reproduce the experimentally deduced displacements and, consequently, the intensity distribution in the absorption spectrum. In this procedure, one first determines the parameters for the ground state to match the equilibrium structure and ground state normal modes. Next, one iteratively imposes changes in the potential parameters for the excited state, calculates the displacements, and compares these with the spectroscopically derived ones. During this process, most of the computer time is spent on calculating the matrix of the force constants, which changes with every change in the potential parameters.

We have used this procedure to find simple, yet good quality surfaces for both the ground and first excited states of stilbene [21]. Fourteen normal modes are used to model the absorption spectrum of *trans*-stilbene and calculate the spectroscopic displacements for each mode. Then, the empirical parameters are adjusted until these displacements match those inferred from the experimental spectrum. The final parameters are given in Table 2.

Table 2(a). Parameters for bond stretching potentials (6)

	Ground state		Excited state	
Bond	k_b, kcal/mole·Å^2	b_0, Å	k_b', kcal/mole·Å^2	b_0', Å
C_{ph}–C_{ph}	993.7	1.397	893.0	1.397
C_{ph}–H	725.6	1.113	725.0	1.113
C_e–C_e	1150.0	1.338	750.0	1.358
C_{ph}–C_e	818.0	1.458	1100.0	1.435
C_e–H	725.6	1.113	720.0	1.114

Table 2(b). Parameters for angle deformation potentials (7)

Angle	Ground state k_φ, kcal/mole·deg^2	Ground state φ_0, deg	Excited state k_φ', kcal/mole·deg^2	Excited state φ_0', deg
C_{ph}–C_{ph}–C_{ph}	0.016	120	0.013	120.1
C_{ph}–C_{ph}–H	0.014	120	0.011	120.4
C_{ph}–C_{ph}–C_e	0.010	120	0.008	121.3
C_{ph}–C_e–H	0.012	120	0.010	120.0
C_{ph}–C_e–C_e	0.014	120	0.008	121.7
C_e–C_e–H	0.015	120	0.009	120.8

Table 2(c). Parameters for torsional potentials (8)

Atoms	Ground state V_2, kcal/mole	Ground state V_4, kcal/mole	Excited state V_2', kcal/mole	Excited state V_4', kcal/mole
C_{ph}–C_{ph}	3.00	0	2.95	0
C_{ph}–C_{ph}–C_e–C_e	6.245	0	6.25	−1.72
C_{ph}–C_{ph}–C_e–H	3.00	0	3.00	0
C_{ph}–C_e–C_e–C_{ph}	29.00	0	−32.90	11.58
C_{ph}–C_e–C_e–H	3.00	0	2.95	0
H–C_e–C_e–H	3.00	0	2.95	0

Table 2(d). Parameters for van der Waals potentials (9)

Atom(s)	f, kcal/mole	r_{vdW}, Å
C–C	0.107	—
C–H	0.067	—
H–H	0.042	—
C	—	1.7
H	—	1.2

Later in this chapter, we will use these surfaces to model the photoinduced isomerization of *cis*-stilbene. These surfaces have also been used recently by Gershinsky and Pollak to study solvent effects on the isomerization of *trans*-stilbene [33, 34] and by Bolton and Nordholm for a study of IVR in *trans*-stilbene [35].

3 Nonadiabatic Dynamics

A typical photochemical reaction involves two electronic transitions: photoexcitation and radiationless transition back to the ground electronic state. The latter process is incompletely understood due to the absence of quantitatively accurate potential energy surfaces for excited states and the conical intersections they make with the ground state. In this section, we consider electronic transitions from two perspectives: *(i)* a quantum-dynamical treatment of a simple 1-D model, and *(ii)* a semiclassical treatment of a multidimensional model.

3.1 Time-Dependent Quantum Mechanical Modeling

The primary photochemical event in vertebrate rhodopsin involves an 11-*cis* to 11-*trans* photoisomerization [36–38]. The reaction coordinate in this case can be associated with the C_{10}–C_{11}–C_{12}–C_{13} torsional angle θ, and both electronic structure calculations and the observed ultrafast timescales for the reaction suggest a barrierless potential energy surface along θ. Here, we study a simple one-dimensional model of photoinduced isomerization, with parameters chosen to be representative of the energetics of isomerization in rhodopsin. The internal conversion process is inherently quantum-mechanical, and one goal of our studies is to determine to what extent a semiclassical treatment can be used to capture the essential dynamics of the process.

The present model consists of two diabatic surfaces,

$$V_{11}(\theta) = -\frac{D_2}{2}\left[1-\cos 2\left(\theta-\frac{\pi}{2}\right)\right] - \frac{D_4}{2}\left[1-\cos 4\left(\theta-\frac{\pi}{2}\right)\right] \quad , \tag{12}$$

$$V_{22}(\theta) = \epsilon + D\left(\theta-\frac{\pi}{2}\right)^2 \quad , \tag{13}$$

with coupling between them given by

$$V_{12}(\theta) = V\exp\left[-\left(\frac{\theta-\frac{\pi}{2}}{a}\right)^2\right] \quad . \tag{14}$$

These equations give a barrierless potential for the upper surface, $V_{22}(\theta)$, with a minimum at $\theta = \frac{\pi}{2}$. The lower surface, $V_{11}(\theta)$, has minima at both planar configurations ($\theta = 0, \pi$) and a maximum at $\theta = \frac{\pi}{2}$. The nonadiabatic transition region is centered at $\theta = \frac{\pi}{2}$, which is typical for polyene isomerization processes, as are the parameters $D = 8\times10^3$ cm^{-1}, $D_2 = 1\times10^4$ cm^{-1}, and $D_4 = 0$ cm^{-1}. The energy difference between V_{11} and V_{22} is ϵ at $\theta = \frac{\pi}{2}$, and is used as one of the variable parameters in the modeling. The two diabatic and adiabatic surfaces are shown in Fig. 2 for $\epsilon = -2000$ cm^{-1}. The barrier height on the lower diabatic state is 1×10^4 cm^{-1}, and the difference between the $\theta = 0$ and $\theta = \frac{\pi}{2}$ positions on the upper state is 1.974×10^4 cm^{-1}.

The dependence of the diabatic matrix elements $V_{12}(\theta)$ is Gaussian with the peak at $\theta = \frac{\pi}{2}$, as suggested by electronic structure calculations [39]. The interaction between the two surfaces is localized to the region of their crossing—

this is achieved by using small values for the parameter a. However, we also examine the dependence of the results on the magnitude of a. The corresponding adiabatic surfaces, obtained by diagonalizing the matrix $V_{ij}(\theta)$ at each point θ, are shown in Fig. 2b, along with the diabatic coupling $V_{12}(\theta)$.

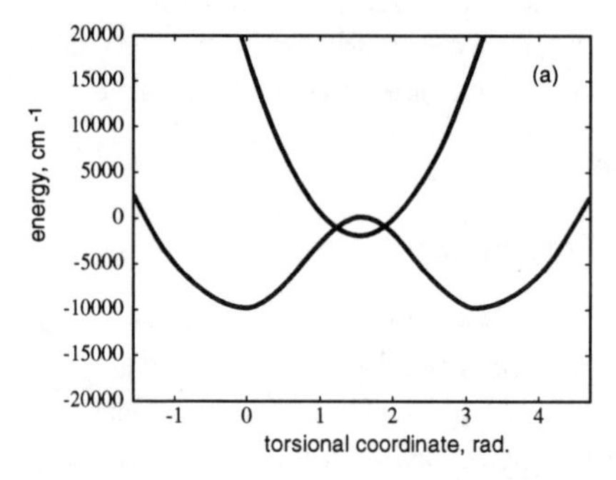

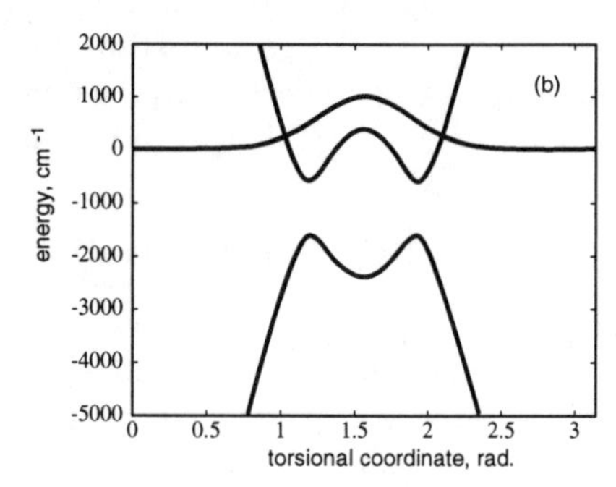

Fig. 2. (a) Diabatic potential energy surfaces used in the 1-D model. (b) Adiabatic potential energy surfaces with Gaussian interaction between them. Note that the horizontal and vertical scales of the two panels are different.

We use these model surfaces to simulate a process that begins with a vertical excitation from the left well of the lower surface. Dynamics are initiated on the upper state for a Gaussian wave packet with initial position $\theta_0 = 0$ and width approximating to the ground vibrational state on the lower surface. The effective mass (moment of inertia for the torsional motion) used in the calculations is 5.0×10^{-40} g·cm^2. The initial Gaussian wave packet is propagated in time by solving the time-dependent Schrödinger equation for the two coupled surfaces, using standard FFT grid methods [40–42] for a spatial grid of 1024 points and a time-step of 0.1 fs.

To illustrate the types of dynamics that accompany internal conversion, we first consider two cases: *(i)* $\epsilon = -1000$ cm^{-1}, $V = 500$ cm^{-1}, and *(ii)* $\epsilon = -2000$ cm^{-1}, $V = 1000$ cm^{-1}. These cases represent both moderate *(i)* and strong *(ii)* coupling between the surfaces, and results obtained for each case are displayed in Figs. 3 and 4, respectively. In panel (a) of each figure, we show the total ground state population as a function of time. For case *(i)*, there is some oscillation of population between the states while the wave packet is in the nonadiabatic crossing region, with a little less than 50% of the population ultimately transferred to the ground state. On the second traversal of the crossing region, approximately 50% of the remaining population on the excited state is again transferred to the ground state. In the strong coupling case *(ii)*, about 95% of the population is transferred with each traversal by the wave packet, and no population oscillation due to recrossing is observed (see Fig. 4a).

Panel (b) of Figs. 3 and 4 shows snapshots of the wave packet at 5 fs time intervals during its propagation. From these snapshots, we see that the wave packet remains localized during and after the nonadiabatic transition, so its dynamics are very classical in appearance. This suggests that it is appropriate to think of the dynamics as a combination of classical trajectory-like evolution on the individual surfaces with "hops" between the surfaces that preserve the position and direction of propagation of the system.

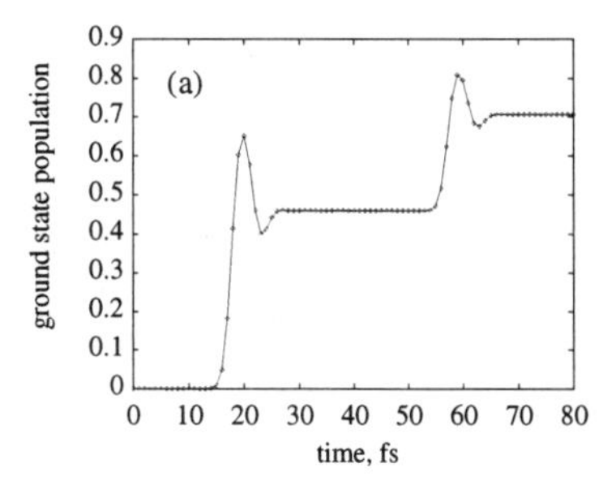

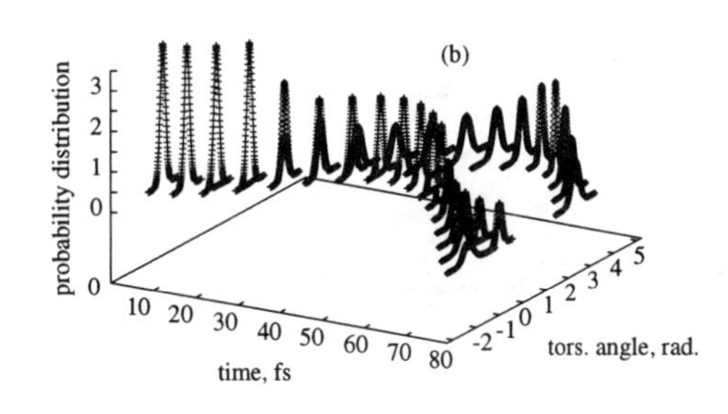

Fig. 3. (a) Ground state population as a function of time; (b) Snapshots of the nonstationary state at various times during propagation. These results correspond to a model system with $\epsilon = -1000$ cm^{-1}and $V = 500$ cm^{-1}. About 50% of the excited state population is transferred to the ground state on each passage through the interaction region.

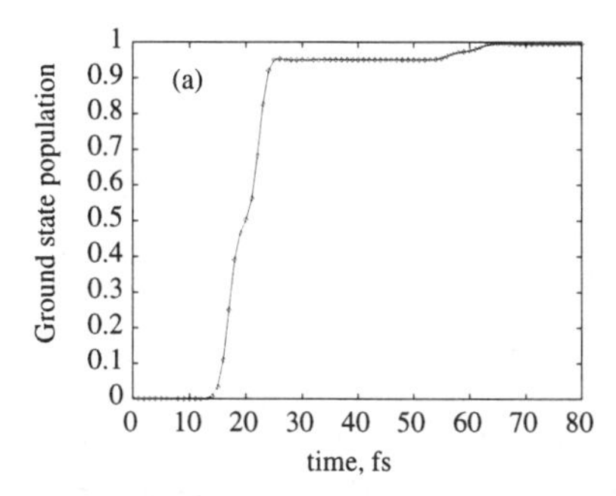

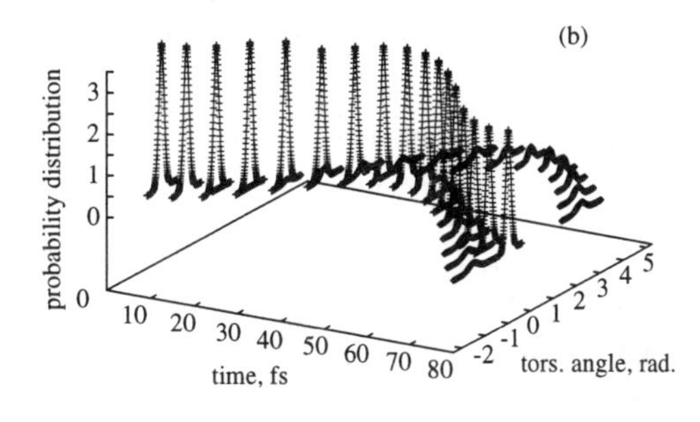

Fig. 4. Same as Fig. 3, but for $\epsilon = -1000$ cm^{-1}and $V = 500$ cm^{-1}. In this case, the interaction between the two electronic surfaces is very strong, and about 95% of the excited state population is transferred to the ground state on each passage through the interaction region.

To survey a range of dynamical behavior, we examine several sets of parameters (ϵ, V) to gain a better understanding of the internal conversion process. In Fig. 5, we summarize results for a wide range of these parameters. For a given interaction strength V, we calculate the fractional population transferred to the ground state on each passage through the interaction region for a number of different values of ϵ. Case *(i)* examined above is represented by one of the points on curve 2, while case *(ii)* is one of the points on curve 5.

For each interaction strength, we observe the same qualitative behavior. As observed before, the initial wave packet splits into two localized wave packets, one continuing on the excited state, the other one appearing on the ground state. Altogether, these results are very promising for the use of semiclassical methods to model internal conversion processes. Classical trajectories follow similar pathways when initiated in the same region as the initial probability distribution of the wave packet.

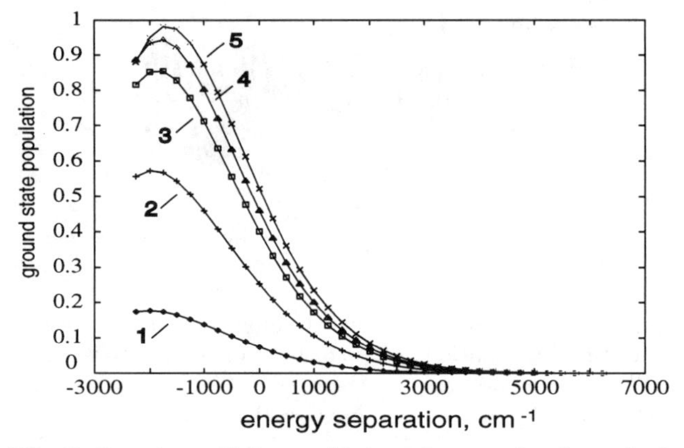

Fig. 5. Quantum efficiency of internal conversion for a single passage through the interaction region. The curves represent the ground state population after a single passage through the interaction region and correspond to different nonadiabatic interaction parameters: *(1)* 250; *(2)* 500; *(3)* 700; *(4)* 900; and *(5)* 1000 cm^{-1}.

The parameter sets leading to very high quantum efficiency for population transfer in Fig. 5 also have interesting implications for a semiclassical simulation of the nonadiabatic dynamics. These sets correspond to a crossing of the two surfaces with the minimum of the excited state lying 1000 to 2500 cm^{-1}lower than the maximum of the ground state and an interaction parameter in the range of 700 to 1000 cm^{-1}. In these ranges of ϵ and V, most of the population of the excited state is transferred to the ground state during a single passage. Therefore, the "surface hopping" of trajectories can be assumed to occur with near 100% efficiency when a trajectory enters the crossing region in these cases. A more important characteristic in this case is how closely the classical trajectories follow the quantum path in phase space. In the case of a multidimensional reaction coordinate and multiple products, a reliable algorithm should correctly reproduce the quantum yields of the products.

We have also investigated the dependence of the results in Fig. 5 on the localization of the interaction between the two states that is dictated by the parameter a in (14). Since very little is known about this parameter, we studied a broad range from 10^{-5} to 1, the smaller values representing a localized interaction, the larger a flat spatial dependence. Interestingly, we find that the quantum efficiency varies only by about 10% over this range of a-values for constant interaction strength V. The qualitative features of the dynamics are not affected. Thus, we expect a quasiclassical treatment to be valid, even for very delocalized coupling between the electronic states.

3.2 Quasiclassical Modeling

In this section, we outline semiclassical methods for simulating both the photoexcitation of a molecule at finite temperature and the nonadiabatic dynamics associated with internal conversion. These methods are used in conjunction with quasiclassical simulations of the dynamics on the individual electronic states. Quasiclassical simulations are simply classical molecular dynamics calculations in which the initial conditions of an ensemble of trajectories are chosen to represent an initial quantum state. From this ensemble of trajectories, one can compute quantities of physical insterest by averaging over the ensemble, thus mimicking the probabilistic aspect of quantum mechanics. For a chemical process with multiple product channels, we can also simulate the branching ratios between the various channels.

We begin by modeling the excitation of a molecule by an ultrashort laser pulse of finite duration. Pulse lengths of 100 fs are now routinely achievable in the laboratory. Since the fastest vibrations have a period of about 10 fs, the finite length of the laser pulse cannot be ignored when simulating the excitation process [43–46].

To generate initial conditions representing excitation with a laser pulse, we consider the dynamical displacements along each normal coordinate during the excitation. For a time interval τ, the displacement of the system is given by [21]

$$\Delta \boldsymbol{Q}(\tau) = -\boldsymbol{\omega}'^{\,-1} \mathbf{s}_2^{-1} (\mathbf{c}_1/\boldsymbol{\omega}'' + \mathbf{c}_2/\boldsymbol{\omega}')^{-1} \boldsymbol{D} \quad , \tag{15}$$

where the various quantities have the same meaning as in (4). Equation (15) is used to sample the phase space of the wave packet on the excited state. It represents the quantum delocalization that occurs due to time evolution during the laser pulse. This supplements the delocalization of the initial state due to the uncertainty spread and the Boltzmann thermal distribution in the ground electronic state. The displacements $\Delta \boldsymbol{Q}(\tau)$ are calculated from (15) for time intervals τ within the total laser pulse duration. Each calculated vector $\Delta \boldsymbol{Q}$ is added to the equilibrium normal coordinate vector and transformed to Cartesian coordinates to obtain a set of initial conditions for a classical trajectory on the excited state surface. The number of initial conditions generated this way determines the size of the trajectory ensemble used to study the excited state dynamics and photochemical reactions.

The excited state dynamics of the system are approximated well by the classical trajectory ensemble until the nonadiabatic crossing region is encountered. Nonadiabatic transitions cannot generally be treated using a purely classical description; however, several useful semiclassical approaches for treating such processes in conjunction with classical dynamics simulations have been proposed [47–54]. Initially developed to treat problems with very localized regions of strong nonadiabatic coupling [50], the "surface hopping" approaches have recently been modified to treat problems with extended regions of coupling and to minimize the occurrence of recrossing nonadiabatic trajectories [51]. The resulting stochastic switching algorithm due to Tully [51] minimizes computational effort, simplifies final averaging, and is easy to incorporate in classical molecular dynamics calculations.

For a two-electronic-state system $|0\rangle$ and $|1\rangle$, the surface hopping method consists in first integrating the classical mechanical equations of motion for the nuclei, then quantum-mechanically propagating the electronic wave function along the calculated piece of trajectory and determining the probabilities a_{jj}^2 for the system to be in each of the electronic states $|j\rangle$ $(j = 0, 1)$. Switching from one surface to the other occurs if the switching probability is larger than a random number in the interval [0, 1].

The simplicity of the method makes it widely applicable; however, for large systems, the electronic part of the calculation becomes impractical. In the spirit of the algorithm, one can use an appropriate semiclassical expression for the transition probability to supplement the classical dynamics. From first-order perturbation theory, the probability of making a transition at moment τ is given by [52]

$$P_{10}(\tau) = \left(\int_0^{\tau} [\langle \psi_0 | \frac{\partial \psi_1}{\partial t} \rangle a_1(t) \exp(-\frac{\mathrm{i}}{\hbar} \int_0^t \Delta V_{10} \mathrm{d}t')] \mathrm{d}t \right)^2 , \tag{16}$$

where ψ_0 and ψ_1 are the electronic wave functions of the ground and excited states and $\Delta V_{10} = V_1 - V_0$ is the time-dependent potential energy difference. Semiclassical evaluation of the time integral gives [52]

$$P_{10}(\tau) = \exp\left\{ -\frac{4\Delta V_{10}(\tau)}{3\hbar} \left[\frac{2\Delta V_{10}(\tau)}{(\frac{\mathrm{d}^2 \Delta V_{10}}{\mathrm{d}t^2})_{t=\tau}} \right]^{\frac{1}{2}} \right\} . \tag{17}$$

In practice, the quasiclassical ensemble of trajectories are propagated until they enter the region of the potential where internal conversion can occur. Within this region, the probability of a nonadiabatic transition is calculated at each time step using (17), with the second derivatives evaluated numerically. If $P_{10}(\tau) > \xi$, a random number in [0, 1], then the trajectory is switched to the ground state surface. When this happens, the momenta are scaled to conserve energy, and the trajectory is continued, now on the ground state potential energy surface. In this way, one can retain the convenience of classical dynamics, while incorporating electronic transitions in the simulation.

4 The Photoisomerization Dynamics of *cis*-Stilbene

To illustrate the methods we have outlined thus far, we now apply them to the $S_1 \leftarrow S_0$ photoisomerization dynamics of *cis*-stilbene using the potential surfaces described in Sect. 2. The *cis–trans* photoisomerization of stilbene (pictured in Fig. 6) has been exhaustively studied under many different conditions [55–58]. The irradiation of either isomer results in an approximate 1:1 ratio of both isomers; however, *cis*-stilbene also can undergo photocyclization to form dihydrophenanthrene (by forming a new bond, C_2–$C_{2'}$, between the two phenyl rings). Our goal in these studies is to understand the dynamical mechanism of photoproduct formation for each of the products and to explore different ways to manipulate the final product yields.

If we consider a one-dimensional reaction coordinate corresponding to the ethylenic torsion θ, then the *cis–trans* isomerization reaction can be pictured in much the same way as the examples of Sect. 3.1. Following excitation to either *cis** or *trans**, the molecule undergoes torsion about θ to an excited state minimum corresponding to a perpendicular configuration ($\theta = \frac{\pi}{2}$) denoted by p^*. From p^* the molecule relaxes to the ground state to form both *cis-* and *trans-*stilbene. In the isolated molecule, the excited state lifetime of *trans** is about 80 ps, while that of *cis** is approximately 300 fs. A barrier of about 1200 cm^{-1}has been inferred experimentally for $trans^*$, while the extremely short lifetime of cis^* has been cited as evidence for a barrierless surface along θ. The lifetime of p^* is believed to be less than 150 fs. Quantum yields have been measured in a variety of solvents, with average values of $\Phi_{cis} \approx 0.5$ and $\Phi_{trans} \approx 0.35$ for excitation of *trans-* and *cis*-stilbene, respectively.

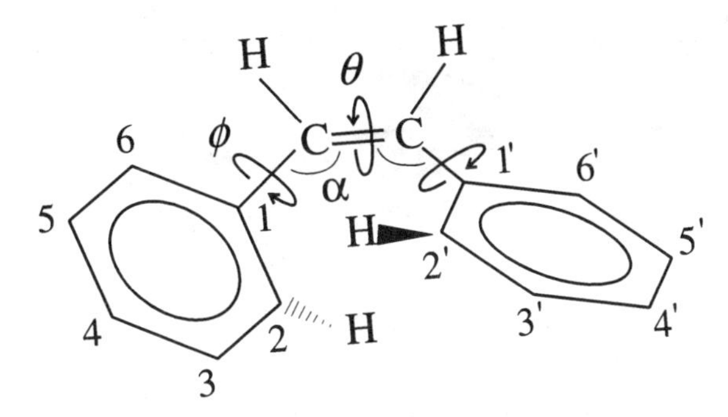

Fig. 6. *cis*-Stilbene ($C_{14}H_{12}$) structure. In the present study, all 78 degrees of freedom are included. The most important coordinates for isomerization are the torsional (θ) and bending (α) angles, as well as the $C_2 - C_{2'}$ distance, which is directly associated with photocyclization.

Several recent studies indicate that this one-dimensional picture of the reaction is an oversimplification in the case of *cis*-stilbene[58, 59]. Upon excitation, substantial changes occur in the region of the ethylenic bridge. The bond order of $C_e = C_e$ decreases and that of $C_1 - C_e$ increases for the $\pi^* \leftarrow \pi$ excitation. The changes in the electronic structure of the π-system also increase the torsion (θ) angle and decrease the bend (α) and twist (ϕ) angles in the S_1 state, leading to several active vibrations upon the vertical transition. These activated vibrations also assist in the formation of DHP, the secondary photoproduct of *cis*-stilbene, which is formed with quantum yields from 0.05 to 0.20 depending on the solvent environment. Finally, a careful analysis of the energetics shows that the energy difference between *cis** and p^* is about 45 kcal/mole. Our calculations show that on a barrierless surface, the transition $cis^* \rightarrow p^*$ would require only about 30 fs, but this contradicts experimental observation, providing evidence for the existence of a small barrier along θ from the *cis-* side and for internal motion along coordinates other than θ [22].

In our studies, we include all 72 vibrational degrees of freedom in a quasi-classical molecular dynamics simulation of the photoisomerization reactions. In addition to the basic potential surfaces given in (6–9), we use a bonding potential to model the DHP channel of the reaction. Since little is known about the shape of the PES along this coordinate, a Morse function is used,

$$U_{C_2C_2'} = D_{CC}\left\{\exp\left[-2\beta_c(R_{CC} - R_{eq})\right] - 2\exp\left[-\beta_c(R_{CC} - R_{eq})\right]\right\} \quad , \qquad (18)$$

with $D_{CC} = 35.0$ kcal/mole, $\beta_c = 4.67$ Å^{-1}, and $R_{eq} = 1.53$ Å for the ground state; and $D_{CC} = 39.0$ kcal/mole, $\beta_c = 2.0$ Å^{-1}, and $R_{eq} = 1.82$ Å for the excited state. A cross-section along the two coordinates θ and R_{CC} is shown in Fig. 7 for the model surfaces used in the present simulation.

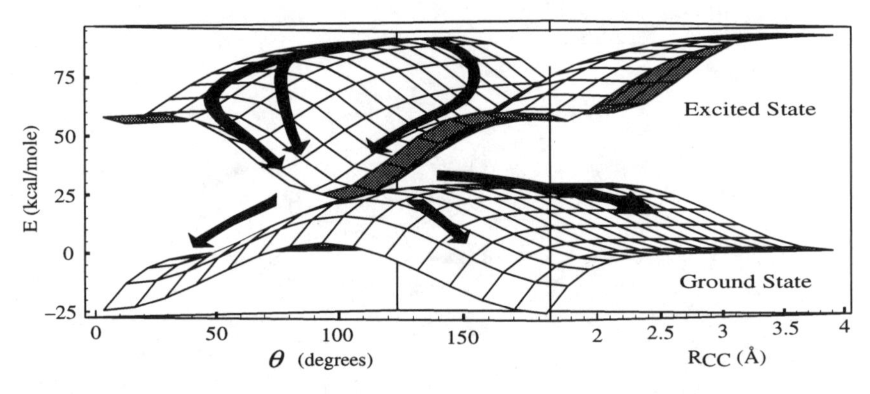

Fig. 7. Potential energy surfaces for the S_0 and S_1 states of stilbene as a function of the ethylenic torsional angle (θ) and the interatomic distance between C_2 and $C_{2'}$ (R_{CC}). Three pathways are indicated, corresponding to the different decay lifetimes observed in the quasiclassical simulations.

4.1 The Excited State Dynamics

First, we present results for the excited state dynamics of an ensemble of 300 trajectories whose initial conditions represent the quantum nonstationary state prepared by excitation of a vibrationally cold harmonic molecule with a 100 fs Gaussian laser pulse having 4000 cm^{-1}excess vibrational energy. Several observations can be made regarding the dynamics of the trajectory ensemble. First, the average lifetime of the ensemble is 280 fs. This lifetime is arbitrarily defined as the time required for trajectories to escape from the Franck–Condon region by overcoming the local barrier V_θ along the ethylenic torsion coordinate θ. In spectroscopic experiments (pump-probe transient absorption or fluorescence decay) this represents the time for the system to leave the observation window. The lifetime depends mainly on the local barrier V_θ. In this case, the potential parameters have been adjusted to reproduce approximately the experimental value of 307 fs. The experimental lifetime is best reproduced by a barrier $V_\theta = 340\ \mathrm{cm}^{-1}$ occurring at $\theta \approx \frac{\pi}{8}$.

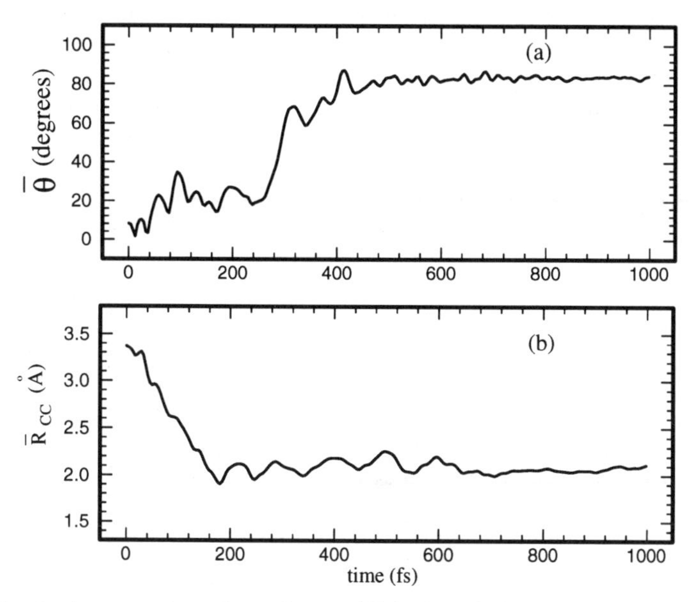

Fig. 8. Average time dependence of (**a**) the ethylenic torsional angle (θ) and (**b**) the interatomic $C_2 - C_{2'}$ distance in the excited state. An ensemble of 300 trajectories is used for averaging, with initial conditions representing excitation with 100 fs Gaussian laser pulse and 4000 cm^{-1}excess vibrational energy.

In Fig. 8, the ensemble average values of the torsional angle (θ) and the distance between the C_2 and $C_{2'}$ atoms (R_{CC}) are shown as functions of time. For the individual trajectories, typical characteristics include initial movement along the photocyclization $C_2 - C_{2'}$ coordinate from 3.4 Å to $\approx$ 2.0 Å in 200–300 fs, as well as oscillations along the ethylenic torsion coordinate around $\theta = 0°$ with an amplitude of $8 - -20°$. Most of the trajectories display a combination of these motions, with torsional oscillation persisting until a time t_θ when the trajectory is able to overcome the local barrier to *cis–trans* isomerization. Once the system passes over this barrier, it rapidly moves within 30 fs toward the excited state potential minimum at $\theta = \frac{\pi}{2}$ forming a vibrationally hot, twisted intermediate. The remainder of the vibrational modes do not appear to contribute significantly to the excited state dynamics, remaining predominantly low-amplitude vibrations; however, they participate in the dynamics by providing an energy bath from which energy can flow in or out of the vibrationally hot large-amplitude motions.

Although the potential parameters are adjusted so that the ensemble average lifetime agrees with the experimental lifetime, a more interesting observation is the distribution of the barrier crossing times t_θ, shown in Fig. 9. Three distinct groups of barrier-crossing times can be distinguished. Almost half of the trajectories (145 of 300) cross the V_θ barrier between t_θ = 260 and 300 fs, close to the average value. However, two other groups of trajectories display markedly different lifetimes: 50 trajectories traverse the barrier in less than 100 fs, while 57 cross between 360 fs and 400 fs after photoexcitation. About 5% of all trajectories have crossing times longer than 400 fs, with a couple as long as 900 fs.

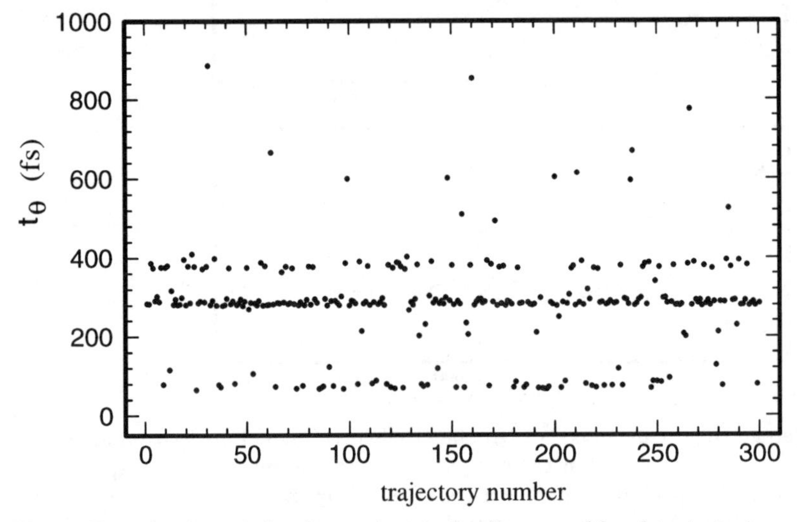

Fig. 9. Distribution of the decay time t_θ in the ensemble of trajectories, where t_θ is determined as the time for leaving the Frank–Condon region by overcoming the local barrier to isomerization.

The clustering of the barrier crossing times at three distinct values suggests that the dynamics follow three distinct pathways leading away from the Franck–Condon region. A closer examination of the trajectories in each of these clusters reveals that the fast group (t_θ < 100 fs) represents trajectories with energy distributed in such a way as to allow direct crossing of the torsional barrier to $\theta \approx \frac{\pi}{2}$, followed by motion to smaller $C_2 - C_{2'}$ distances. The slow group (360 fs < t_θ < 400 fs) does not have sufficient energy in θ to surmount the barrier and instead migrates to smaller R_{CC} before finally turning toward the excited state minimum and undergoing torsion to $\theta \approx \frac{\pi}{2}$. The majority cluster of trajectories (with 260 fs < t_θ < 300 fs) follows an intermediate path in which the ensemble begins to migrate to smaller R_{CC}, then undergoes torsion to $\theta \approx \frac{\pi}{2}$ earlier than the slow group. All three pathways lead to the same nonadiabatic crossing region near the minimum of the excited state potential (characterized by $\theta = 90 \pm 10°$

and $R_{CC} = 2.0 \pm 0.2$ Å). These pathways are schematically drawn on the excited state surface shown in Fig. 7.

Once back in the ground electronic state, the system relaxes via IVR to form one of three photoproducts: *trans*-stilbene (20.3%), *cis*-stilbene (71.7%), or DHP (8.0%). The quantum yields in gas phase are not known experimentally, and we have adjusted our model, specifically the energy gap between S_0 and S_1, to obtain product yields similar to those found experimentally for *cis*-stilbene photoisomerization in methylcyclohexane–isohexane [58]. Since each excited state pathway crosses to the ground state in the same region of the S_1 potential, it would appear that the differentiation between the photoproducts does *not* occur in the excited state, but is the result of statistical partitioning of the vibrationally hot ground state dynamics. However, careful analysis reveals that the momentum of the system leading into its nonadiabatic crossing is important in determining its final destination on the ground state potential. In other words, different excited state pathways *do* correlate with different final photoproducts, as we discuss in the following sections.

4.2 The DHP Channel

The photocyclization of *cis*-stilbene to form dihydrophenanthrene (DHP) arises from two predominant pathways in the excited state. About 65% of the trajectories leading to this photoproduct belong to the fast decay pathway ($t_\theta < 100$ fs). This pathway enters the nonadiabatic crossing region, moving rapidly to smaller $R_{C_2C_{2'}}$ distances. Once these trajectories "hop" to the ground state surface, they continue moving in the direction of smaller C_2–$C_{2'}$ distances, thus favoring the photocyclization channel. After the system crosses to the ground state, it requires another 500 fs to relax before $R_{C_2C_{2'}} \approx 1.53$ Å and the new bond creating the fused ring system is formed.

The remaining 35% of the DHP-forming trajectories follow an extremely slow pathway to the nonadiabatic crossing region ($t_\theta > 600$ fs). At the time of torsional barrier crossing in these trajectories, the $R_{C_2C_{2'}}$ distance has already shortened to about 1.95 ± 0.05 Å, suggesting that the initial motion of the system in this case is along the $R_{C_2C_{2'}}$ coordinate. In contrast to the fast group of trajectories, the formation of DHP for this group is dictated by the potential in the ground state rather than the momentum of the system. The subsequent ground state IVR occurs in about 100 fs after internal conversion to form the final DHP product. Both of the pathways that lead to DHP formation exhibit very short lifetimes in the nonadiabatic crossing region corresponding to the perpendicular configuration (p^*), ranging from 15–45 fs with an average of about 38 fs.

4.3 The *trans*-Stilbene Channel

Two different pathways also contribute to the formation of *trans*-stilbene, although they are much more similar to one another than the two pathways leading to DHP formation. The average behavior of these pathways is characterized by torsional barrier crossing at t_θ=336 fs, with internal conversion occurring 355 fs after photoexcitation. Most (66%) of these trajectories belong to the slow cluster, characterized by 370 fs $< t_\theta <$ 390 fs. After overcoming the local torsion barrier in the excited state, all trajectories in this group move directly along θ to form *trans*-stilbene. For most trajectories, internal conversion occurs at $\theta < \frac{\pi}{2}$; therefore, the isomerization is initially dictated by the system's momentum after crossing to the ground state, and relaxation to the final product occurs in about 100 fs.

Since the slow excited state pathway corresponds to initial motion toward smaller R_{CC}, followed by rapid torsional motion in θ toward $\theta = \frac{\pi}{2}$, the mechanism for *trans*-stilbene formation is thematically similar to that for the fast pathway to DHP formation. Ironically, in each case, the initial motion is along a coordinate associated with the formation of the *other* photoproduct. The important point is that the momentum leading into the nonadiabatic crossing region generally plays a pivotal role in determining the final outcome of the reaction.

Finally, we note that the major product of the photoinduced dynamics of *cis*-stilbene is *cis*-stilbene itself, with all of the intermediate pathway trajectories (260 fs $< t_\theta <$ 300 fs) leading to re-formation of the starting material. Upon internal conversion to the ground state, these trajectories have insufficient energy to surmount the potential barrier to *trans*-stilbene formation and generally reflect off the barrier and decay via IVR into the *cis*-stilbene potential minimum on the ground state.

5 Chemical Control via Two-Photon Excitation

5.1 Simulation of Two-Photon (IR + UV) Excitation

The excited state dynamics of stilbene suggests the possibility of chemical control due to the correlation between the dynamical paths followed and the final photoproducts formed. Even though all paths lead to the same excited state minimum from which the system internally converts to the ground state, the momentum the system carries with it into the crossing region guides its subsequent ground state dynamics. Thus we are led to the supposition that if a means can be found to restrict the dynamics to a single pathway leading out of the Franck–Condon region, then it may be possible to manipulate the final product yields. In this section, we examine one way in which this might be accomplished.

Several ingenious methods have been proposed for manipulating the dynamics of a molecule, including preparing a coherent superposition of states, using a shaped pulse train to construct a desired initial state, using an ultrafast "pump–dump" excitation to prepare the initial state, and preparing a vibrationally excited molecule prior to electronic excitation [1–7]. Of these schemes, the last is perhaps the simplest to achieve experimentally and, consequently, has been applied successfully to a number of small systems [2, 6]. The basic approach involves using infrared radiation to prepare a vibrationally excited state (a normal mode eigenstate), then promoting a vertical transition to an electronic excited state with an ultraviolet photon. In most of these cases, the initial normal mode excitation is localized in a bond (or bonds) where one desires the eventual chemical change to occur.

This scheme becomes more difficult to apply in larger molecular systems such as stilbene. The higher density of vibrational states makes it more difficult to excite a single normal mode, even at modest levels of excitation. This means that IVR processes will begin to compete with the desired photochemical dynamics. Moreover, the low intensity of normal mode overtones makes the preparation of vibrational states with $\nu > 1$ virtually impossible for all but hydride bonds. The nature of the normal modes themselves is less likely to be localized in the molecule, and so the connection between the vibrationally prepared initial state and a desired chemical change is more subtle.

In designing a scheme to alter the photoproduct branching ratio in *cis*-stilbene, our initial goal is to examine the effects of preliminary normal mode excitation on the excited state dynamics. To simplify our treatment, we model this process by first adding two quanta of energy to a specified normal mode. The initial conditions for this mode are then represented by a uniform sample of the relative phases of the normal mode oscillator at this fixed energy. All other normal modes are prepared as in the previous section. After converting the normal mode initial conditions to Cartesian coordinates and momenta, a vertical transition is then made to the electronic excited state, and the classical molecular dynamics of these initial conditions are simulated as before.

5.2 Selected Results

There are 72 normal modes of vibration in stilbene, and a rich variety of reaction outcomes can be obtained from two-stage excitation schemes in which one first prepares a vibrational excited state. Here, we survey some selected results from our preliminary studies. A more detailed account will be presented in a forthcoming publication.

In our simulations, we first add vibrational energy to the molecule, then promote the system to the S_1 excited state with a UV photon of frequency 302 nm. This is the same wavelength used in the one-photon excitation simulations described in Sect. 4. Thus, one effect of the two-stage excitation is to provide additional energy to the molecule during its excited state dynamics. The excited state dynamics are not ergodic on the time scale of the reaction, however, so the results remain sensitive to the manner in which the energy is allocated among the vibrations. As a benchmark for the two-photon simulations given below, we show the effect of excess energy in a one-photon excitation in Fig. 10. The average barrier-crossing time, t_θ, decays exponentially with excess energy (Fig. 10a), but the product yields are only mildly affected by the additional energy (Fig. 10b). The faster reaction rates at higher energy may be indicative of more trajectories having sufficient torsional energy to surmount the V_θ barrier upon arriving on the S_1 surface, and thus following the fast reaction pathway on the excited state. This partly explains why DHP formation is slightly enhanced at the expense of *trans*-stilbene for the highest energies observed.

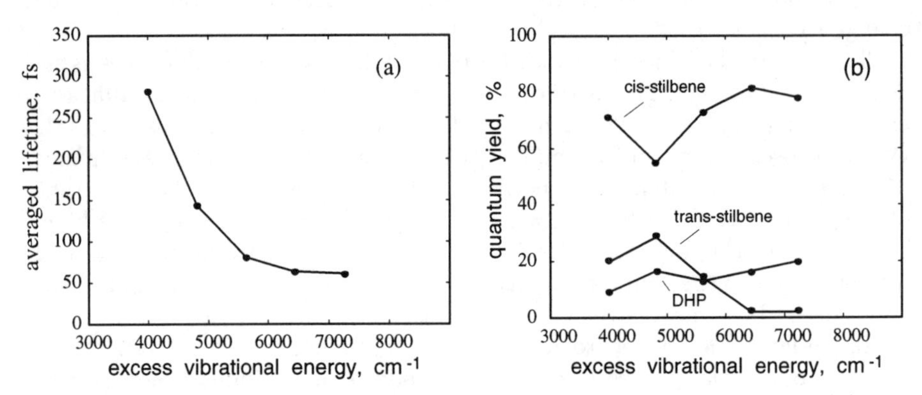

Fig. 10. Effect of excess energy on (**a**) torsional barrier crossing time and (**b**) final photoproduct yield for one-photon excitation of *cis*-stilbene.

Since the three excited state pathways observed in one-photon excitation of *cis*-stilbene represent large-amplitude motions that lead to structural changes and photoisomerization, a natural supposition would be that vibrational excitation of the low-frequency vibrations would have the greatest effect on the final product yields. This is not the case, however. For vibrational excitation of the seven lowest-frequency normal modes, ranging from 40 to 168 cm^{-1}, the product yields all have the same profile as the one-photon yields, with DHP < *trans*-stilbene < *cis*-stilbene. In particular, the yield for *cis*-stilbene is always greater than 59% and that of *trans*-stilbene is always less than half that of *cis*-stilbene.

Much greater effects are observed for normal modes 13–46, whose frequencies lie in the middle range (400–2000 cm^{-1}). As an example, in Fig. 11 we show the final product yields obtained from initially placing two quanta of energy in modes 13–19, then promoting the system to the excited state. Dramatic changes in the yields are evident, with *trans*-stilbene constituting the major product in three of the simulations (modes 13, 14, and 15), and DHP attaining yields in excess of 35% in two (modes 15 and 16). Similar observations can be made for simulations of two quanta of vibrational excitation placed in other normal modes. Final yields of *trans*-stilbene range from 1% (mode 17) to 90% (mode 36), while yields of DHP range from $< 1\%$ (mode 14) to $> 50\%$ (mode 42). These results suggest that it is possible to greatly enhance or suppress the formation of a given product simply by placing energy in different normal modes prior to the electronic excitation.

Interestingly, we find that the distribution of barrier-crossing times on the S_1 state is much narrower for systems that are first vibrationally excited. For example, all 200 trajectories used to simulate the excited state dynamics of a system with two quanta of energy initially placed in mode 14 cross the torsional barrier in 220 ± 5 fs after promotion to the excited state. Although the dynamics appear to have collapsed to a single pathway on the excited state, the relationship between the nature of the normal mode initially excited and the pathway eventually followed is not clear and bears further investigation. Mode 14 corresponds to a skeletal C–C bond stretch that is delocalized over most of the molecule—this mode does not appear to have any relationship to the ethylenic torsion and R_{CC} coordinates that are featured in the photoisomerization dynamics of *cis*-stilbene.

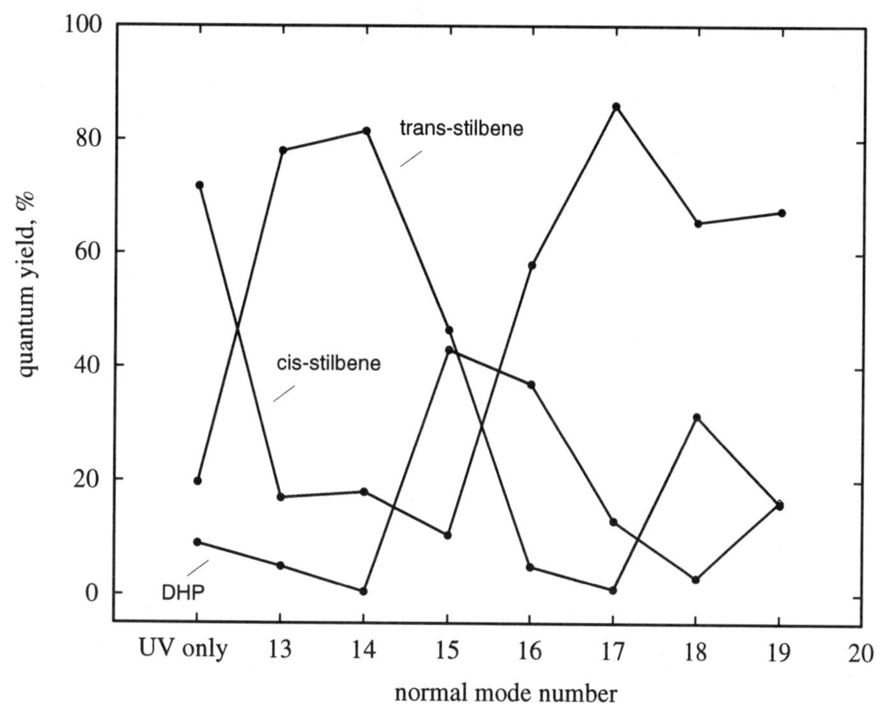

Fig. 11. Effect of initial vibrational excitation on the final photoproduct yields of *cis*-stilbene. For each of the modes indicated, two quanta of vibrational energy were added. The yields for one-photon excitation of *cis*-stilbene are included on the left side for comparison.

Finally, we examine the effect of adding varying amounts of energy to a given normal mode before the system is promoted to the electronic excited state. In Fig. 12, we show the effect of varying the energy placed in modes 14 and 16 prior to UV photoexcitation. In both cases, the final product yields vary substantially depending on the degree of vibrational excitation. Although spectroscopically based experiments can only add integer numbers of vibrational quanta to the system, these plots suggest that a greater variety of different product yields might be accessed if one can devise efficient ways to place varying amounts of energy in a specific normal mode.

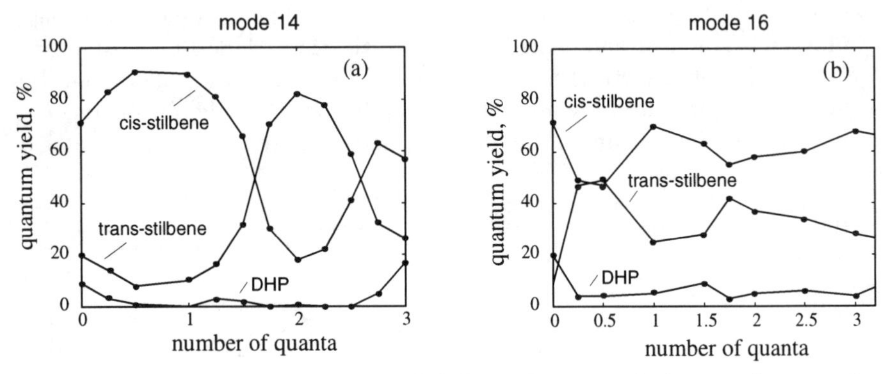

Fig. 12. Variation of product yields with different amounts of vibrational energy placed in (**a**) mode 14 (1714 cm^{-1}), and (**b**) mode 16 (1621 cm^{-1}).

We have reviewed the techniques used to construct quantitative empirical models of potential energy surfaces for the ground and excited electronic states of large molecules and shown how these can form the basis for studying the molecular dynamics of photoisomerization processes. With sufficient knowledge about the structure and dynamics of molecules, one can begin to design control schemes that allow the experimentalist to manipulate the final product yields of photochemical reactions. We believe that the synthetic chemist and materials scientist of the future will find great utility in simulating the photochemical dynamics of molecular systems as a routine part of designing photochromic molecular devices, much as today's chemists routinely use electronic structure calculations of the ground electronic state to simulate and predict molecular properties. These studies represent a first step toward this future.

Acknowledgments. We are grateful to Ms. Kathleen Nojima and Ms. Rilun Tang for performing some of the computations presented in Sect. 5. We owe much to our many collaborators over the years, including Dr. Hrvoje Petek, Prof. Keitaro Yoshihara, Prof. Victor Zadkov, and Prof. Boris Grishanin. We thank the National Science Foundation for financial support of this research. Acknowledgment is also made to the Donors of the Petroleum Research Fund, administered by the American Chemical Society, for partial support of this work.

References

1 P. Avouris, Accts. Chem. Res. **28**, 95 (1995).
2 F. F. Crim, Ann. Rev. Phys. Chem. **44**, 397 (1993).
3 Z.-M. Lu and H. Rabitz, J. Phys. Chem. **99**, 13731 (1995).
4 B. Kohler, J. L. Krause, F. Raksi, K. R. Wilson, R. M. Whitnell, and Y. J. Yan, Accts. Chem. Res. **28**, 133 (1995).
5 P. Brumer and M. Shapiro, Ann. Rev. Phys. Chem. **43**, 257 (1992).
6 J. D. Thoemke, J. M. Pfeiffer, R. M. Metz, and F. F. Crim, J. Phys. Chem. **99**, 13748 (1995).
7 H. Kawashima, M. M. Wefers, and K. A. Nelson, Ann. Rev. Phys. Chem. **46**, 627 (1995).
8 L. E. Fried and S. Mukamel, Adv. Chem. Phys. **84**, 435 (1993).
9 C. S. Parmenter, Far. Disc. Chem. Soc. **75**, 7 (1983).
10 R. Moore, F. E. Doany, E. J. Heilweil, and R. M. Hochstrasser, Far. Disc. Chem. Soc. **75**, 331 (1983).
11 M. Quack, Ann. Rev. Phys. Chem. **41**, 839 (1990).
12 D. H. Levy, Ann. Rev. Phys. Chem. **31**, 197 (1980).
13 M. Ito, T. Ebata, and N. Mikami, Ann. Rev. Phys. Chem. **39**, 123 (1988).
14 J. Manz and L. Wöste, L. (Eds.), *Femtosecond Chemistry* (VCH, Weinheim, 1995).
15 M. Klessinger, Angew. Chem. Int. Ed. Engl. **34**, 549 (1995).
16 F. Bernardi, M. Olivucci, M. A. Robb, Pure Appl. Chem. **67**, 17 (1995).
17 J. Michl, and V. Bonačić-Koutecký, *Electronic Aspects of Organic Photochemistry* (Wiley, New York, 1990).
18 E. J. Heller, Accts. Chem. Res. **14**, 368 (1981).
19 Y. J. Yan and S. Mukamel, J. Chem. Phys. **85**, 5908 (1986).
20 K. Shan, Y. J. Yan, and S. Mukamel, J. Chem. Phys. **87**, 2021 (1987).
21 V. D. Vachev, J. H. Frederick, B. A. Grishanin, V. N. Zadkov, and N. I. Koroteev, J. Phys. Chem. **99**, 5247 (1995).
22 V. D. Vachev, J. H. Frederick, B. A. Grishanin, V. N. Zadkov, and N. I. Koroteev, Chem. Phys. Lett. **215**, 306 (1993).
23 J. H. Frederick, E. J. Heller, J. Ozment, and D. W. Pratt, J. Chem. Phys. **88**, 2169 (1988).
24 J. H. Frederick and J. E. Hadder, in *Time-Resolved Vibrational Spectroscopy V, Springer Proceedings in Physics 68* (Springer-Verlag, Berlin, 1992).
25 J. H. Frederick, in *Structures and Conformations of Non-Rigid Molecules, NATO ASI Series C, Vol. 410*, ed. by J. Laane, M. Dakkouri, B., van der Veken, and H. Oberhammer (Kluwer, Dordrecht, 1993).
26 J. Svitak, Z. Li, J. Rose, and M. E. Kellman, J. Chem. Phys. **102**, 4340 (1995).
27 H. Abe, S. Kamei, N. Mikami, and M. Ito, Chem. Phys. Lett. **109**, 217 (1984).
28 S. Kamei, T. Sato, N. Mikami, and M. Ito, J. Phys. Chem. **90**, 5615 (1986).
29 K. W. Holtzlaw and D. W. Pratt, J. Chem. Phys. **84**, 4713 (1986).
30 V. D. Vachev and J. H. Frederick, Chem. Phys. Lett. **249**, 476 (1996).
31 V. Burkert and N. L. Allinger, *Molecular Mechanics* (ACS Press, Vol. 177, Washington D.C., 1982).
32 N. L. Allinger, Y. H. Yuh, and J.-H. Lii, J. Am. Chem. Soc. **111**, 8551 (1989).
33 G. Gershinsky and E. Pollak, J. Chem. Phys. **105**, 4388 (1996).
34 G. Gershinsky and E. Pollak, J. Chem. Phys. **107**, 812 (1997).
35 K. Bolton, S. Nordholm, Chem. Phys. **203**, 101 (1996).
36 R. R. Birge, Biochem. Biophys. Acta **1016**, 293 (1990).

37 Q. Wang, R. W. Schoenlein, L. A. Peteanu, R. A. Mathies, and C. V. Shank, Science **266**, 422 (1994).
38 A. Warshel, Z. T. Chu, and J.-K. Hwang, Chem. Phys. **158**, 303 (1991).
39 J. M. Dormans, G. C. Groenenboom, and H. M. Buck, J. Chem. Phys. **86**,4895 (1987).
40 H. Metiu and J. Alvarellos, J. Chem. Phys. **88**, 4957 (1988).
41 M. D. Feit, J. A. Fleck, Jr., and A. Steiger, J. Comp. Phys. **47**, 412 (1982).
42 R. Kosloff, J. Phys. Chem. **92**, 2087 (1988).
43 R. D. Levine and R. B. Bernstein, *Molecular Reaction Dynamics and Chemical Reactivity* (Oxford University Press, Oxford, 1987).
44 R. D. Taylor and P. Brumer, Far. Disc. Chem. Soc. **75**, 117 (1983).
45 K.-A. Suominen, B. M. Garraway, and S. Stenholm, Phys. Rev. A **45**, 3060 (1992).
46 B. M. Garraway and S. Stenholm, Opt. Commun. **83**, 349 (1991).
47 L. D. Landau and E. M. Lifshitz, *Quantum Mechanics: Nonrelativistic Theory* (Pergamon, New York, 1987).
48 H. Köppel, W. Domcke, and L. S. Cederbaum, Adv. Chem. Phys. **57**, 59 (1984).
49 L. Seidner and W. Domcke, Chem. Phys. **186**, 27 (1994).
50 J. C. Tully and R. K. Preston, J. Chem. Phys. **55**, 562 (1971).
51 J. C. Tully, J. Chem. Phys. **93**, 1061 (1990).
52 W. H. Miller and T. F. George, J. Chem. Phys. **56**, 5637 (1972).
53 S. Chapman, Adv. Chem. Phys. **82**, 423 (1992).
54 D. F. Coker and L. Xiao, J. Chem. Phys. **102**, 496 (1995).
55 H. Görner and H. J. Kuhn, in *Advances in Photochemistry*, vol. 19 (Wiley, New York, 1995).
56 D. Waldeck, Chem. Revs. **91**, 415 (1991).
57 J. S. Baskin, S. Pedersen, L. Bañares, and A. H. Zewail, J. Chem. Phys. **98**, 6291 (1993).
58 R. J. Sension, S. T. Repinec, A. Z. Szarka, and R. M. Hochstrasser, J. Chem. Phys. **98**, 6291 (1993).
59 J. H. Frederick, Y. Fujiwara, J. H. Penn, K. Yoshihara, and H. Petek, J. Phys. Chem. **95**, 2845 (1991).
60 K. A. Muszkat and E. Fischer, J. Chem. Soc. B. **1967**, 662.

Structure and Dynamics of Electronically Excited Diphenylacetylene in Different Environments

Hiro-o Hamaguchi and Taka-aki Ishibashi

Department of Chemistry, School of Science, The University of Tokyo, 3-8-1 Hongo, Bunkyo-ku, Tokyo 113, Japan E-mail: hhama@chem.s.u-tokyo.ac.jp

Abstract. Recent spectroscopic as well as quantum-mechanical studies on the structure and dynamics of electronically excited diphenylacetylene are critically reviewed with an emphasis placed on the effect of the environment.

1 Introduction

Electronically excited singlet states interact strongly with their environment. In case more than two excited singlet states are lying close to one another, the state ordering may change with the environment, and if so, the structure and dynamics associated with these states will be influenced profoundly. In order to understand the behavior of electronically excited molecules in condensed phases, we need to elucidate the mechanisms of such interactions. Diphenylacetylene (hereafter DPA, also known as tolan) serves as an excellent probe for the interaction of the excited singlet states with the environment. A few low-lying excited singlet states are known to exist for this molecule, but their assignments have been controversial over the past five years, owing to the complicated dependence of these states on the environment. DPA is a prototype conjugated molecule having a C≡C triple bond, and the change in the molecular structure with the environment is also of considerable interest. The change in the bond order of the central CC bond will be manifested as the change in the CC stretch vibrational frequency.

In the present article, we first review recent spectroscopic studies on the excited singlet states of DPA in different environments; fluorescence excitation and multiphoton ionization spectroscopy in a supersonic jet [1], one- and two-photon fluorescence excitation spectroscopy in low-temperature rare-gas matrices [2], and picosecond time-resolved absorption and fluorescence spectroscopy in solution [3]. We then summarize our picosecond time-resolved CARS studies on the S_1 and the S_2 states of DPA in solution [4, 5]. Finally, we discuss the structure and dynamics in the S_1 and S_2 states of DPA with reference to a quantum-chemical calculation [6].

An x-ray diffraction study [7–9] showed that DPA in a single crystal takes a planar D_{2h} structure in the ground electronic state. Electron diffraction data were consistent with a D_{2h} structure [10]. Polarized ultraviolet absorption spectra in a single crystal [11] and in stretched polymer films [12, 13] indicated that observed electronic transition is long-axis polarized. Electronic [14] and vibrational [15–17] spectra of DPA in solution have been interpreted soundly in terms of a planar D_{2h} structure. The prominent ultraviolet absorption of DPA is assignable to a $B_{1u} \leftarrow A_g$ transition that is long-axis polarized under the D_{2h} symmetry. We should note here that there is confusion in the literature about notation of irreducible representations of the D_{2h} point group. In this article, as shown in Fig. 1, we take the principal twofold axis as the long molecular axis (z). The other twofold axes are the in-plane short molecular axis (y) and the out-of-plane axis (x). According to this definition, the long-axis polarized transition is designated as $B_{1u} \leftarrow A_g$, the short-axis polarized transition as $B_{2u} \leftarrow A_g$, and the out-of-plane transition as $B_{3u} \leftarrow A_g$. In References [3] and [12], the principal axis is taken as the short molecular axis (y), and therefore the long-axis polarized transition is designated as $B_{2u} \leftarrow A_g$, which is referred to as $B_{1u} \leftarrow A_g$ in this article.

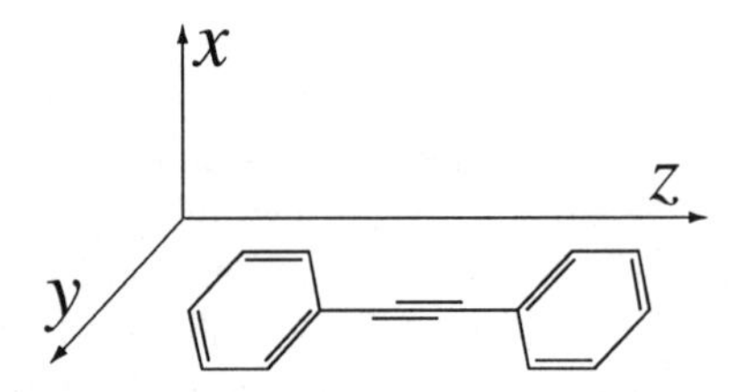

Fig. 1. Definition of molecular axes of DPA.

2 The Supersonic-Jet Study

Fluorescence excitation, SVL dispersed fluorescence, and multiphoton ionization spectra of DPA in a supersonic free jet were reported by Okuyama et al. [1]. Two series of bands were found in the fluorescence excitation spectrum: Series A starting from 35248 cm^{-1} and Series B starting from 35051 cm^{-1}. The intensity of the band origin in Series A was about five times stronger than that in Series B. The strongest two bands in Series A coincided in their positions with the two prominent peaks in the gas-phase absorption spectrum, which showed a typical vibrational structure of an allowed electronic transition. On the basis of these findings, Okuyama et al. assigned Series A to the B_{1u} state. The remaining Series B bands were tentatively assigned to the weakly allowed B_{2u} state. In the two-photon resonant four-photon ionization spectrum, another series of bands starting from 34960 cm^{-1} were observed. They were assigned to the two-photon allowed A_g state. Thus, three different electronic states were found, the 34960 cm^{-1} state (assigned to A_g), the 35051 cm^{-1} state (B_{2u}), and the 35248 cm^{-1} state (B_{1u}). They are located close to one another in a wave-number range of only three hundred wave-numbers. From the mutual exclusion relation between the one- and two-photon absorption spectra, Okuyama et al. suggested D_{2h} or C_{2h} structures for the ground and the excited electronic states.

Another point of interest in the jet study is the extraordinary dependence of the observed fluorescence intensity on the excitation energy. No fluorescence excitation band was observed above 700 cm^{-1} from the band origin of the 35248 cm^{-1} state. This sharply contrasts with the vapor-phase absorption spectrum, in which extensive vibrational structure is observed up to 2500 cm^{-1}. Disagreement between the fluorescence excitation and absorption spectra means that the quantum yield of fluorescence depends strongly on the excess energy. Okuyama et al. [1] ascribed this finding to the onset of an efficient nonradiative decay channel around excess energy of 700 cm^{-1} in the manifold of the 35248 cm^{-1} state.

3 The Matrix Isolation Study

Site-selected one- and two-photon fluorescence excitation spectra were measured by Gutmann et al. [2] for DPA in three different matrices (N_2, Ar, Kr) at 12 K. The observed one-photon fluorescence excitation spectra did not resemble the corresponding supersonic-jet spectrum. The spectra showed extensive vibrational structures starting from the band origin (33996 cm^{-1} in Ar). An excellent mirror relation was found between the dispersed fluorescence and fluorescence excitation spectra. The band origins of the dispersed fluorescence (33998 cm^{-1} in Ar) and fluorescence excitation spectra agreed excellently with each other, showing that the electronic transition involved is an allowed transition. Gutmann et al. naturally assigned the 33996 cm^{-1} state to the B_{1u} state. Instead of the abrupt disappearance of the vibrational structure at 700 cm^{-1} in the jet spectrum, marked band broadenings were observed in matrices for bands with excess energies higher than 1300 cm^{-1}. The band broadening increased with increasing excess energy. No extra bands assignable to other electronic states were found in the one-photon fluorescence excitation spectra.

Several features starting from 34408 cm^{-1} were found in the two-photon fluorescence excitation spectrum in Ar. The observed two-photon polarization indicated that the final vibronic states of these transitions are of B_g symmetry. Only the B_{3g} symmetry is the suitable candidate, because the in-plane two-photon absorption from the ground state is forbidden for the B_{1g} and B_{2g} states. The 34408 cm^{-1} band was therefore assigned to a vibronic band that borrowed two-photon absorption activity from a nearby B_{3g} state. Since no two-photon absorption bands were found below 34408 cm^{-1}, this B_{3g} state is likely to be located higher in energy than the 33996 cm^{-1} state. Existence of the third low-lying singlet excited state, the A_u state, was suggested by a CNDO/S-CI calculation. Though this state was not observed in the one- and two-photon fluorescence excitation spectra in the region below 33996 cm^{-1}, it may well be located just above the B_{1u} state and cause the band broadening of the vibronic levels of the B_{1u} manifold. Gutmann et al. presumed that this A_u state corresponds to the 34960 cm^{-1} state found by Okuyama et al. [1] in the two-photon resonant four-photon ionization spectrum in jet. The one-photon allowed B_{1u} state shows a large matrix effect on going from the gaseous state (35248 cm^{-1}) to the Ar-matrix state (33996 cm^{-1}) and that this may reverse the state ordering in the free molecule and make the A_u state lower in energy than the B_{1u} state.

4 The Solution-Phase Study

Picosecond time-resolved absorption and fluorescence studies of DPA in various solvents were carried out by Hirata et al. [3]. Three transient absorption bands with different lifetimes were found at 415 nm (lifetime $\gg$ 2 ns), at 500 nm ($\approx$200 ps), and at 435 nm ($\approx$10 ps). The 415 nm band was identical with the known $T_n \leftarrow T_1$ absorption band. The rise of this $T_n \leftarrow T_1$ absorption agreed with the decay of the 500 nm band, and the rise of the 500 nm band agreed with the decay of the 435 nm band. On the basis of these findings, Hirata et al. assigned the 500 nm band to the S_1 state and the 435 nm band to the S_2 state. The lifetime $\approx$200 ps corresponds to the $S_1 \rightarrow T_1$ intersystem crossing, and the lifetime $\approx$10 ps corresponds to the $S_2 \rightarrow S_1$ internal conversion.

The measured fluorescence lifetime agreed with the decay time of the 500 nm transient absorption. This means that the fluorescence is emitted not from the S_1 state but from the S_2 state. The fluorescence spectrum showed a good mirror-image relation with the ultraviolet absorption spectrum. Therefore, the principal activity in the ultraviolet absorption originates from the $S_2 \leftarrow S_0$ transition. Then it is straightforward to identify the S_2 state as the strongly allowed B_{1u} state.

The lifetime of the S_2 state showed marked temperature dependence. Arrhenius plots produced straight lines giving activation energies in the range of 800–1100 cm^{-1} depending on the solvent. These values should be regarded as the activation energies for the $S_2 \rightarrow S_1$ internal conversion. It is of great interest how this activation energy is associated with the process of internal conversion, which is most likely to be an intramolecular process. Hirata et al. [3] explained the observed activation energy in terms of the turnover from the intermediate coupling case to the statistical limit of the $S_2 \rightarrow S_1$ internal conversion. A similar value of activation energy, 1200 cm^{-1}, was also estimated by Ferrante et al. [6] from the measurement of the temperature-dependent fluorescence quantum yield in solution.

5 Picosecond CARS Spectroscopy in Solution

We recently developed a new technique of picosecond time-resolved CARS spectroscopy that we call 2-D multiplex CARS spectroscopy [18, 19]. A subpicosecond actinic laser and nanosecond CARS probing lasers were used with a streak camera and a polychromator to obtain a time–frequency two-dimensional image of transient CARS signals. This technique has enabled us to obtain picosecond Raman spectra of strongly fluorescent excited states like the singlet states of DPA, for which spontaneous Raman spectroscopy was not applicable [20]. Figure 2 shows the 2-D CARS image of photoexcited DPA in cyclohexane. The ordinate corresponds to time increasing from the top to the bottom, and the abscissa to the wave-number shift increasing from the right to the left. The vertical lines existing throughout the time course are the solvent CARS bands. In addition to these signals, we see two transient CARS signals with different time behaviors. The long-lived signals ($\approx$1100 cm^{-1}, $\approx$980 cm^{-1}; lifetime $\approx$200 ps) are assigned to the S_1 state, and the short-lived signal ($\approx$980 cm^{-1}; lifetime < 10 ps) to the S_2 state, as indicated in Fig. 2.

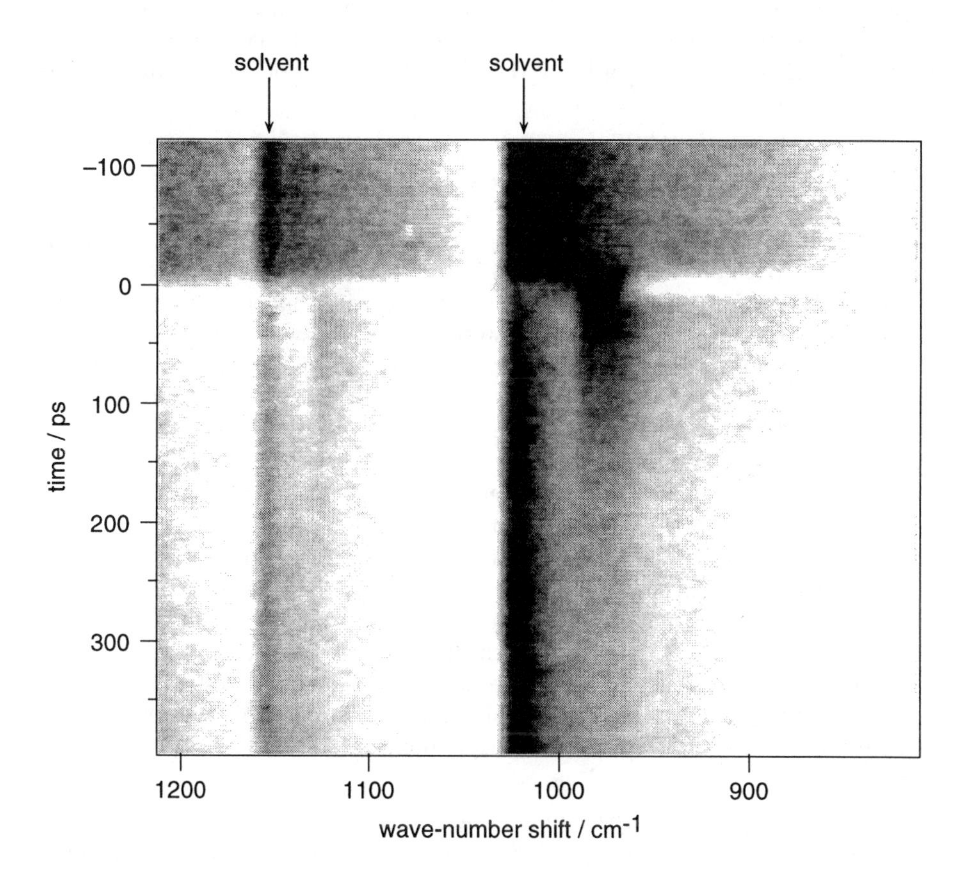

Fig. 2. A 2-D CARS image of DPA.

The CARS spectrum averaged over 40–60 ps is shown in Fig. 3. Four weak bands are observed in the 2300–1900 cm^{-1} region . Though this time range should be dominated by the S_1 CARS signals, the time behaviors of these four bands are totally different form that expected for the S_1 state. With reference to the known spontaneous Raman frequencies [20], we assign these bands to the C≡C stretch vibrations in the S_0 state ($2215\,cm^{-1}$), the cation radical ($2142\,cm^{-1}$), the anion radical ($2091\,cm^{-1}$) and the T_1 state ($1974\,cm^{-1}$). It is of interest how the radical ions are photogenerated in a nonpolar solvent like cyclohexane. This issue will be treated separately in a future publication.

No CARS band assignable to the S_1 state is found in the 1900–2300 cm^{-1} region. Two possibilities are conceivable. First, the CC stretch frequency is located in the 1900–2300 cm^{-1} region but the corresponding CARS band is too weak to be observed. Second, the bond order of the central CC bond is decreased so much that the CC stretch vibration has a frequency lower than $1900\,cm^{-1}$. The highest two frequencies observed in the S_1 CARS spectrum are $1577\,cm^{-1}$ and $1557\,cm^{-1}$. If the second possibility is the case, at least one of these two frequencies is affected significantly by the ^{13}C substitution of the central carbons of DPA.

Figure 4 shows the CARS spectra of S_1 DPA in the frequency region of 1900-1400 cm^{-1}. The upper spectrum is for the normal DPA and the lower spectrum for ^{13}C-DPA, in which one of the two central carbons is substituted with ^{13}C. The spectral pattern changes greatly on going from normal DPA to ^{13}C-DPA, indicating that both of the two bands at $1577\,cm^{-1}$ and $1557\,cm^{-1}$ are contributed by the central CC stretch vibration. In order to clarify this point quantitatively, we carried out a normal coordinate analysis based on a two-mode model. In this model, two vibrations are assumed to interact with each other. One is the central CC stretch vibration and the other is a phenyl-ring stretch vibration whose vibrational frequency falls in the region of 1500–1600 cm^{-1} (most probably the Wilson's 8a type mode). Four parameters, one diagonal element of the G-matrix and two diagonal and one off-diagonal elements of the F-matrix, were used. The diagonal element of the G-matrix for the CC stretch vibration was calculated from the atomic mass and was fixed. The calculation reproduced the observed four frequencies within an error of $2\,cm^{-1}$. The change in the intensity pattern was also very well accounted for [5]. We therefore conclude that the central CC bond order of S_1 DPA is decreased markedly and that the CC stretch frequency is lowered to the frequency region 1600–1500 cm^{-1}. The intrinsic CC stretch frequency obtained by the two-mode analysis is $1567\,cm^{-1}$. This means that the central CC bond of S_1 DPA is not a triple bond any more but is practically a double bond.

The CARS spectrum in the time range -5 to 5 ps is shown in Fig. 5. This time range is dominated by the CARS spectrum of the S_2 state. Contamination from strong S_1 bands are designated as S_1. A weak and broad feature is observed at $2099\,cm^{-1}$. The corresponding band in the ^{13}C-DPA spectrum is found at $2060\,cm^{-1}$. This isotopic frequency shift ($-39\,cm^{-1}$) is very close to the theoretical value for the pure CC stretch vibration ($-41\,cm^{-1}$). We therefore assign the $2099\,cm^{-1}$ band to the central CC stretch vibration of S_2 DPA. In contrast to what is found for the S_1 state, the central CC bonding in the S_2 state retains the character of a triple bond. The width of the $2099\,cm^{-1}$ band is determined as $40\,cm^{-1}$ by a curve-fitting analysis. The bandwidth of $40\,cm^{-1}$ is unusually large for a vibrational band. In fact, bandwidths determined for the

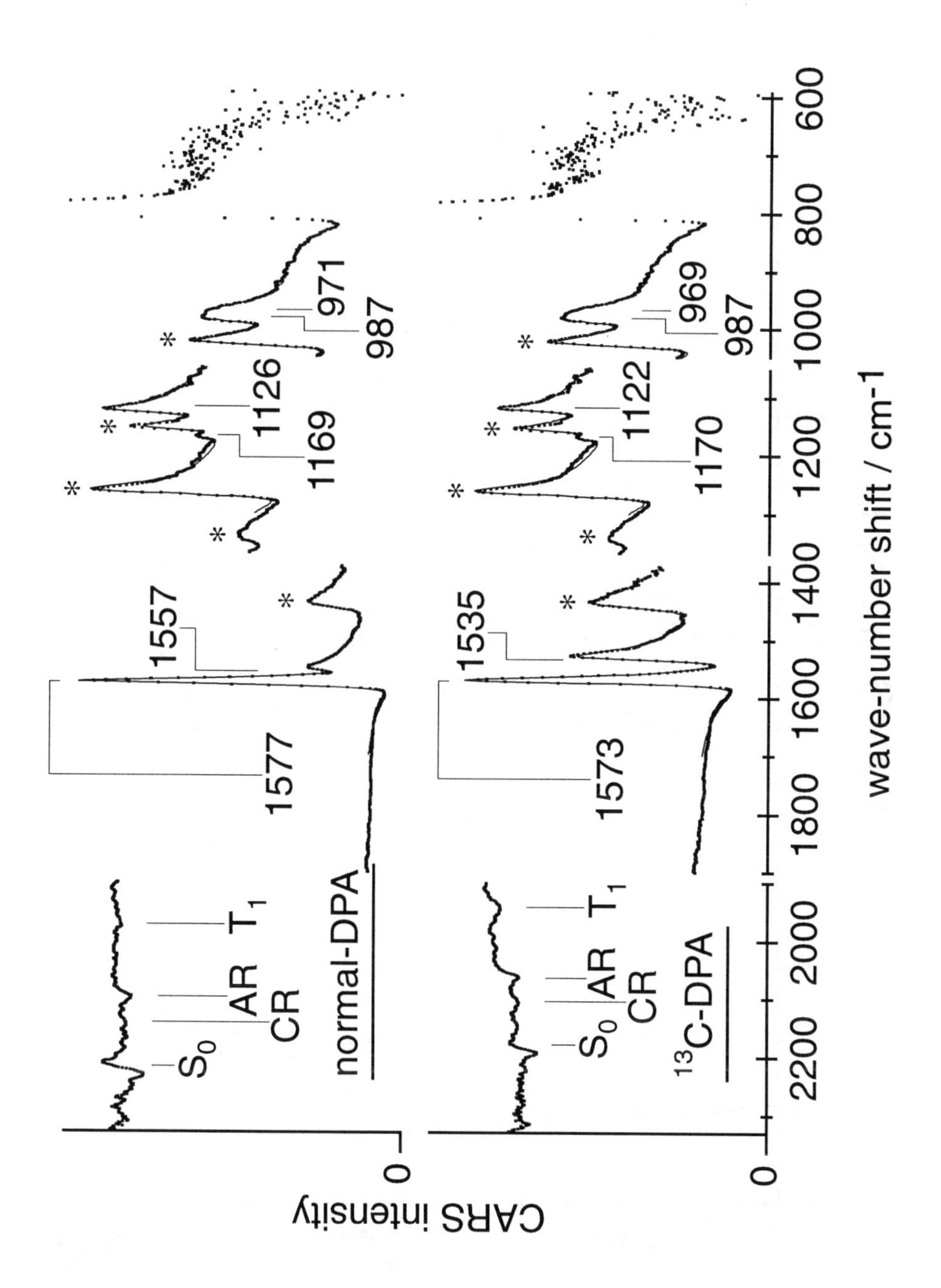

Fig. 3. CARS spectra of S_1 DPA.

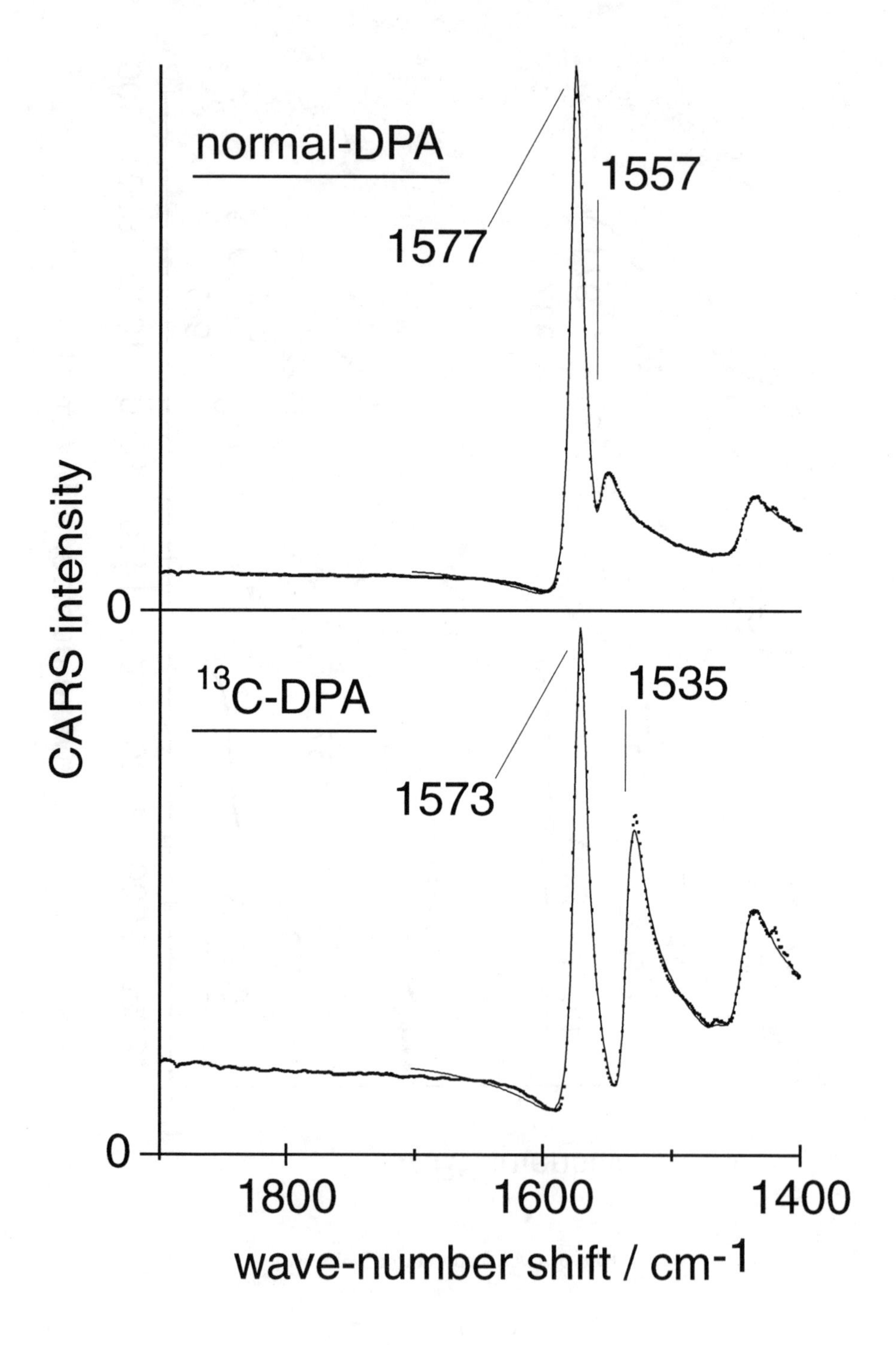

Fig. 4. CARS spectra of S_1 DPA (1900–1400 cm^{-1} region).

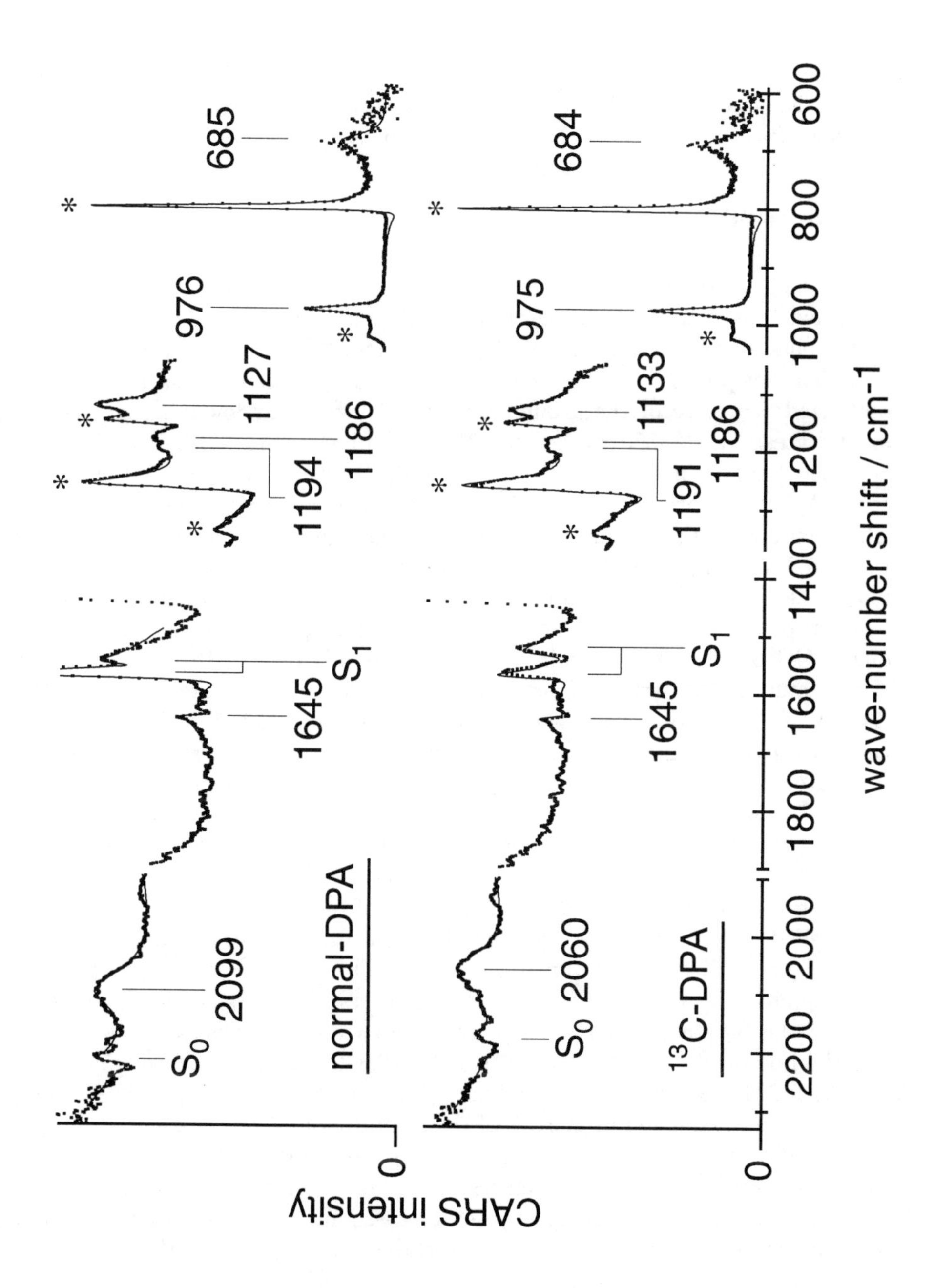

Fig. 5. CARS spectra of S_2 DPA.

other S_2 CARS bands are all smaller than $20\,cm^{-1}$. This finding suggest that there exists a mode-specific mechanism of vibrational dephasing that selectively broadens the width of the CC stretch band to as much as $40\,cm^{-1}$.

6 Discussion

Ferrante et al. [6] carried out empirical molecular orbital calculations of DPA and tried to explain the apparently contradictory observations in different environments, in the supersonic jet [1], in matrices [2], and in solution [3], from a unified viewpoint. They first performed an INDO/S calculation at the AM1 optimized structure of the ground electronic state. Four low-lying singlet excited states, B_{1u} ($33360\,cm^{-1}$ above the ground state A_g), B_{2u} ($34618\,cm^{-1}$), B_{3g} ($34642\,cm^{-1}$), and A_u ($34779\,cm^{-1}$) were obtained. The fifth excited singlet state, A_g, was located more than $5000\,cm^{-1}$ higher in energy than the A_u state. The accuracy of the calculation is not high enough to predict the ordering of the four low-lying states but is good enough to confine our attention to these four states in the explanation of the DPA photophysics. A population analysis in terms of Cohen's all-valence-electron bond order was carried out, and changes of the CC bond orders on going from the ground state to the excited states were examined. A marked change was obtained for the A_u state, for which the central CC bond order was calculated to be as small as 2.145. This means that the central CC bond is practically a double bond in the A_u state, as far as the calculated results are concerned. Appreciable change was also found for the B_{1u} state. The calculated CC bond order for this state was 2.514 and was 0.335 smaller than the ground state value (2.849). Only small changes, less than 0.05, were calculated for the other two states, B_{2u} and B_{3g}.

At this stage, we find good correspondence of the calculated results with the established experimental observations. First of all, the vibrational frequencies of the S_1 and S_2 states obtained by picosecond CARS spectroscopy [4, 5] are well explained by the calculated CC bond orders, if we assign the S_1 state to A_u and the S_2 state to B_{1u}. The correspondence of the observed frequencies and the bond orders is excellent; $2217\,cm^{-1}$ (2.849) for the S_0 state, $1567\,cm^{-1}$ (2.145) for the S_1 state, and $2099\,cm^{-1}$ (2.514) for the S_2 state. The strong absorption and emission properties of the S_2 state found by Hirata et al. [3] are fully consistent with this assignment of the S_2 state to B_{1u}. The proximity of the four singlet excited states may well result in different state orderings in different environments and leave uncertainty for the assignments of the other observed low-lying states. It seems certain, however, that the B_{1u} state is the S_3 state in jet, the S_1 state in matrices, and the S_2 state in solution. We note the possibility that the reported S_1 state in matrices may not be the real S_1 state, since the two-photon absorption was detected by one-photon fluorescence in the matrix experiments [2] and the levels lying below the 0-0 band of the emitting B_{1u} state were less likely to be detected.

The mutual exclusion relation between the one- and two-photon activities in jet was used by Okuyama et al. [1] to evidence D_{2h} or C_{2h} structures of DPA in the ground and excited electronic states. No evidence for a further distorted structure was found in matrices [2] nor in solution [3]. Thus, it seems reasonable to assume D_{2h}/C_{2h} structures for all the excited states observed. As for the S_2 (B_{1u}) state, observed vibrational frequency is as high as 2100 cm^{-1} both in matrices [2] and in solution [4, 5], and the central CC bond can be regarded as a triple bond. It is highly likely that S_2 DPA retains a D_{2h} structure as it is in the ground state, regardless of what environment it is in. The vibrational frequency of the S_1 (A_u) state was for the first time observed by CARS spectroscopy [4, 5]. This frequency, 1567 cm^{-1}, indicates a double-bond character of the central CC bond. The structure of DPA in the S_1 state may well be a trans bent structure with C_{2h} symmetry. It is well known that acetylene undergoes such a structural change on going from the linear ground state to the bent S_1 state [21]. The S_1 state of acetylene belongs to the A_u representation and corresponds to the A_u state of DPA with regard to the electronic structure in the central triple bond. In this regard, reexamination of the two-photon spectrum in jet is of vital importance. The 34960 cm^{-1} state in jet was assigned by Okuyama et al. [1] to the $2A_g$ state but this assignment is not consistent with recent molecular orbital calculations [2, 6]. If this state is re-assigned to A_u, which borrows two-photon activity from the nearby B_{3g} state by vibronic coupling, the analysis of the vibrational structure will afford detailed information on the structure of the S_1 (A_u) state.

Next, the dynamics related to the B_{1u} state are considered. Sudden disappearance of the vibrational structure above 700 cm^{-1} was found for the S_3 (B_{1u}) state in jet [1]. Broadenings of the vibrational bands above 1300 cm^{-1} were found for the S_1 (B_{1u}) state in matrices [2]. Solvent-dependent activation energies of about 1000 cm^{-1} were obtained for the $S_2(B_{1u}) \rightarrow S_1(A_u)$ internal conversion in solution [3, 6]. A mode-specific band broadening was observed in the CARS spectrum of S_2 (B_{1u}) state in solution [4, 5]. These observations suggest that some dynamic processes dependent on the excess vibrational energy and/or vibrational mode are operating in the vibrational manifold of the B_{1u} state of DPA. However, we need to note the fact that the electronic state interacting with the B_{1u} state might be quite different in different environments. Here we confine our discussion to the $S_2 \rightarrow S_1$ internal conversion process in solution, which is well established by time-resolved spectroscopies [3].

In order to explain the observed activation energy for the $S_2 \rightarrow S_1$ internal conversion, Ferrante et al. [6] extended their calculation to obtain potential energy curves of the four low-lying excited singlet states along the CC stretch coordinate. Their results are reproduced in Fig. 6. At the equilibrium CC distance in the ground state, the B_{1u} state has the lowest energy. The calculated potential minimum in the B_{1u} state is located at 1.24 Å, which is slightly longer than that of the ground state (1.20 Å). On the other hand, the A_u state has a potential minimum at far longer CC distance (1.28 Å), and the energy at this minimum is more than 1000 cm^{-1} lower than that of the B_{1u} minimum. In other words, the B_{1u} state is the lowest excited singlet state at the CC bond length of the ground state, but the A_u state is the real S_1 state whose potential minimum has the lowest electronic energy. The calculated B_{1u} and A_u potential curves cross each other at around 1.25 Å. Ferrante et al [6] argued that this potential crossing may result in a potential barrier that causes the observed activation energy for the $B_{1u} \rightarrow A_u$ internal conversion. Though this argument looks quite attractive, the following two points must be made clear before it is finally accepted. First, purely electronic interaction never occurs between two states with different symmetries, and therefore the B_{1u} and A_u potentials never make a barrier by an avoided crossing. Distortion along the CC stretch vibration (that belong to the a_g representation) cannot mix the B_{1u} and A_u states. The only possibility of the B_{1u}-A_u mixing is the vibronic coupling via the b_{1g} vibrations. In this case, however, the simple picture of barrier generation by an avoided potential crossing must be abandoned. Second, a large shift of the 0-0 transition of the B_{1u} state was observed on going from the jet (35248 cm^{-1}) to the solution (33577 cm^{-1} [22]). If the calculated B_{1u} potential curve in Fig. 6 is shifted toward lower energy by 1600 cm^{-1}, there will be no barrier generation by potential crossing. Note that the strongly one-photon allowed B_{1u} state will be significantly affected by the environment and the one-photon-forbidden A_u state will be much less affected [23]. We need to be very careful when we discuss the dynamics in solution on the basis of the calculations performed for the free molecule.

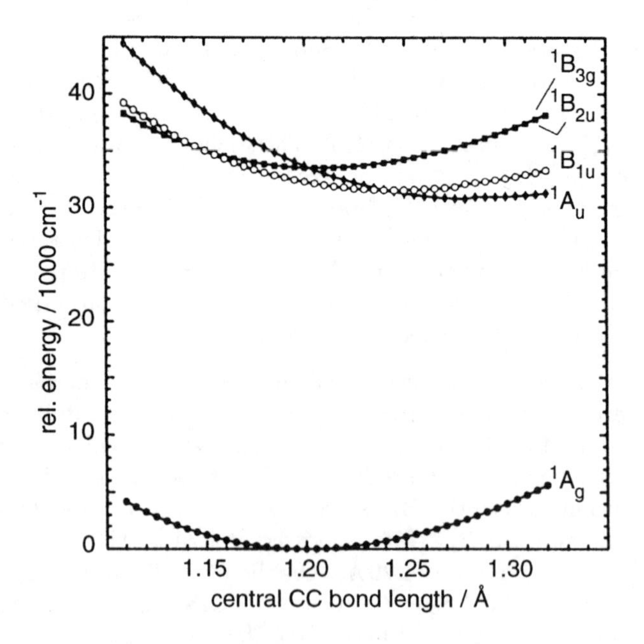

Fig. 6. Calculated potentials of excited singlet states of DPA [6].

Finally, we present an alternative view to account for the $S_2 \rightarrow S_1$ internal conversion dynamics of DPA in solution. This view is based on the model of solvent-induced dynamic polarization [24–26]. The model was first proposed by us to explain the newly observed relationship between the solvent-dependent Raman bandwidth of the S_1 state of *trans*-stilbene (S_1 *t*SB) and the solvent-dependent rate of isomerization; the CC stretch Raman band of S_1 *t*SB exhibits a mode-selective broadening the magnitude of which depends markedly on the solvent and shows a linear relationship with the rate of isomerization. The model assumes the solvent-induced dynamic polarization of the central CC bond of S_1 *t*SB which can be regarded as time-dependent state mixing between the S_1 state and a nearby zwitterionic excited state, $\mathcal{Z}$, whose potential minimum is located at the twisted configuration. Such solvent-induced state mixing causes the time-dependent change of the vibrational frequency and hence the dephasing of the CC stretch vibration. The same state mixing can cause the isomerization along the $\mathcal{Z}$ potential surface toward the twisted structure. The relationship between the vibrational band width and the rate of isomerization is thus rationalized by the dynamic polarization model. The model is readily applicable to the mode selective band broadening ($\approx 40\,\mathrm{cm}^{-1}$) of the CC stretch band of DPA with regard to the $S_2 \rightarrow S_1$ internal conversion. The solvent fluctuation occasionally mixes the S_1 state character into the S_2 state, causing the dephasing of the CC stretch vibration and the band broadening. Concomitantly, the mixing can result in the relaxation to the solvent-stabilized S_1 state to accomplish the $S_2 \rightarrow S_1$ internal conversion. The solvent fluctuation may well lower the local symmetry of the DPA molecule and make the mixing of the B_{1u} and A_u states possible. This thought is yet to be tested quantitatively, as was done for S_1 *t*SB [24–26]. Accurate measurements of the solvent and/or temperature dependence of the CC stretch band width of S_2 DPA are required as well as the those of the solvent and/or time dependence of the $S_2 \rightarrow S_1$ internal conversion rate.

References

1 K. Okuyama, T. Hasegawa, M. Ito, and N. Mikami, J. Phys. Chem. **88**, 1711 (1984).
2 M. Gutmann, M. Gudipati, P.-F. Schönzart, and G. Hohlneicher, J. Phys. Chem. **96**, 2433 (1992).
3 Y. Hirata, T. Okada, N. Mataga, and T. Nomoto, J. Phys. Chem. **96**, 6559 (1992).
4 T. Ishibashi and H. Hamaguchi, Chem. Phys. Lett. **264**, 551 (1997).
5 T. Ishibashi and H. Hamaguchi, submitted for publication.
6 C. Ferrante, U. Kensy, and B. Dick, J. Phys. Chem. **97**, 13457 (1993).
7 J. M. Robertson and I. Woodward, Proc. Roy. Soc. A **162**, 436 (1937).
8 A. Mavridis and I. Moustakali-Mavridis, Acta Crystallogr., Sect. B **33**, 3612 (1977).
9 A. A. Espiritu and J. G. White, Z. Krit **147**, 177 (1978).
10 A. V. Abramenkov, A. Almenningen, B. N. Cyvin, S. J. Cyvin, T. Jonvik, L. S. Khaikin, C. Rømming, and L. V. Vilkov, Acta Chem. Scand. A **42**, 674 (1988).
11 S. C. Chakravorty and S. C. Ganguly, Z. Phys. Chem. (Munich) **72**, 34 (1970).
12 Y. Tanizaki, H. Inoue, T. Hoshi, and J. Shiraishi, Z. Phys. Chem. (Munich) **74**, 45 (1971).

13 E. W. Thulstrup and J. Michl, J. Am. Chem. Soc. **104**, 5594 (1982).
14 H. Suzuki, Bull. Chem. Soc. Japan, **33**, 389 (1960); **33**, 944 (1960).
15 B. Kellerer, H. H. Hacker, and J. Brandmüller, Ind. J. Pure & Applied Phys. **9**, 903 (1971).
16 B. Kellerer and H. H. Hacker, J. Mol. Struct. **13**, 79 (1972).
17 G. Balanović, L. Colombo, and D. Skare, J. Mol. Struct. **147**, 275 (1986).
18 T. Tahara, B. N. Toleutaev, and H. Hamaguchi, J. Chem. Phys. **100**, 786 (1994).
19 T. Tahara and H. Hamaguchi, Rev. Sci. Instru. **65**, 3332 (1994).
20 H. Hiura and H. Takahashi, J. Phys. Chem. **96**, 8909 (1992).
21 C. K. Ingold and G. W. King, J. Chem. Soc., 2725 (1953).
22 Unpublished data.
23 W. Liptay, Z. Naturforsch. **20a**, 1441 (1965).
24 H. Hamaguchi and K. Iwata, Chem. Phys. Lett. **208**, 465 (1993).
25 H. Hamaguchi, Mol. Phys. **88**, 463 (1996).
26 V. Deckert, K. Iwata, and H. Hamaguchi, J. Photochem. Photobiol. A **102**, 35 (1996).

Laser Control of Electrons in Molecules

André D. Bandrauk, Hengtai Yu
Laboratoire de Chimie Théorique, Faculté des sciences,
Université de Sherbrooke, Sherbrooke, Qué, J1K 2R1, Canada
E-mail :bandrauk@gauss.chimie.usherb.ca

Abstract

Coherent superposition of electronic states can be achieved by simultaneous laser excitation at different frequencies and phases. A three level excitation scheme is analyzed and applied to phase control of electrons or equivalently charge transfer in complex molecules. *Ab initio* molecular orbitals are used to demonstrate the principle in the charge transfer molecule DMABN, 4-(N,N-dimethylamino)benzonitrile.

1 Introduction

One of the most active areas of investigation in the search for new photochemical reactions has been electron transfer reactions initiated by photoexcitation of donor-acceptor pairs. The ensuing radical cation of the donor usually undergoes efficient bond fragmentation, thus offering novel reactions, triggered by electron transfer, ET [1, 2]. Lately , much interest has been focused on laser control of the dynamics of photochemical processes via the relative phases of laser beams or pulses available with current laser technology [3]. Recent theoretical and experimental investigations have shown the feasibility of controlling electrons during ionization processes [4, 5], and in quantum wells [6]. Recently we have shown that one can control ET in the photodissociating Cl_2 molecule by using a symmetry-breaking excitation scheme with photons of frequency ω and 2ω [7]. Such a scheme has now been demonstrated experimentally to provide a useful means of controlling current in quantum wells [6], which have strong analogies to molecular systems.

In the present chapter we investigate the use of optically induced quantum coherence effects that can be achieved using moderately intense laser fields, well below intensities $I = 10^{13}$ W/cm^2 above which ionization begins to occur on picosecond time scales [8]. Coherent superposition of excited states results in linear combinations of waves called wave packets [9]. With appropriate phase, one should be able to shape such wave packets and guide them. We have shown in a previous work how coherent excitation of a three-level system can be applied to manipulate and control ET in the charge transfer molecule DMABN [10] (Fig. 1) using previous results of *ab initio* calculations on this molecule [11]. We reexamine here this problem in detail and conclude with a discussion of the feasibility of laser control of electrons in complex molecules in general .

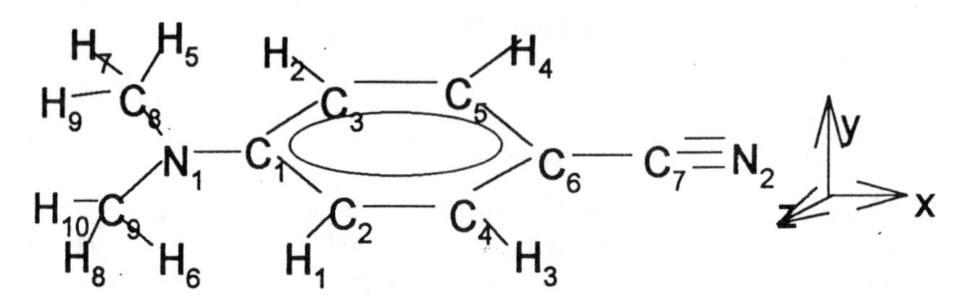

Fig. 1. Coordinate system of DMABN

2 Coherent excitation of a three-level system

Three-level systems offer the possibility of investigating analytically coherence properties initiated by laser excitation, and therefore been examined in the atomic case in detail. The analysis shows the occurrence of laser-induced dark states that are completely uncoupled from the initial ground states as a consequence of radiation nonperturbative coherent properties of such systems [9]. In particular, when excitations occur in a Λ-scheme (pump-probe or Raman), then these dark states are stable (long-lived) coherent superpositions of the initial and final states. We examine here a scheme more appropriate for photochemistry, the V-scheme (Fig. 2), where one excites simultaneously from the ground state $|0\rangle$ to two excited states $|1\rangle$ and $|2\rangle$ with lasers of frequency ω_1 and ω_2 by fields or pulses of amplitude E_1 and E_2. The corresponding radiative couplings are expressed in terms of Rabi frequencies Ω such that $\hbar\Omega_1 = \mu_{01}E_1$, $\hbar\Omega_2 = \mu_{02}E_2$, where the μ's are transition dipole moments. For real time-dependent fields, $E(t) = E\cos(\omega t + \phi) = \frac{E}{2}[\exp(i\omega t + \phi) + \text{cc.}]$, retaining only one exponential term, the resonant term, leads to the rotating wave approximation (RWA) [8, 9].

2.1 **Resonant Case**

RWA leads to a time-independent Hamiltonian in the dressed state representation [8, 9] for resonant excitation ($\hbar\omega_1 = \mu_{01}E_1, \hbar\omega_2 = \mu_{02}E_2$, solid arrows, Fig. 2)

$$H = \frac{\hbar}{2}\begin{pmatrix} 0 & \Omega_1 e^{i\phi_1} & \Omega_2 e^{i\phi_2} \\ \Omega_1 e^{-i\phi_1} & 0 & 0 \\ \Omega_2 e^{-i\phi_2} & 0 & 0 \end{pmatrix}, \tag{1}$$

where ϕ_1 and ϕ_2 are the phases of the two laser beams E_1 and E_2. Defining an effective 3-level Rabi frequency

$$\Omega = (\Omega_1^2 + \Omega_2^2)^{\frac{1}{2}}, \tag{2}$$

one obtains the following three dressed states, or eigenmodes, of the three level laser-molecule system (n = photon number) ,

$$\begin{aligned}
|\alpha\rangle &= 2^{-\frac{1}{2}}[|0, n_1, n_2\rangle + \frac{\Omega_1}{\Omega}e^{-i\phi_1}|1, n_1 - 1, n_2\rangle + \frac{\Omega_2}{\Omega}e^{-i\phi_2}|2, n_1, n_2 - 1\rangle], && (3)\\
E_\alpha &= \frac{\hbar\Omega}{2}, && (4)\\
|\beta\rangle &= 2^{-\frac{1}{2}}[-|0_1, n_1, n_2\rangle + \frac{\Omega_1}{\Omega}e^{-i\phi_1}|1, n_1 - 1, n_2\rangle \\
&\quad + \frac{\Omega_2}{\Omega}e^{-i\phi_2}|2, n_1, n_2 - 1\rangle], && (5)\\
E_\beta &= -\frac{\hbar\Omega}{2}, && (6)\\
|\gamma\rangle &= -\frac{\Omega_2}{\Omega}e^{i\phi_2}|1, n_1 - 1, n_2\rangle + \frac{\Omega_1}{\Omega}e^{i\phi_2}|2, n_1, n_2 - 1\rangle, && (7)\\
E_\gamma &= 0, && (8)
\end{aligned}$$

where $E_{\alpha(\beta,\gamma)}$ are the dressed state energies of the laser-molecule system. We note that the states $|\alpha\rangle$ and $|\beta\rangle$ are Stark shifted by $\pm\hbar\frac{\Omega}{2}$, so that the frequency separation between these two dressed states is the effective Rabi frequency Ω defined in (2). Both states reflect the laser-induced superposition of all three electronic states that are phaser-ependent. The state $|\gamma\rangle$ is the dark state, which remains unshifted and does not contain the initial state $|0\rangle$. Clearly the state $|\gamma\rangle$ will never be populated from the ground state $|0\rangle$ in the V-configuration. As a consequence the total wave function can be written as combinations of $|\alpha\rangle$ and $|\beta\rangle$ only:

$$|\psi(t)\rangle = 2^{-\frac{1}{2}}(|\alpha\rangle e^{\frac{-i\Omega t}{2}} - |\beta\rangle e^{\frac{i\Omega t}{2}}), \tag{9}$$

satisfying the initial condition $|\psi(0)\rangle = |0, n_1, n_2\rangle$, i.e., the ground state. Thus at times $\tau = \frac{\pi}{\Omega}$, one has the superpositions

$$\begin{aligned}
|\psi(\tau)\rangle &= -i2^{-\frac{1}{2}}(|\alpha\rangle + |\beta\rangle) \\
&= -i(\frac{\Omega_1}{\Omega}e^{-i\phi_1}|1, n_1 - 1, n_2\rangle + \frac{\Omega_2}{\Omega}e^{-i\phi_2}|2, n_1, n_2 - 1\rangle). && (10)
\end{aligned}$$

One observes that at such times, where $\tau = \frac{\pi}{\Omega}$, the laser induced coherent state is a pure superposition of the excited states $|1\rangle$ and $|2\rangle$ only, and the relative sign of these amplitudes can be controlled by the relative laser phase difference

$(\phi_2 - \phi_1)$. In order to see this more succinctly, we specialize further to the particular case, i.e., the equal radiative coupling case

$$\Omega_1 = \Omega_2 = 2^{-\frac{1}{2}}\Omega. \tag{11}$$

This can be achieved by varying the laser field amplitudes E_1 and E_2 so that $\mu_{01}E_1 = \mu_{02}E_2$. In this case one recovers the simply transparent result

$$|\psi(\tau)\rangle = -i2^{-\frac{1}{2}}(e^{-i\phi_1}|1, n_1 - 1, n_2\rangle + e^{-i\phi_2}|2, n_1, n_2 - 1\rangle). \tag{12}$$

Thus if $\phi_1 = \phi_2 = \phi$, one obtains

$$|\psi^+(\tau)\rangle = -i2^{-\frac{1}{2}}e^{-i\phi}(|1, n_1 - 1, n_2\rangle + |2, n_1, n_2 - 1\rangle). \tag{13}$$

If $\phi_1 - \phi_2 = \pi$, one now has

$$|\psi^-(\tau)\rangle = -i2^{-\frac{1}{2}}e^{-i\phi}(|1, n_1 - 1, n_2\rangle - |2, n_1, n_2 - 1\rangle). \tag{14}$$

We observe that at times $\tau = \frac{\pi}{\Omega}$, one can create new excited state amplitudes that are simple sums or differences of the original excited states. For molecules, this implies that one should be able to prepare simple superpositions of delocalized molecular orbitals and thus create more localized electronic states leading to laser induced CT.

We designate the states $|\psi^\pm(\tau)\rangle$ as our target states and examine their time evolution from the occupation probability

$$P^\pm(t) = |\langle\psi^\pm(\tau)|\psi(t)\rangle|^2. \tag{15}$$

The resulting time dependent occupation probabilities of the target states $|\psi^\pm(\tau)\rangle$ are readily obtained from equations (13-15):

$$P^\pm(t) = \frac{1}{2}\sin^2(\frac{\Omega t}{2})[1 \pm \cos(\phi_2 - \phi_1)]. \tag{16}$$

We conclude by observing that in order to obtain complete selectivity of one state or another, i.e., $P^\pm(t) = 1$, then

(i) the phases of the two lasers must be locked, i.e., $\phi_2 - \phi_1 = 0$ or $= \pi$;

(ii) the recurence time of the selectivity is $t_n = (2n + 1)\tau = \frac{(2n+1)\pi}{\Omega}$.

Thus two essential parameters enter into the *controllability* of the system: the Rabi frequency Ω, which depends on transition moments μ and field amplitudes E and finally the phase difference ϕ between the two laser beams or pulses. We observe furthermore from (10) that a special feature of the three-level V system is that in the general case the total system oscillates between the ground state $|0\rangle$ and the coherent linear superpositions of $|1\rangle$ and $|2\rangle$ defined by $|\psi(\tau)\rangle$ (10). In fact, the general state (9) can be written as

$$|\psi(t)\rangle = |0\rangle\cos(\frac{\Omega t}{2}) - i|\psi(\tau)\rangle\sin(\frac{\Omega t}{2}), \tag{17}$$

thus showing that at recursion times $t_n = (2n+1)\frac{\pi}{\Omega}$, one is indeed in the general coherent state $|\psi(\tau)\rangle$ (10), whereas at times $t_m = 2m(\frac{\pi}{\Omega})$, one is in the ground state $|0\rangle$.

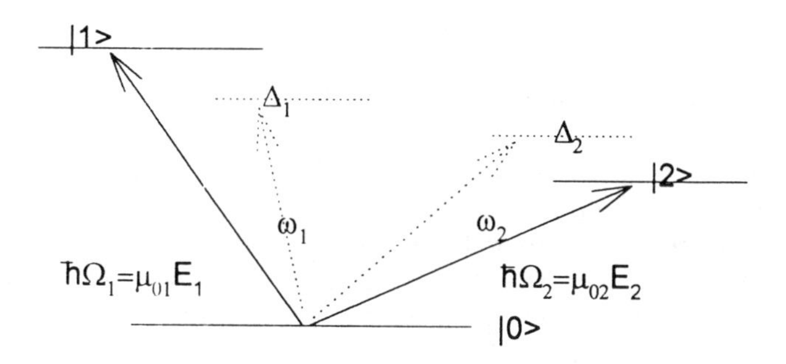

Fig. 2. Three level V system with appropriate Rabi frequencies Ω , transition moments μ and laser field amplitudes E. resonant: solid arrows; nonresonant : broken arrows ($\Delta_1 < 0; \Delta_2 > 0$)

2.2 Non Resonant Case

In general it is not possible to define the absolute phase difference $\phi = \phi_1 - \phi_2$ between two different laser fields E_1 and E_2 unless the frequencies are multiples of each other. Thus one must consider the nonresonant case, $\hbar\omega_1 - E_{10} = \Delta_1 \langle 0$; $\hbar\omega_2 - E_{20} = \Delta_2 \rangle 0$ (Fig. 2b, broken arrows). The nonresonant RWA Hamiltonian now becomes

$$H = \frac{1}{2}\begin{pmatrix} 0 & \hbar\Omega_1 e^{i\phi_1} & \hbar\Omega_2 e^{i\phi_2} \\ \hbar\Omega_1 e^{-i\phi_1} & 2\Delta_1 & 0 \\ \hbar\Omega_2 e^{-i\phi_2} & 0 & 2\Delta_2 \end{pmatrix}. \tag{18}$$

The new dressed states now become

$$|\alpha\rangle = C_\alpha[|0, n_1, n_2\rangle + \frac{\hbar\Omega_1 e^{-i\phi_1}}{2(E_\alpha - \Delta_1)}|1, n_1 - 1, n_2\rangle + \frac{\hbar\Omega_2 e^{-i\phi_2}}{2(E_\alpha - \Delta_2)}|2, n_1, n_2 - 1\rangle]. \tag{19}$$

with similar expressions for $|\beta\rangle$ and $|\gamma\rangle$. Simple analytical solutions for the dressed energies exist when [12]

$$\Omega_1^2\Delta_2 = -\Omega_2^2\Delta_1, \tag{20}$$

$$E_{\alpha(\beta)} = \frac{1}{2}(\Delta_1 + \Delta_2) \pm \{(\Delta_1 - \Delta_2)^2 + (\Delta_1)^2 + (\Delta_2)^2\}^{\frac{1}{2}}, \quad E_\gamma = 0, \tag{21}$$

with for example

$$|\gamma\rangle = C_\gamma[|0, n_1, n_2\rangle - \frac{\hbar\Omega_1 e^{-i\phi_1}}{2\Delta_1}|1, n_1 - 1, n_2\rangle - \frac{\hbar\Omega_2 e^{-i\phi_2}}{2\Delta_2}|2, n_1, n_2 - 1\rangle], \quad (22)$$

and the C's in equations (19) and (22) are normalizing factors. We note that as $\Omega_1, \Omega_2 \longrightarrow 0$ at t=0 or ∞, one now obtains in the adiabatic limit $|\alpha\rangle \longrightarrow |1\rangle, |\beta\rangle \longrightarrow |2\rangle$ and $|\gamma\rangle \longrightarrow |0\rangle$. This is different from the resonant case, (3 7) where $|\alpha\rangle$ and $|\beta\rangle$ evolve adiabatically into the ground state $|0\rangle$; whereas the dark state $|\gamma\rangle$ evolves adiabatically from $|1\rangle$ at $t = -\infty$ to $|3\rangle$ at t=$+\infty$. The latter is the basis of the Stimulated Raman Adiabatic Passage inversion scheme, STIRAP, whereby one can transfer efficiently from state $|1\rangle$ to state $|3\rangle$ in a Λ-scheme [13]. Clearly for strong fields such that $\hbar\Omega/\Delta >> 1$, the radiative interaction Ω is larger than the detuning Δ, then one obtains in the nonresonant case essentially time-independent coherent states when $|\Omega_1/\Delta_1| = |\Omega_2/\Delta_2| >> 1$, or

$$|\gamma\rangle = C_\gamma[|1, n_1 - 1, n_2\rangle \pm e^{i\phi_1}|2, n_1, n_2 - 1\rangle]. \quad (23)$$

This is illustrated in Fig. 3 for a hypothetical three level system with $E_{10} = 0.31, E_{20} = 0.17, \omega_1 = 0.32, \omega_2 = 0.16$ such that $\Delta_1 = 0.01, \Delta_2 = -0.01$ and $\Omega_1 = \Omega_2 = 0.05$. We use atomic units (E (a.u.) = 27.2 eV, t(a.u.) = 2.47×10^{-17}s). For $\Omega_1/\Delta_1 = 5 = -\Omega_2/\Delta_2$, results are shown for the phase $\phi = \pi$ where $|\gamma\rangle \approx (|1\rangle + |2\rangle)/\sqrt{2}$. Calculations were done by solving the exact time-dependent Schrödinger equation for 1000 cycles of laser 1(2). We note the quick saturation of the coherent state $|\gamma\rangle = (|1\rangle + |2\rangle)/\sqrt{2}$ after 800 cycles, i.e., transient effects from the nascent population of the initial $|0\rangle$ state disappear after that time.

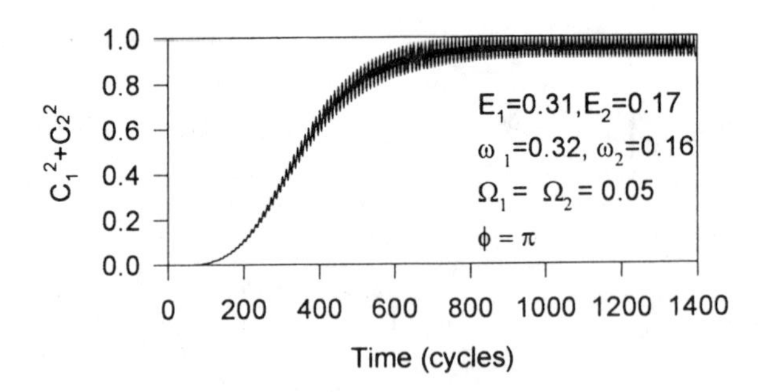

Fig. 3. Three level system with $E_{10} = 0.31, E_{20} = 0.17, \omega_1 = 0.32, \omega_2 = 0.16$, and $\Omega_1 = \Omega_2 = 0.05$ a.u.; 1000 cycle rising time

In conclusion we have shown that resonant excitation, (1) and non resonant excitation (18) lead in each case to phase-locked coherent superposition of the two excited states $|1\rangle$ and $|2\rangle$, (13 14) and equivalently, (23). Such coherent electronic states depend only on the phase difference $\phi = \phi_1 - \phi_2$ between the two laser fields of amplitudes E_1 and E_2. In both resonant and nonresonant $(\Omega/\Delta\rangle 1)$ cases, the maximum state difference is obtained between $\phi = 0$ and π, since the three-level V-scheme (Fig. 1) involves coherences between the two excited states $|1\rangle$ and $|2\rangle$ only .

3 *Ab Initio* Simulations

DMABN (Fig. 1) is the prototype of an ET, molecule, in which the formation of a twisted intramolecular ET or equivalently charge transfer (CT), state is well documented from fluorescence studies. Thus from the red shifts of the fluorescence in polar solvents, it is inferred that the dimethylamine (DMA) group undergoes a 90 degree internal rotation with respect to the aromatic ring, and an electron is completely transferred from the amine group to the benzonitrile moiety because the conjugation between the amine lone pair orbital n and π-orbitals in the aromatic ring is then absent [14, 15]. The time for such a rotation has been estimated to be at least 40 picoseconds [14].

We have previously calculated static electronic properties of DMABN by *ab initio* CI methods using the GAMESS quantum chemistry program [16]. It is found to be in agreement with other calculations [11] that the first excited state, S_1, has different symmetry from the ground state, as it is obtained from an $n\pi^*$ transition. Efficient ET from the amino group to the triple bond can be achieved by considering the excited S_2 and S_3 states only. To construct an appropriate coherent electronic, we assume that the S_2 and S_3 state wave functions Ψ_I and Ψ_{II} can be expressed as linear combinations of three main electronic configuration state functions, CSF, Φ_A, Φ_B, and Φ_C [11],

$$
\begin{aligned}
|\Psi_I\rangle &= a_1\Phi_A + b_1\Phi_B + c_1\Phi_C, \\
|\Psi_{II}\rangle &= a_2\Phi_A + b_2\Phi_B + c_2\Phi_C,
\end{aligned}
\tag{24}
$$

where $a_i, b_i, c_i(\mathrm{i} = 1, 2)$ are the coefficients of the CSF in the CI calculations. The calculated values are $a_1 = 0.9635, b_1 = 0.1088, c_1 = -0.2447$ and $a_2 = -0.3405, b_2 = 0.9125, c_2 = -0.2270$ after renormalization to eliminate smaller CSF contributions. Thus $|\Psi_I\rangle$ and $|\Psi_{II}\rangle$ correspond to the dressed states $|1, n_1 - 1, n_2\rangle$ and $|2, n_1, n_2 - 1\rangle$ defined in section 2 (see (7)).

We therefore represent a general coherent state as

$$
|\Psi_c\rangle = C_I\Psi_I + C_{II}e^{i\phi}\Psi_{II},
\tag{25}
$$

where C_I and C_{II} are coefficients depending on transition moments from the ground state and laser amplitudes E_1, E_2 respectively, and ϕ is the laser phase difference between the two laser fields. Substituting (24) into (25) results in the coherent state function expressed as

$$|\Psi_c\rangle = A\Phi_A + B\Phi_B + C\Phi_C, \tag{26}$$

$$\begin{aligned} A &= a_1 C_I + a_2 C_{II} e^{i\phi}, \\ B &= b_1 C_I + b_2 C_{II} e^{i\phi}, \\ C &= c_1 C_I + c_2 C_{II} e^{i\phi}. \end{aligned} \tag{27}$$

One can write the coherent state density as

$$|\Psi_c|^2 = (A^*, B^*) \begin{pmatrix} \rho_{AA} & \rho_{AB} \\ \rho_{BA} & \rho_{BB} \end{pmatrix} \begin{pmatrix} A \\ B \end{pmatrix} + C^2 \rho_{CC}. \tag{28}$$

where $\rho_{AA} = |\Phi_A|^2$, $\rho_{BB} = |\Phi_B|^2$, $\rho_{CC} = |\Phi_C|^2$, and $\rho_{AC} = \rho_{BC} = 0$ [11].

We next introduce a unitary matrix U to remove the nondiagonal density matrix ρ_{AB} in order to localize the electron density; i.e.,

$$\begin{aligned} |\Psi_c|^2 &= (A^*, B^*) UU^+ \begin{pmatrix} \rho_{AA} & \rho_{AB} \\ \rho_{BA} & \rho_{BB} \end{pmatrix} UU^+ \begin{pmatrix} A \\ B \end{pmatrix} + C^2 \rho_{CC} \\ &= (A'^*, B'^*) \begin{pmatrix} \rho_{11} & \rho_{12} \\ \rho_{21} & \rho_{22} \end{pmatrix} \begin{pmatrix} A' \\ B' \end{pmatrix} + C^2 \rho_{CC} \end{aligned} \tag{29}$$

such that ρ_{12} is minimized. As suggested earlier by Kato et al. [11], a useful method is to diagonalize the dipole matrix between the two CSFs Φ_A and Φ_B,

$$\begin{aligned} D &= \begin{pmatrix} D_{AA} & D_{AB} \\ D_{BA} & D_{BB} \end{pmatrix}, \\ D_{AB} &= \sum_{i,j} c_i^{A*} c_j^B \langle \phi_i | z | \phi_j \rangle = \int\int z \rho_{AB} \mathrm{d}v, \end{aligned} \tag{30}$$

where D_{AA} and D_{BB} correspond to the dipole moments in diabatic states. Thus in our case we obtained (unit: Debye)

$$\begin{aligned} D &= \begin{pmatrix} -13.663 & -2.242 \\ -2.242 & -5.204 \end{pmatrix}, \\ D' &= U^+ D U = \begin{pmatrix} -14.220 & 0 \\ 0 & -4.647 \end{pmatrix}. \end{aligned} \tag{31}$$

From the above results it can be seen that the dipole moments in diabatic states obtained from GAMESS are very similar to those from larger CI calculations [11]. This procedure resulted in small off-diagonal elements ρ_{12}, ρ_{21}, as anticipated by Kato et al.[11], as such a procedure corresponds to separation of charge. The total coherent density is thus a sum of simply separated charge densities ρ_{11} and ρ_{22}, as we shall see below.

$$|\Psi_c|^2 = A'^2\rho_{11} + B'^2\rho_{22} + C^2\rho_{CC}. \tag{32}$$

From equation (29) we have

$$\begin{pmatrix} A' \\ B' \end{pmatrix} = U^+ \begin{pmatrix} A \\ B \end{pmatrix} = \begin{pmatrix} U_{11} & U_{21} \\ U_{12} & U_{22} \end{pmatrix} \begin{pmatrix} A \\ B \end{pmatrix}. \tag{33}$$

In order to localize the charge density via the laser dependent parameters C_I, C_{II}, we examine separately the three case: $|A'|^2 = 0, |B'|^2 = 0$, and $|C''|^2 = 0$, i.e. we eliminate linear combinations of the CSFs. As an example, one can set $B' = 0$ under the condition:

$$\begin{aligned} (a_2U_{12} + b_2U_{22})C_{II}\sin\phi &= 0, \\ \frac{C_I}{C_{II}} &= -\frac{a_2U_{12} + b_2U_{22}}{a_1U_{12} + b_1U_{22}}\cos\phi. \end{aligned} \tag{34}$$

Two simple solutions are possible, $\phi = 0$ or π. Setting $|B'|^2 = 0$ would give other solutions which are more complicated [17]. The simple procedure with $\phi = 0$ gives

$$|\Psi_c|^2 = 0.997\rho_{11} + 0.084\rho_{CC}. \tag{35}$$

Finding the condition for which $A' = 0$ with $\phi = 0$ gives

$$|\Psi_c|^2 = 0.997\rho_{22} + 0.071\rho_{CC}, \tag{36}$$

and $C = 0$ with $\phi = 0$ resulted in

$$|\Psi_c|^2 = 0.649\rho_{11} + 0.761\rho_{22}. \tag{37}$$

Clearly, in each case, $A' = 0$ and $B' = 0$, the contribution from the CSF Φ_C was always found to be negligible.

We illustrate in Fig. 4 our calculations for the densities $\rho_{11}, \rho_{12}, \rho_{22}, \rho_{CC}$, which were then used to obtain the new coherent densities $|\Psi(A' = 0)|^2, |\Psi(B' = 0)|^2, |\Psi(C = 0)|^2$ in Fig. 5. We note first that ρ_{11}, ρ_{22} are indeed well separated in the molecule and correspond to opposite charges due to ET; ρ_{CC} is more delocalized but is a minor component. Thus, eliminating each component separately using the laser parameters results in an electron distribution always well localized in one end of the molecule or the other. Other localizations are obtained for $\phi = \pi$, as found earlier [10], and for laser parameter values giving $|A'|^2 = 0, |B'|^2 = 0$ [17]. The density difference figures are given in Fig. 6, so we can see more clearly the resulting electron localizations.

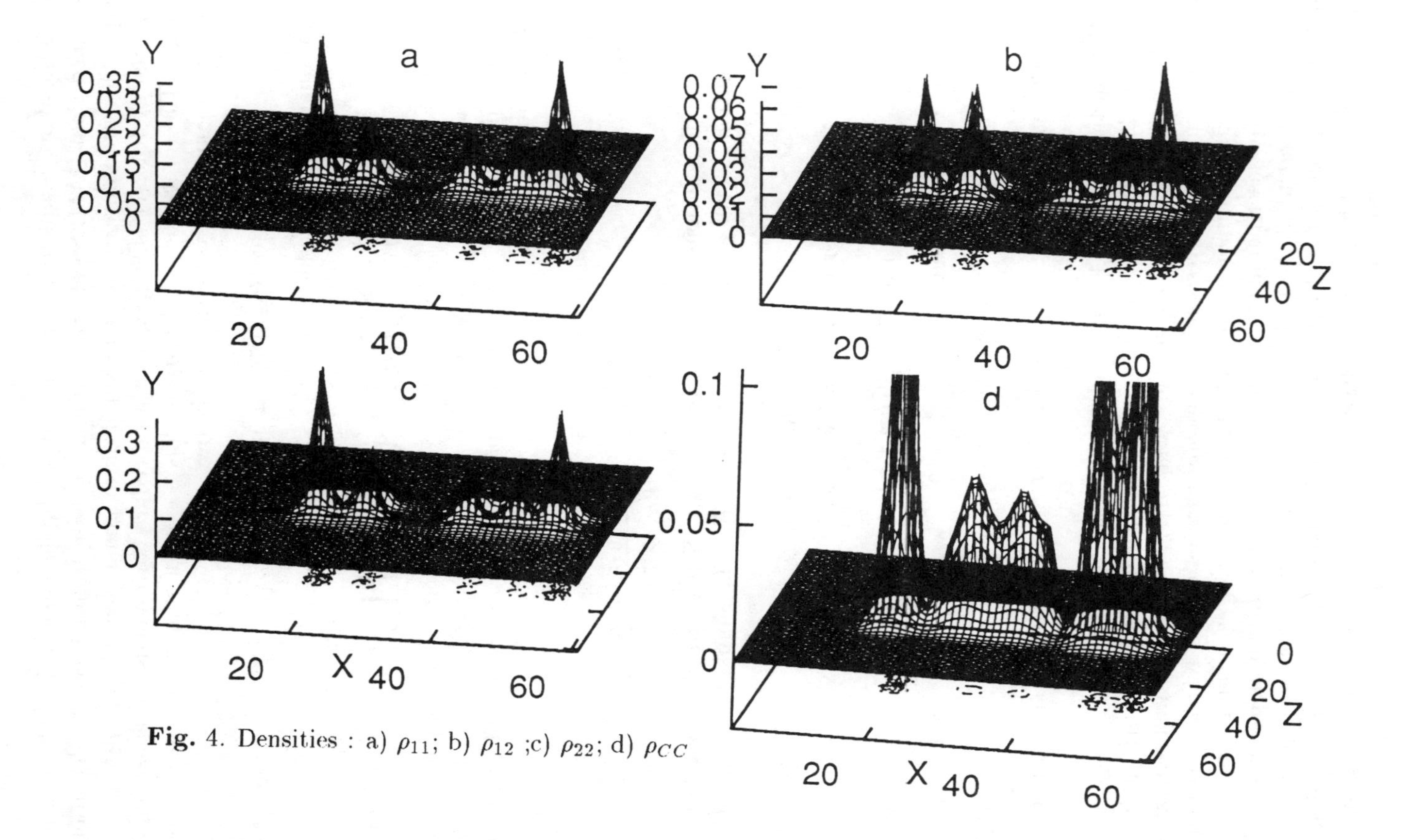

Fig. 4. Densities : a) ρ_{11}; b) ρ_{12} ;c) ρ_{22}; d) ρ_{CC}

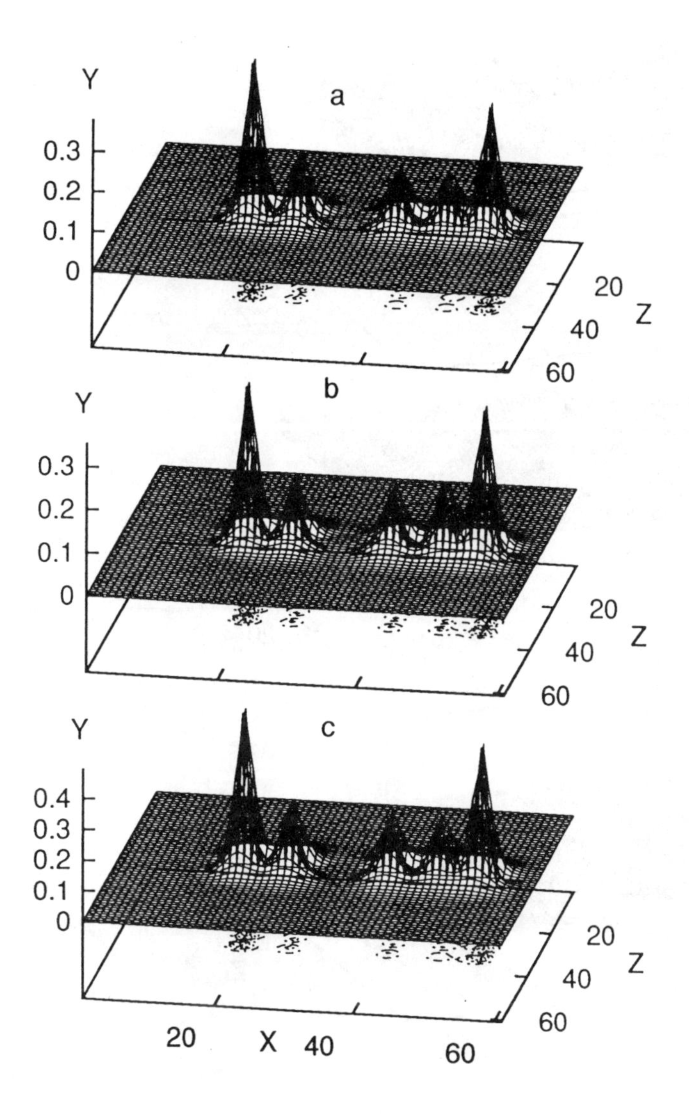

Fig. 5. Coherent density: a) $|Psi(A' = 0)|^2$; b) $|Psi(B' = 0)|^2$; c) $|Psi(C = 0)|^2$

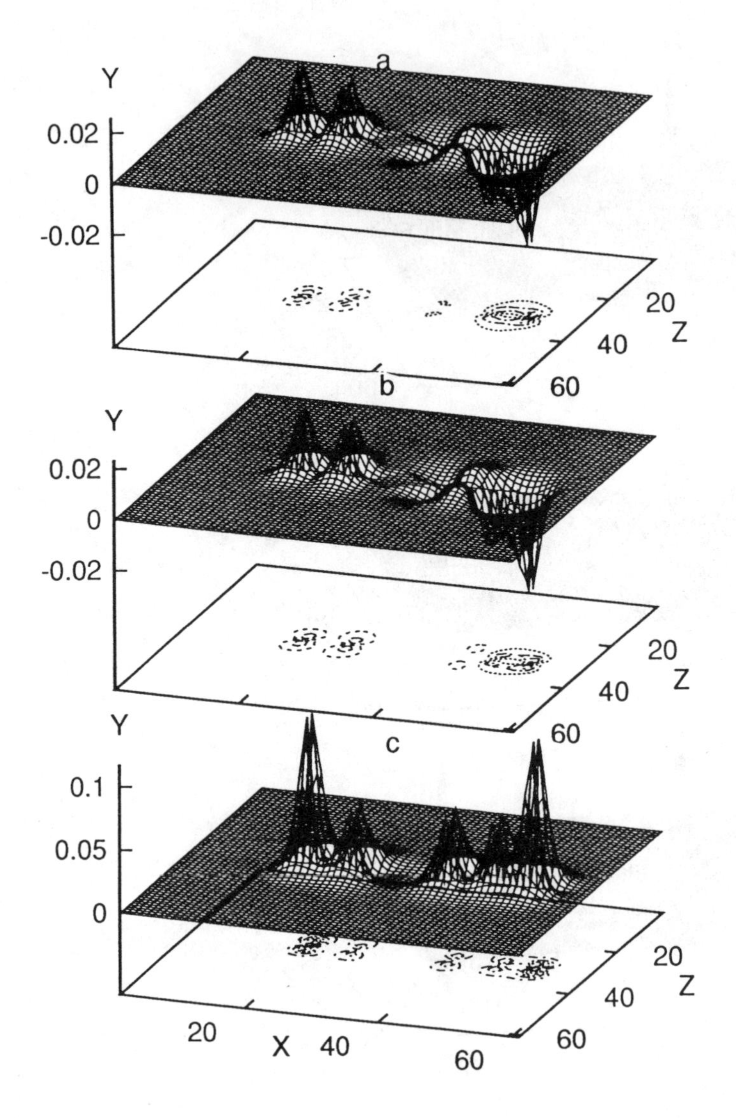

Fig. 6. Density differences: a) ρ_{22} - ρ_{11}; b) $|Psi(A' = 0)|^2$ - $|Psi(B' = 0)|^2$; c) $|Psi(C = 0)|^2$ - $|Psi(A' = 0)|^2$

4 Conclusions

Coherence and decay of electronic wave packets has been studied in detail for Rydberg atomic states [18] where dephasing due to nuclear motion is of no concern. We have shown above that simultaneous laser excitation of different electron states in CT systems can be used to transfer electrons between different functional groups, e.g., amino to nitrile (Fig. 1). This is achieved by varying appropriately configuration state function coefficients, (26) through the laser field amplitude E and relative phase ϕ. As shown from (17), the laser induced ET or CT only persists during the laser illumination at the recursion time $\tau_n = (2n+1)(\pi/\Omega)$, where Ω is the Rabi frequency, (11). Since $\Omega(\mathrm{cm}^{-1}) = 10^{-3} I(\mathrm{W/cm}^2)\mu$ (a.u.) [8], then for intensities $I = 10^8$ W/cm^2 and $\mu = 10$ a.u., we obtain $\Omega = 100$ cm^{-1} and $\tau = \pi/\Omega c \approx 10^{-12} s = 1$ ps. Thus since the internal rotation requires 40 ps to stabilize CT in solution, the above calculation shows that laser-induced CT should be feasible with moderate intensity (I $\approx 10^8$ W/cm^2) picosecond pulses before vibrational relaxation will occur, e.g., bending of the cyano group instead of twisting of the amino group [19]. Finally, the large radiative coupling implied by the Rabi frequency Ω ensures that large molecules such as DMABN will easily be aligned by the laser field [20] in order to enhance the laser-induced CT proposed in this work.

5 Acknowledgments

We thank the Natural Sciences and Engineering Research Council of Canada for support of the present research and Dr. M. W. Schmidt (Iowa State University) for useful discussions about transition moment calculations with GAMESS.

References

1. E. R. Gaillard and D. G. Whitten, Acc. Chem. Res. **29** , 292 (1996).
2. Y. Wang, L. A. Lucia, and K. S. Schanze, J. Phys. Chem. **99**, 1961 (1995).
3. P. Brumer and M. Shapiro, in *Molecules in Laser Fields*, ed. A. D. Bandrauk, (M. Dekker Pub., New York, 1993), Chap. 6.
4. T. Zuo, A. D. Bandrauk, M. Ivanov, and P. B. Corkum, Phys. Rev. **A 51**, 3991 (1995).
5. T. Zuo and A. D. Bandrauk, Phys. Rev. **A 54**, 3254 (1996) .
6. F. Dupont, P. B. Corkum, Phys. Rev. Lett. **74**, 3596 (1995).
7. E. E. Aubanel, A. D. Bandrauk, Chem. Phys. Lett. **229**, 169 (1994); Chem. Phys. **198**, 195 (1995).
8. A. D. Bandrauk, *Molecules in Laser Fields* , ed. A. D. Bandrauk, (M. Dekker Pub., New York, 1993), Chaps. 1, 4.
9. B. W. Shore, *The Theory of Coherent Atomic Excitation* (J. Wiley, New York, 1990). .
10. A. D. Bandrauk, H. Yu, E. E. Aubanel, Can. J. Chem. **74**, 988 (1996).
11. S. Kato and Y. Amatatsu, J. Chem. Phys. **92**, 7241 (1990).
12. D. A. Cardimona, Phys. Rev., **A 41**, 5016 (1990).
13. S. Schiemann, A. Kuhn and K. Bergmann, Phys. Rev. Lett. **71**, 3637 (1993).
14. J. Hicks, M. Vandersall, Z. Babarogic, and K. Eisenthal, Chem. Phys. Lett. **116**, 18 (1981).
15. O. Kajimoto, M. Futakami, T. Kobayashi, and K. Yamasaki, J. Phys. Chem. **92**, 18 (1981).
16. M. W. Schmidt, K. K. Baldridge, J. A. Boatz, J. H. Jensen, S. Koseki, M. S. Gordon, K. A. Nguyen, T. L. Windus, and S. T. Elbert, QCPE Bulletin, version 1992.
17. H. Yu, A. D. Bandrauk, in preparation.
18. J. Parker, C. R. Stroud, Phys. Rev. Lett. **56**, 716 (1986).
19. A. L. Sobolewski, W. Domcke, Chem. Phys. Lett. **250**, 428 (1996).
20. P. Dietrich, D. T. Strickland, M. Laberge, and P. B. Corkum, in *Molecules in Laser Fields*, Ed. A. D. Bandrauk (M. Dekker Pub., New York, 1993), Chap. 4.

Coherent Control of Molecular Dynamics

Paul Brumer[1] and Moshe Shapiro[2]

[1] Chemical Physics Theory Group, Department of Chemistry, University of Toronto, Toronto, Canada M5S 3H6
[2] Chemical Physics Department, The Weizmann Institute of Science, Rehovot, Israel 76100

Abstract. Coherent control provides a quantum interference based method for controlling molecular dynamics. This theory is reviewed, as are applications to a variety of processes including photodissociation, asymmetric synthesis, and the control of bimolecular processes. Control scenarios and computations on the control of IBr, H_2O, DOH, and Na_2 photodissociation reactions are discussed. We show that a wide range of yield control is possible under suitable laboratory conditions. Recent experiments on the control of photocurrent directionality in semiconductors and of the Na_2 photodissociation to yield Na atoms in different excited states are shown to confirm the theory.

1 Introduction

Product selectivity is at the heart of chemistry, and the control of reactions using lasers has been a goal for decades. Recently, we [1] - [17] and other groups [18] - [27] have demonstrated theoretically that one can achieve this goal by using quantum interference phenomena. We showed that phases acquired by a quantum system while excited by lasers enable one to control quantum interferences, and hence the outcome, of many dynamical processes. Experimental tests [28] - [38] of our approach, termed coherent control (CC), have confirmed many of the theoretical predictions and proven the viability of the method.

The purpose of this article is to provide an introduction to the concepts [39] underlying CC and to discuss its current status in both chemistry and physics. We focus on the essential principles, and provide references to specific computational and experimental results on particular cases.

1.1 Aspects of Scattering Theory and Reaction Dynamics

The processes we wish to control include branching "half" collisions,

$$AB + C \leftarrow ABC \rightarrow A + BC \ , \tag{1}$$

and "full" collisions,

$$AB(m'') + C \leftarrow A + BC(m) \rightarrow A + BC(m') \ . \tag{2}$$

In the above, A, B, C are either atoms, groups of atoms, electrons, or photons, m, m' denote the internal (vibrational, rotational, photon occupation) quantum numbers of the reactants or products.

Given $\Psi(t=0)$, the system wave function at an initial time, the evolution of the system is determined by the time-dependent material Schrödinger equation,

$$H_M \Psi(t) = i\hbar \partial \Psi(t)/\partial t \,, \tag{3}$$

where H_M is the system Hamiltonian. The long-term behavior of $\Psi(t)$ is intimately connected with the nature of the time-independent continuum energy eigenstates. For every continuum energy value E, each of the possible outcomes observed in the product region is represented by an independent wave function. This "boundary" condition is expressed more precisely by denoting the different possible chemical products of the breakup of ABC in Eq. (1) by an index q (e.g., $q = 1$ denotes the $A + BC$ products) and all additional identifying state labels by m. The set of continuum eigenfunctions of the material Hamiltonian,

$$H_M | E, m, q^- \rangle = E | E, m, q^- \rangle, \tag{4}$$

is now defined via the requirement that asymptotically every $| E, m, q^- \rangle$ state goes over to a state of the separated products, denoted by $| E, m, q^0 \rangle$ [40] that is of energy E, chemical identity q, and remaining quantum numbers m.

The description of the system in terms of $| E, m, q^- \rangle$ has an important advantage: Expressing the state of the system in the present in terms of these states, i.e., writing an initial continuum state as

$$\Psi(t=0) = \sum_{q,m} \int dE c_{q,m}(E) | E, m, q^- \rangle, \tag{5}$$

means that we know the fate of the system in the future. Since each of the $| E, m, q^- \rangle$ states correlates with a *single* product state, the probability of observing each $| E, m, q^0 \rangle$ product state is simply given by $|c_{q,m}(E)|^2$, the *preparation* probabilities. The probability of producing a chemical product q in the future is therefore given as

$$P_q = \sum_m \int dE |c_{q,m}(E)|^2. \tag{6}$$

Below we demonstrate that the key to laser control is to change one $c_{q,m}(E)$ coefficient relative to another $c_{q',m}(E)$ coefficient *at the same energy*. In order to understand how this can be done we discuss now the process of preparation.

1.2 Perturbation Theory, System Preparation, and Coherence

Consider the effect of an electric field $\epsilon(t)$ on an initially bound eigenstate $| E_g \rangle$ of the radiation-free Hamiltonian H_M. The overall Hamiltonian is then given by

$$H = H_M - \mathsf{d}[\overline{\epsilon}(t) + \overline{\epsilon}^*(t)], \tag{7}$$

where d is the component of the dipole moment along the electric field.

If the impinging photon is energetic enough to dissociate the molecule, it is then necessary to expand $|\Psi(t)\rangle$ in the bound and scattering eigenstates of the radiation-free Hamiltonian,

$$|\Psi(t)\rangle = \sum_i c_i(t)|E_i\rangle \exp(-iE_it/\hbar) + \sum_{m,q}\int dE c_{E,m,q}(t)|E,m,q^-\rangle \exp(-iEt/\hbar). \tag{8}$$

Insertion of Eq. (8) into the time dependent Schrödinger equation results in a set of first-order differential equations for the $c_\nu(t)$ coefficients, where ν represents either the bound (i) or scattering (E, m, q) indices.

For weak fields, the use of first-order perturbation theory gives, for the post-pulse preparation coefficient,

$$c_{E,m,q}(t \gg \Gamma) = (\sqrt{2\pi}/i\hbar)\epsilon(\omega_{E,E_g})\langle E,m,q^-|\mathsf{d}|E_g\rangle \tag{9}$$

where Γ is the pulse duration and

$$\epsilon(\omega) = \frac{1}{\sqrt{2\pi}}\int_{-\infty}^{\infty} \exp(i\omega t)\,\overline{\epsilon}(t)\,dt. \tag{10}$$

It follows from Eq. (6) and Eq. (9) that the probability $P(E,q)$ of forming an asymptotic product in arrangement q is

$$P(E,q) = \sum_m |c_{E,m,q}(t \gg \Gamma)|^2 = (2\pi/\hbar^2)\sum_m |\epsilon(\omega_{E,E_g})\langle E_g|\mathsf{d}|E,m,q^-\rangle|^2 \tag{11}$$

and that the branching ratio $R(1,2;E)$ between the $q = 1$ products and the $q = 2$ products at energy E is given by

$$R(1,2;E) = \frac{\sum_m |\langle E_g|\mathsf{d}|E,m,1^-\rangle|^2}{\sum_m |\langle E_g|\mathsf{d}|E,m,2^-\rangle|^2}\,. \tag{12}$$

1.3 Coherent Control of Chemical Reactions

We now address the issue of how to alter the above yield ratio $R(1,2;E)$ in a *systematic* fashion. Equation (12) makes clear that (at least in the weak field regime) this can not be achieved by altering the laser intensity, since the field strength cancels out in the expression for R. Quantum interference phenomena can, however, alter the numerator or denominator of R in an independent and controlled way. This can be achieved by accessing the final continuum state via two or more interfering pathways. One of the first examples that we studied [1] involves preparing a molecule in a superposition $c_1|\phi_1\rangle + c_2|\phi_2\rangle$ state and exciting the two components to the same final continuum energy E by using two CW sources. The field employed is of the form

$$\overline{\epsilon}(t) = \epsilon_1 e^{-i\omega_1 t + i\chi_1} + \epsilon_2 e^{-i\omega_2 t + i\chi_2}, \tag{13}$$

where $\hbar\omega_i = E - E_i$. A straightforward computation [1] yields that

$$R(1,2;E) = \frac{\sum_m |\langle \tilde{\epsilon}_1 c_1 \phi_1 + \tilde{\epsilon}_2 c_2 \phi_2 | \mathrm{d} | E, m, 1^- \rangle|^2}{\sum_m |\langle \tilde{\epsilon}_1 c_1 \phi_1 + \tilde{\epsilon}_2 c_2 \phi_2 | \mathrm{d} | E, m, 2^- \rangle|^2}, \tag{14}$$

where $\tilde{\epsilon}_i = \epsilon_i \exp(i\chi_i)$. Expanding the square gives

$$R(1,2;E) =$$

$$\frac{\sum_m [|\tilde{\epsilon}_1 c_1 \langle \phi_1 | \mathrm{d} | E, m, 1^- \rangle|^2 + |\tilde{\epsilon}_2 c_2 \langle \phi_2 | \mathrm{d} | E, m, 1^- \rangle|^2 + 2Re[c_1 c_2^* \tilde{\epsilon}_1 \tilde{\epsilon}_2^* \langle \phi_1 | \mathrm{d} | E, m, 1^- \rangle]}{\sum_m [|\tilde{\epsilon}_1 c_1 \langle \phi_1 | \mathrm{d} | E, m, 2^- \rangle|^2 + |\tilde{\epsilon}_2 c_2 \langle \phi_2 | \mathrm{d} | E, m, 2^- \rangle|^2 + 2Re[c_1 c_2^* \tilde{\epsilon}_1 \tilde{\epsilon}_2^* \langle \phi_1 | \mathrm{d} | E, m, 2^- \rangle]} \tag{15}$$

The structure of the numerator and denominator of Eq. (15) is of the type desired: i.e., each has a term associated with the excitation of the $|\phi_1\rangle$ state, a term associated with the excitation of the $|\phi_2\rangle$ state, and a term corresponding to the interference between the two excitation routes. The interference term, which can be either constructive or destructive, is in general different for the two product channels. What makes Eq. (15) so important *in practice* is that the interference term has coefficients whose magnitude and sign depend upon *experimentally controllable* parameters. In the case of Eq. (15), the experimental parameters that alter the yield [1] are contained in the complex quantity $A = \tilde{\epsilon}_2 c_2 / \tilde{\epsilon}_1 c_1$. Both $x \equiv |A|$ and $\theta_1 - \theta_2 \equiv arg(A)$ can be controlled separately in the experiment.

The "real time" analogue of the above two CW frequencies scenario, in which the superposition state preparation is affected by a single broadband pulse and the dissociation by a second pulse, is discussed in detail in Sect. 2.2 below.

2 Representative Control Scenarios

The two-step approach is but one particular implementation of coherent control; numerous other scenarios may be designed. They all rely upon the same "coherent control principle," that *in order to achieve control one must drive a state through multiple independent optical excitation routes to the same final state.*

It would seem that laser incoherence would lead to loss of control, since incoherence implies that the phases of $\tilde{\epsilon}_1$ and $\tilde{\epsilon}_2$ in Eq. (15) are random. An ensemble average of these phases is expected to lead to the disappearance of the interference term. This is true, however, only in the fully chaotic limit. Control can persist in the presence of some laser incoherence [16], or when the initial state is described by a *mixed*, as distinct from *pure*, state [7]. Most surprising is the fact, described below, that by utilizing strong laser fields one can attain quantum interference control with completely *incoherent* sources [41].

We now describe in more detail three control scenarios.

2.1 1-Photon, 3-Photon Interference

So far, we have exploited quantum interference phenomena by dissociating a superposition of several energy eigenstates with a single type (one photon absorption) process. It is possible instead to start with a *single* energy eigenstate and employ interference between optical routes of *different* types. Such is the interference between two multiphoton processes of different multiplicities. In order to satisfy the coherent control principle, which requires that we reach the same final energy E, we must use photons of commensurate frequencies, i.e., frequencies that satisfy an $m\omega_1 = n\omega_2$ relation, with integer m and n. Selection rules dictate the acceptable n, m pairs.

As the simplest example, we examine a one photon process interfering with a three photon process ("3 vs. 1" control). Let H_g and H_e be the nuclear Hamiltonians for a ground and an excited electronic state. H_g is assumed to have a discrete spectrum and H_e to possess a continuous spectrum. The molecule, initially in an eigenstate $|\, E_i \,\rangle$ of H_g, is subjected to two electric fields given by

$$\epsilon(t) = \epsilon_1 \cos(\omega_1 t + \mathbf{k}_1 \cdot \mathbf{R} + \theta_1) + \epsilon_3 \cos(\omega_3 t + \mathbf{k}_3 \cdot \mathbf{R} + \theta_3). \tag{16}$$

Here $\omega_3 = 3\omega_1$, $\epsilon_l = \epsilon_l \hat{\epsilon}_l$, $l = 1, 3$; ϵ_l is the magnitude and $\hat{\epsilon}_l$ is the polarization of the electric fields, $\mathbf{k}_i$ are the wavevectors. The two fields are chosen parallel, with $\mathbf{k}_3 = 3\mathbf{k}_1$.

The probability $P(E, q; E_i)$ of producing product with energy E in arrangement q from a state $|\, E_i \,\rangle$ is given by

$$P(E,q;E_i) = P_3(E,q;E_i) + P_{13}(E,q;E_i) + P_1(E,q;E_i), \tag{17}$$

where $P_1(E,q;E_i)$ and $P_3(E,q;E_i)$ are the probabilities of dissociation due to the ω_1 and ω_3 excitation, and $P_{13}(E,q;E_i)$ is the term due to interference between the two excitation routes.

In the weak field limit, $P_3(E,q;E_i)$ is given by

$$P_3(E,q;E_i) = (\frac{\pi}{\hbar})^2 \epsilon_3^2 F_3^{(q)}, \tag{18}$$

where

$$F_3^{(q)} = \sum_n |\langle\, E, n, q^- \,|\, (\hat{\epsilon}_3 \cdot \mathsf{d})_{e,g} \,|\, E_i \,\rangle|^2 \,, \tag{19}$$

d is the electric dipole operator, and

$$(\hat{\epsilon}_3 \cdot \mathsf{d})_{e,g} = \langle\, e \,|\, \hat{\epsilon}_3 \cdot \mathsf{d} \,|\, g \,\rangle \,, \tag{20}$$

with $|\, g \,\rangle$ and $|\, e \,\rangle$ denoting the ground and excited electronic states, respectively. $P_1(E,q;E_i)$ is given in third-order perturbation theory by [6]

$$P_1(E,q;E_i) = (\frac{\pi}{\hbar})^2 \epsilon_1^6 F_1^{(q)}, \tag{21}$$

where

$$F_1^{(q)} = \sum_n |\langle\, E, n, q^- \,|\, T \,|\, E_i \,\rangle|^2 \,, \tag{22}$$

with

$$T = (\hat{\epsilon}_1 \cdot \mathsf{d})_{e,g}(E_i - H_g + 2\hbar\omega_1)^{-1}(\hat{\epsilon}_1 \cdot \mathsf{d})_{g,e}(E_i - H_e + \hbar\omega_1)^{-1}(\hat{\epsilon}_1 \cdot \mathsf{d})_{e,g} \, . \quad (23)$$

We assumed that $E_i + 2\hbar\omega_1$ is below the dissociation threshold and that dissociation occurs from the excited electronic state only.

A similar derivation [6] gives the cross term in Eq. (17) as

$$P_{13}(E, q; E_i) = -2(\frac{\pi}{\hbar})^2 \epsilon_3 \epsilon_1^3 \cos(\theta_3 - 3\theta_1 + \delta_{13}^{(q)})|F_{13}^{(q)}| \quad (24)$$

with the amplitude $|F_{13}^{(q)}|$ and phase $\delta_{13}^{(q)}$ defined by

$$|F_{13}^{(q)}| \exp(i\delta_{13}^{(q)}) = \sum_n \langle E_i \,|\, T \,|\, E, n, q^- \rangle \langle E, n, q^- \,|\, (\hat{\epsilon}_3 \cdot \mathsf{d})_{e,g} \,|\, E_i \rangle \, . \quad (25)$$

The branching ratio $R_{qq'}$ between the q and q' products can then be written as

$$R_{qq'} = \frac{F_3^{(q)} - 2x\cos(\theta_3 - 3\theta_1 + \delta_{13}^{(q)})|F_{13}^{(q)}| + x^2 F_1^{(q)}}{F_3^{(q')} - 2x\cos(\theta_3 - 3\theta_1 + \delta_{13}^{(q')})|F_{13}^{(q')}| + x^2 F_1^{(q')}} \quad (26)$$

, where x is defined as

$$x = \epsilon_1^3 / \epsilon_3 . \quad (27)$$

The numerator and denominator of Eq. (26) contain contributions from two independent routes and an interference term. Since the interference term is controllable through variation of laboratory parameters, so too is the product ratio $R_{qq'}$. Thus the principle upon which this control scenario is based is the same as in the first example above, although the interference is introduced in an entirely different way.

Experimental control over $R_{qq'}$ is obtained by varying the difference $(\theta_3 - 3\theta_1)$ and the parameter x. The former is the phase difference between the ω_3 and the ω_1 laser fields, and the latter, via Eq. (27), incorporates the ratio of the two lasers' amplitudes. Experimentally, one envisions using "tripling" to produce ω_3 from ω_1; the subsequent variation of the phase of one of these beams provides a straightforward method of altering $\theta_3 - 3\theta_1$. Indeed, generating ω_3 from ω_1 allows for compensation of any phase jumps in the two laser sources. Thus the relative phase $\omega_3 - 3\omega$ is well-defined.

As pointed out above, "3 vs. 1" is not necessarily the only viable control scenario in the "*nvs.m*" family. It has the advantage that one may generate one of the frequencies (the tripled photon) from the other. This is indeed the reason why the "3 vs. 1" route was the first control scenario to be implemented experimentally (see the discussion below).

Control of *integral* (in contrast to *differential*) cross-sections requires that the $|\, E, n, q^- \rangle$ continuum states be made up of equal parity $|\, J, M \,\rangle$ angular momentum states. This means that in the "*mvs.n*" control scheme, the integer n must have the same parity as the integer m. Thus, studies of a "2 vs. 2" scheme for the control of the Na_2 photodissociation [15, 42] (discussed in detail in Sect. 2.3) and of a "2 vs. 4" scenario for the control of the Cl_2 photodissociation [43], have been published. In addition, studies of "3 vs. 1" control with strong fields, have also appeared [44, 45, 46]. These studies and others [47] have verified that "$nvs.m$" control is viable even when strong fields are used, although the dependence on the x amplitude and the $\theta_n - 3\theta_m$ phase factors is no longer as transparent as in the weak field case, discussed above.

The weak field "3 vs. 1" scenario has now been experimentally implemented in part in REMPI type experiments. The experiments demonstrated control of the total ionization rate, first in Hg [28], and then in HCl and CO [29]. In the case of HCl [29], the molecule was excited to an intermediate $^3\Sigma^-(\Omega^+)$ vib-rotational resonance, using a combination of three ω_1 ($\lambda_1 = 336$ nm) photons and one ω_3 ($\lambda_3 = 112$ nm) photon. The ω_3 beam was generated from an ω_1 beam by tripling in a Kr gas cell. Ionization of the intermediate state takes place by absorption of one additional ω_1 photon.

The relative phase of the light fields was varied by passing the ω_1 and ω_2 beams through a second Ar or H_2 ("tuning") gas cell of variable pressure. The HCl REMPI experiments verified the prediction of a sinusoidal dependence of the ionization rates on the relative phase of the two exciting lasers of Eq. (26). The HCl experiment also verified the prediction of Eq. (26) of the dependence of the strength of the sinusoidal modulation of the ionization current on the x amplitude factor. More recently, control over branching processes such as dissociation vs. ionization in HI was demonstrated by Gordon et al. [30].

If one is content with controlling angular distributions, one can lift the equal parity restriction. Absorption of two photons of perpendicular polarizations [5, 8], or of two photons interfering with their second-harmonic photon ("2 vs. 1" scenario) [8, 33, 34], result in states of different parities. Though such processes do not lead to control of integral quantities, they do allow for control of differential cross-sections. The "1 vs. 2" scenario (discussed in Sect. 4) has been implemented experimentally for the control of photo-current directionality in semiconductors using no bias voltage [33, 35] and for the control of the orientation of the HD^+ photodissociation [36].

2.2 The Pump-Dump Scheme

An alternative version of the scenario outlined in Sect. 1.3 is a "pump-dump" scheme [18, 19], in which an initial superposition of bound states is prepared with one laser pulse and subsequently dissociated with another. The pump and dump steps are assumed to be temporally separated by a time delay τ. The analysis below shows that under these circumstances the control parameters are the central frequency of the pump pulse and the time delay between the two pulses.

Consider a molecule, initially ($t = 0$) in eigenstate $| E_g \rangle$ of Hamiltonian H_M, subjected to two transform limited light pulses. The field $\bar{\epsilon}(t)$ consists of two temporally separated pulses $\bar{\epsilon}(t) = \bar{\epsilon}_x(t) + \bar{\epsilon}_d(t)$, with the Fourier transform of $\bar{\epsilon}_x(t)$ denoted by $\epsilon_x(\omega)$, etc. For convenience, we have chosen Gaussian pulses peaking at $t = t_x$ and t_d respectively. As discussed in Sect. 1.2, the $\bar{\epsilon}_x(t)$ pulse induces a transition to a linear combination of two excited bound electronic state with nuclear eigenfunctions $| E_1 \rangle$ and $| E_2 \rangle$, and the $\overline{\epsilon}_d(t)$ pulse dissociates the molecule by further exciting it to the continuous part of the spectrum. Both fields are chosen sufficiently weak for perturbation theory to be valid [48].

The superposition state prepared by the $\overline{\epsilon}_x(t)$ pulse, whose width is chosen to encompass just the two E_1 and E_2 levels, is given in first-order perturbation theory as

$$| \phi(t) \rangle = | E_g \rangle e^{-iE_g t/\hbar} + c_1 | E_1 \rangle e^{-iE_1 t/\hbar} + c_2 | E_2 \rangle e^{-iE_2 t/\hbar}, \tag{28}$$

where

$$c_k = (\sqrt{2\pi}/i\hbar) \langle E_k | \mathsf{d} | E_g \rangle \epsilon_x(\omega_{kg}), \; k = 1, 2, \tag{29}$$

with $\omega_{kg} \equiv (E_k - E_g)/\hbar$.

After a delay time of $\tau \equiv t_d - t_x$ the system is subjected to the $\overline{\epsilon}_d(t)$ pulse. It follows from Eq. (28) that after this delay time, each preparation coefficient has picked up an extra phase factor of $e^{-iE_k\tau/\hbar}$, $k = 1, 2$. Hence, the phase of c_1 relative to c_2 at that time increases by $[-(E_1 - E_2)\tau/\hbar = \omega_{2,1}\tau]$. Thus the natural two-state time evolution replaces the relative laser phase of the two-frequency control scenario of Sect. 1.3.

After the decay of the $\overline{\epsilon}_d(t)$ pulse, the system wave function is given as

$$| \psi(t) \rangle = | \phi(t) \rangle + \sum_{n,q} \int dE B(E, n, q|t) | E, n, q^- \rangle e^{-iEt/\hbar}. \tag{30}$$

The probability of observing the q fragments at total energy E in the remote future is therefore given as

$$\begin{aligned} P(E, q) &= \sum_n |B(E, n, q|t = \infty)|^2 \\ &= (2\pi/\hbar^2) \sum_n | \sum_{k=1,2} c_k \, \langle E, n, q^- | \mathsf{d} | E_k \rangle \epsilon_d(\omega_{EE_k})|^2 \,, \end{aligned} \tag{31}$$

where $\omega_{EE_k} = (E - E_k)/\hbar$, and c_k is given by Eq. (29).

Expanding the square and using the Gaussian pulse shape gives

$$\begin{aligned} P(E, q) = (2\pi/\hbar^2)[&|c_1|^2 \mathsf{d}_{1,1}^{(q)} \epsilon_1^2 + |c_2|^2 \mathsf{d}_{2,2}^{(q)} \epsilon_2^2 \\ &+ 2|c_1 c_2^* \epsilon_1 \epsilon_2 \mathsf{d}_{1,2}^{(q)}| \cos(\omega_{2,1}(t_d - t_x) + \alpha_{1,2}^{(q)}(E) + \phi)] \,, \end{aligned} \tag{32}$$

where $\epsilon_i = |\epsilon_d(\omega_{EE_i})|$, $\omega_{2,1} = (E_2 - E_1)/\hbar$, and the phases ϕ, $\alpha_{1,2}^{(q)}(E)$ are defined by

$$\begin{aligned} \langle E_1 | \mathsf{d} | E_g \rangle \langle E_g | \mathsf{d} | E_2 \rangle &\equiv |\langle E_1 | \mathsf{d} | E_g \rangle \langle E_g | \mathsf{d} | E_2 \rangle| e^{i\phi} \\ \mathsf{d}_{i,k}^{(q)}(E) \equiv |\mathsf{d}_{i,k}^{(q)}(E)| e^{i\alpha_{i,k}^{(q)}(E)} &= \sum_n \langle E, n, q^- | \mathsf{d} | E_i \rangle \langle E_k | \mathsf{d} | E, n, q^- \rangle \,. \end{aligned} \tag{33}$$

Integrating over E to encompass the full width of the second pulse and forming the ratio $Y = P(q)/[\sum_q P(q)]$ gives the ratio of products in each of the two arrangement channels, i.e. the quantity we wish to control. Once again it is the sum of two direct photodissociation contributions plus an interference term.

Examination of Eq. (32) makes clear that the product ratio Y can be varied by changing the delay time $\tau = (t_d - t_x)$ or ratio $x = |c_1/c_2|$; the latter is most conveniently done by detuning the initial excitation pulse.

An example of this type of pump-dump control [11] is provided by the example of IBr photodissociation. Specifically, we showed computationally that it is possible to control the Br* vs. Br yield in this process, using two conveniently chosen picosecond pulses. The first pulse was chosen to prepare a linear superposition of two bound states that arise from mixing of the X and A states. A subsequent pulse pumps this superposition to dissociation where the relative yields of Br and Br* are examined. The results show the vast range of control that is possible with this relatively simple experimental setup.

Theoretical work on similar pump-dump scenarios for the control of the

$$\mathrm{D + OH \leftarrow HOD \rightarrow H + OD}$$

dissociation via the B-state [49] of HOD and the A-state [50] of HOD have recently been published. Work on pump-dump control of Li_2 photodissociation is in progress [51]

2.3 Resonantly Enhanced "2 vs. 2" Control of a Thermal Ensemble

In practice, there are a number of sources of incoherence that tend to diminish control. Prominent amongst these are effects due to an initial thermal distribution of states and effects due to partial coherence of the laser source. Below we describe one approach, based upon a resonant "2 vs. 2" scenario, that deals effectively with both problems. An alternative method, in which coherence is retained in the presence of collisions, is discussed elsewhere [7].

The specific scheme we advocate is the particular case of Na_2 photodissociation . Here the molecule is lifted from an initial bound state $|E_i, J_i, M_i\rangle$ to energy E via two independent two photon routes. To introduce notation, first consider a single such two-photon route. Absorption of the first photon of frequency ω_1 lifts the system to a region close to an intermediate bound state $|E_m J_m M_m\rangle$, and a second photon of frequency ω_2 carries the system to the dissociating states $|E, \hat{\mathbf{k}}, q^-\rangle$, where the scattering angles are specified by $\hat{\mathbf{k}} = (\theta_k, \phi_k)$. Here the J's are the angular momentum, the M's are their projections along the z-axis, and the values of energy, E_i and E_m, include specification of the vibrational quantum numbers. Specifically, if we denote the phases of the coherent states by ϕ_1 and ϕ_2, the wavevectors by $\mathbf{k}_1$ and $\mathbf{k}_2$ with overall phases $\theta_i = \mathbf{k}_i \cdot \mathbf{R} + \phi_i$ ($i = 1, 2$), and the electric field amplitudes by ϵ_1 and ϵ_2, then the probability amplitude for resonant two-photon ($\omega_1 + \omega_2$) photodissociation is given [15, 42] by

$$T_{\hat{\mathbf{k}}q,i}(E, E_i J_i M_i, \omega_2, \omega_1) =$$

$$\sum_{E_m,J_m} \frac{\langle E,\hat{\mathbf{k}},q^-|\mathrm{d}_2\epsilon_2|E_mJ_mM_i\rangle\langle E_mJ_mM_i|\mathrm{d}_1\epsilon_1|E_iJ_iM_i\rangle}{\omega_1-(E_m+\delta_m-E_i)+i\Gamma_m}\exp[i(\theta_1+\theta_2)] =$$

$$\frac{\sqrt{2\mathrm{m}k_q}}{h}\sum_{J,p,\lambda\geq 0}\sum_{E_m,J_m}\sqrt{2J+1}\begin{pmatrix} J & 1 & J_m \\ -M_i & 0 & M_i \end{pmatrix}\begin{pmatrix} J_m & 1 & J_i \\ -M_i & 0 & M_i \end{pmatrix}$$

$$\cdot D^{Jp}_{\lambda,M_i}(\theta_k,\phi_k,0)t(E,E_iJ_i,\omega_2,\omega_1,q|Jp\lambda,E_mJ_m)\exp[i(\theta_1+\theta_2)] \tag{34}$$

Here d_i is the component of the dipole moment along the electric-field vector of the ith laser mode, $E = E_i + (\omega_1 + \omega_2)$, δ_m and Γ_m are respectively the radiative shift and width of the intermediate state, m is the reduced mass, and k_q is the relative momentum of the dissociated product in the q-channel. The D^{Jp}_{λ,M_i} is the parity-adapted rotation matrix [52] with λ the magnitude of the projection on the internuclear axis of the electronic angular momentum and $(-1)^J p$ the parity of the rotation matrix. We have set $\hbar \equiv 1$ and assumed for simplicity lasers that are linearly polarized and with parallel electric-field vectors. Note that the T-matrix element in Eq. (34) is a complex quantity, whose phase is the sum of the laser phase $\theta_1 + \theta_2$ and the molecular phase, i.e., the phase of $\mathbf{t}$ matrix.

Because the t-matrix element contains a factor of $[\omega_1-(E_m+\delta_m-E_i)+i\Gamma_m]^{-1}$ the probability is greatly enhanced by the approximate inverse square of the detuning $\Delta = \omega_1 - (E_m + \delta_m - E_i)$ as long as the line width Γ_m is less than Δ. Hence only the levels closest to the resonance $\Delta = 0$ contribute significantly to the dissociation probability. *This allows us to photodissociate only a select number of states (preferably one) from a thermal bath.*

Consider then the following coherent control scenario. A molecule is irradiated with three interrelated frequencies $\omega_0, \omega_+, \omega_-$ where photodissociation occurs at $E = E_i + 2\omega_0 = E_i + (\omega_+ + \omega_-)$ and where ω_0 and ω_+ are chosen resonant with intermediate bound state levels. The probability of photodissociation at energy E into arrangement channel q is then given by the square of the sum of the T matrix elements from pathway "a" $(\omega_0 + \omega_0)$ and pathway "b" $(\omega_+ + \omega_-)$. That is, the probability into channel q is given by

$$P_q(E, E_iJ_iM_i; \omega_0, \omega_+, \omega_-) \equiv$$

$$\int d\hat{\mathbf{k}} \left| T_{\hat{\mathbf{k}}_{q,i}}(E, E_iJ_iM_i, \omega_0, \omega_0) + T_{\hat{\mathbf{k}}_{q,i}}(E, E_iJ_iM_i, \omega_+, \omega_-)\right|^2$$

$$\equiv P^{(q)}(a) + P^{(q)}(b) + P^{(q)}(ab). \tag{35}$$

Here $P^{(q)}(a)$ and $P^{(q)}(b)$ are the independent photodissociation probabilities associated with routes a and b respectively, and $P^{(q)}(ab)$ is the interference term between them, discussed below. Note that the two T matrix elements in Eq. (35) are associated with different lasers and as such contain different laser phases. Specifically, the overall phases of the three laser fields are $\theta_0 = \mathbf{k}_0 \cdot \mathbf{R} + \phi_0, \theta_+ = \mathbf{k}_+ \cdot \mathbf{R} + \phi_+$, and $\theta_- = \mathbf{k}_- \cdot \mathbf{R} + \phi_-$, where ϕ_0, ϕ_+, and ϕ_- are the photon phases, and $\mathbf{k}_0$, $\mathbf{k}_+$, and $\mathbf{k}_-$ are the wavevectors of the laser modes ω_0, ω_+, and ω_-, whose electric field strengths are $\epsilon_0, \epsilon_+, \epsilon_-$ and intensities I_0, I_+, I_-.

The optical path-path interference term $P^{(q)}(ab)$ is given by

$$P^{(q)}(ab) = 2\epsilon_0^2\epsilon_+\epsilon_-|\mu_{ab}^{(q)}| \exp[i(\delta_a^q - \delta_b^q)] \cos(\alpha_a^q - \alpha_b^q) \tag{36}$$

with relative phase

$$\alpha_a^q - \alpha_b^q = (\delta_a^q - \delta_b^q) + (2\theta_0 - \theta_+ - \theta_-), \tag{37}$$

where the amplitude $|\mu_{ab}^{(q)}|$ and the molecular phase difference $(\delta_a^q - \delta_b^q)$ are defined by

$$\epsilon_0^2\epsilon_+\epsilon_-|\mu_{ab}^{(q)}| \exp[i(\delta_a^q - \delta_b^q)] = \frac{8\pi \mathrm{m} k_q}{h^2} \sum_{J,p,\lambda\geq 0} \sum_{E_m,J_m;E_m',J_m'}$$
$$\begin{pmatrix} J & 1 & J_m \\ -M_i & 0 & M_i \end{pmatrix} \begin{pmatrix} J_m & 1 & J_i \\ -M_i & 0 & M_i \end{pmatrix} \begin{pmatrix} J & 1 & J_m' \\ -M_i & 0 & M_i \end{pmatrix} \begin{pmatrix} J_m' & 1 & J_i \\ -M_i & 0 & M_i \end{pmatrix}$$
$$\cdot t(E, E_iJ_i, \omega_0, \omega_0, q|Jp\lambda, E_mJ_m) t^*(E, E_iJ_i, \omega_-, \omega_+, q|Jp\lambda, E_m'J_m'). \tag{38}$$

Consider now the quantity of interest $R_{qq'}$, the branching ratio of the product in the q-channel to that in the q'-channel. Noting that in the weak field case $P^{(q)}(a)$ is proportional to ϵ_0^4, $P^{(q)}(b)$ to $\epsilon_+^2\epsilon_-^2$, and $P^{(q)}(ab)$ to $\epsilon_0^2\epsilon_+\epsilon_-$, we can write

$$R_{qq'} = \frac{\mu_{aa}^{(q)} + x^2\mu_{bb}^{(q)} + 2x|\mu_{ab}^{(q)}| \cos(\alpha_a^q - \alpha_b^q) + (B^{(q)}/\epsilon_0^4)}{\mu_{aa}^{(q')} + x^2\mu_{bb}^{(q')} + 2x|\mu_{ab}^{(q')}| \cos(\alpha_a^{q'} - \alpha_b^{q'}) + (B^{(q')}/\epsilon_0^4)}, \tag{39}$$

where $\mu_{aa}^{(q)} = P^{(q)}(a)/\epsilon_0^4$, $\mu_{bb}^{(q)} = P^{(q)}(b)/(\epsilon_+^2\epsilon_-^2)$, and $x = \epsilon_+\epsilon_-/\epsilon_0^2 = \sqrt{I_+I_-}/I_0$. The terms with $B^{(q)}, B^{(q')}$, described below, correspond to resonant photodissociation routes to energies other than $E = E_i + 2\hbar\omega_0$, and hence [4] to terms that do not coherently interfere with the a and b pathways. Minimization of these terms, due to absorption of $(\omega_0 + \omega_-)$, $(\omega_0 + \omega_+)$, $(\omega_+ + \omega_0)$, or $(\omega_+ + \omega_+)$, is discussed elsewhere [15, 42]. Here we just emphasize that the product ratio in Eq. (39) depends upon both the laser intensities and relative laser phase. Hence manipulating these laboratory parameters allows for control over the relative cross section between channels.

The proposed scenario, embodied in Eq. (39), also provides a means by which control can be improved by eliminating effects due to laser jitter. Specifically, the term $2\phi_0 - \phi_+ - \phi_-$ contained in the relative phase $\alpha_a^q - \alpha_b^q$ can be subject to the phase fluctuations arising from laser instabilities. If such fluctuations are sufficiently large, then the interference term in Eq. (39), and hence control, disappears [16]. We can eliminate this problem by generating ω_+ as $\omega_+ = 2\omega_0 - \omega_-$ via frequency doubling of ω_0 and frequency differencing of ω_- from the resulting beam. It is easy to see that in this case the quantity $2\phi_0 - \phi_+ - \phi_-$ vanishes, irrespective of the phase jitter and drift of either source! Control is then attained by introducing an extra, perfectly controlled phase, χ, through the addition of a delay line in, say, the ω_- beam.

Typical computational results for Na_2 are provided elsewhere [15, 42]. Note also that control is not limited to two-product channels, such as those discussed above. Recent computations [42] on higher energy Na_2 photodissociation , where more product arrangement channels are available, show equally large ranges of control for the three-channel case.

3 Control of Symmetry Breaking

Weak field phase interference has one remarkable property; it can lead to controlled *symmetry breaking* [12]. Below we show that the pump–dump scheme described above (Sect. 2.2) can lead to symmetry breaking in systems with three-dimensional spherical symmetry and to the generation of chirality, provided that magnetic quantum state selection is performed. Other mechanisms for collinear symmetry breaking in *strong* fields have recently been proposed [53, 54]. There, for example, it was shown that one can generate *even* high harmonics by exciting a symmetric double quantum-well. However, in contrast to the symmetry breaking scenario described below, the generation of even harmonics is not expected to exist in systems with three dimensional spherical symmetry.

In general, symmetry breaking occurs whenever a system executes a transition to a *nonsymmetric* eigenstate of the Hamiltonian. Strictly speaking, nonsymmetric eigenstates (i.e., states that do not belong to any of the symmetry group representations) can occur if there exist several degenerate eigenstates, each belonging to a different irreducible representation. Any linear combination of such eigenstates is nonsymmetric.

Nonsymmetric eigenstates of a symmetric Hamiltonian also occur in the continuous spectrum of a BAB type molecule. It is clear that the $|\,E, n, R^-\,\rangle$ state, which correlates asymptotically with the dissociation of the right B group, must be degenerate with the $|\,E, n, L^-\,\rangle$ state, giving rise to the departure of the left B group. It is also possible to form a *symmetric* $|\,E, n, s^-\,\rangle$ and an *antisymmetric* $|\,E, n, a^-\,\rangle$ eigenstate of the same Hamiltonian by taking the $\pm$ combination of these states. There is an important physical distinction between the nonsymmetric states and states that are symmetric/antisymmetric: Any experiment performed in the asymptotic $B--AB$ or $BA--B$ regions must, by necessity, measure the probability of populating a nonsymmetric state. This follows because when the $B--AB$ distance or the $BA--B$ distance is large, a given group B is either far away from *or* close to group A. Thus symmetric and antisymmetric states are not directly observable in the asymptotic regime.

We conclude that the very act of observation of the dissociated molecule entails the collapse of the system to one of the nonsymmetric states. As long as the probability of collapse to the $|\,E, n, R^-\,\rangle$ state is equal to the probability of collapse to the $|\,E, n, L^-\,\rangle$ state, the collapse to a nonsymmetric state does not lead to a preference of R over L in an *ensemble* of molecules. This is the case when the above collapse is *spontaneous*, i.e., occurring due to some (random) factors not in our control. Coherent control techniques allow us to influence these probabilities. In this case, symmetry breaking is stimulated rather than spontaneous. This has far-reaching physical and practical significance.

One of the most important cases of symmetry breaking arises when the two B groups (now denoted by B and B') are not identical, but are enantiomers of one another. (Two groups of atoms are said to be enantiomers of one another if one is the mirror image of the other. If these groups are also "chiral," i.e., they lack a center of inversion symmetry, then the two enantiomers are distinguishable and can be detected through the distinctive direction of rotation of linearly polarized light.)

The existence and role of enantiomers is recognized as one of the fundamental broken symmetries in nature [55]. It has motivated a long-standing interest in asymmetric synthesis, i.e., a process that preferentially produces a specific chiral species. Contrary to the prevailing belief [56] that asymmetric synthesis must necessarily involve either chiral reactants or chiral external system conditions such as chiral crystalline surfaces, we show below that preferential production of a chiral photofragment can occur even though the parent molecule is not chiral. In particular, two results are demonstrated: (1) Ordinary photodissociation, using linearly polarized light, of a BAB' "pro-chiral" molecule may yield different cross- sections for the production of right-handed (B) and left-handed (B') products, when the direction of the angular momentum (m_j) of the products is selected; and (2) this natural symmetry breaking may be enhanced and controlled using coherent lasers.

To treat this problem we return to the formulation of the pump–dump scenario described above, with attention focused on control of the relative yield of two product arrangement channels, but with angular momentum projection m_j fixed. Explicitly considering the dissociation of BAB' into right- (R) and left- (L) hand products, we have

$$Y = P(L, m_j)/P(L, m_j) + P(R, m_j). \tag{40}$$

As above, the product ratio Y is a function of the delay time $\tau = (t_d - t_x)$ and ratio $x = |c_1/c_2|$, the latter by detuning the initial excitation pulse. Active control over the products $B + AB'$ vs. $B' + AB$, i.e., a variation of Y with τ and x, and hence control over left- vs. right-handed products, will result only if $P(R, m_j)$ and $P(L, m_j)$ have different functional dependences on x and τ.

We first note that this molecule belongs to the C_s point group, which is a group possessing only one symmetry plane. This plane, denoted by σ, is defined as the collection of the C_{2v} points, i.e., points satisfying the $B - -A = A - -B'$ condition, where $B - -A$ designates the distance between the B and A groups. We now show that $P(R, m_j)$ may be different from $P(L, m_j)$ for the $B'AB$ case. In order to do that, we choose the intermediate state $| E_1 \rangle$ to be *symmetric* and the state $| E_2 \rangle$ to be *antisymmetric* with respect to reflection in the σ plane. Furthermore, we shall focus upon transitions between electronic states of the same representations, e.g., A' to A' or A'' to A'' (where A' denotes the symmetric representation and A'' the antisymmetric representation of the C_s group). We further assume that the ground vibronic state belongs to the A' representation.

The first thing to demonstrate is that it is possible to excite simultaneously, by optical means, both the symmetric $|E_1\rangle$ and antisymmetric $|E_2\rangle$ states. Using Eq. (29) we see that this requires the existence of both a symmetric d component, denoted by d_s, and an antisymmetric d component, denoted by d_a, because by the symmetry properties of $|E_1\rangle$ and $|E_2\rangle$,

$$\langle E_1|\mathsf{d}|E_g\rangle = \langle E_1|\mathsf{d}_s|E_g\rangle, \quad \langle E_2|\mathsf{d}|E_g\rangle = \langle E_2|\mathsf{d}_a|E_g\rangle. \tag{41}$$

The existence of both dipole-moment components occurs in $A' \to A'$ electronic transitions whenever a bent $B'--A--B$ molecule deviates considerably from the equidistant C_{2v} geometries (where $\mathsf{d}_a = 0$). The effect is nonFranck–Condon in nature, because we no longer assume that the dipole moment does not vary with the nuclear configurations. (In the theory of vibronic-transitions terminology the existence of both d_s and d_a is due to a Herzberg–Teller intensity borrowing [57] mechanism).

We conclude that the excitation pulse *can* create a $|E_1\rangle, |E_2\rangle$ superposition consisting of two states of different reflection symmetry, which is therefore nonsymmetric. We now wish to show that this nonsymmetry, established by exciting *nondegenerate bound* states, translates to a nonsymmetry in the probability of populating the two *degenerate* $|E,n,R^-\rangle, |E,n,L^-\rangle$ *continuum* states. We proceed by examining the properties of the bound-free transition matrix elements of Eq. (33) entering the probability expression of Eq. (32).

Although the continuum states of interest $|E,n,q^-\rangle$ are nonsymmetric, they satisfy a closure relation, since $\sigma|E,n,R^-\rangle = |E,n,L^-\rangle$ and vice versa. Working with the symmetric and antisymmetric continuum eigenfunctions

$$|E,n,R^-\rangle \equiv (|E,n,s^-\rangle + |E,n,a^-\rangle)/\sqrt{2}, \tag{42}$$

$$|E,n,L^-\rangle \equiv (|E,n,s^-\rangle - |E,n,a^-\rangle)/\sqrt{2}, \tag{43}$$

using the fact that $|E_1\rangle$ is symmetric and $|E_2\rangle$ antisymmetric, and adopting the notation $A_{s2} \equiv \langle E,n,s^-|\mathsf{d}_a|E_2\rangle$, $S_{a1} \equiv \langle E,n,a^-|\mathsf{d}_s|E_1\rangle$, etc. we have

$$\mathsf{d}_{11}^{(q)} = \sum{}' \left[|S_{s1}|^2 + |A_{a1}|^2 \pm 2Re(A_{a1}S_{s1}^*)\right] \tag{44}$$

$$\mathsf{d}_{22}^{(q)} = \sum{}' \left[|A_{s2}|^2 + |S_{a2}|^2 \pm 2Re(A_{s2}S_{a2}^*)\right] \tag{45}$$

$$\mathsf{d}_{12}^{(q)} = \sum{}' \left[S_{s1}A_{s2}^* + A_{a1}S_{a2}^* \pm S_{s1}S_{a2}^* \pm A_{a1}A_{s2}^*\right], \tag{46}$$

where the plus sign applies for $q = R$ and the minus sign for $q = L$.

Equation (46) displays two noteworthy features:

(1) $\mathsf{d}_{kk}^{(R)} \neq \mathsf{d}_{kk}^{(L)}$, $k = 1,2$. That is, the system displays *natural symmetry breaking* in photodissociation from state $|E_1\rangle$ or state $|E_2\rangle$, with right- and left-handed product probabilities differing by $4\sum{}'\,\mathrm{Re}(S_{s1}^*A_{a1})$ for excitation from $|E_1\rangle$, and $4\sum{}'\,\mathrm{Re}(A_{s2}S_{a2}^*)$ for excitation from $|E_2\rangle$. Note that these symmetry breaking terms may be relatively small, since they rely upon nonFranck–Condon contributions. However, even in the Franck–Condon approximation,

(2) $\mathsf{d}_{12}^{(R)} \neq \mathsf{d}_{12}^{(L)}$. Thus laser-controlled symmetry breaking, which depends upon $\mathsf{d}_{12}^{(q)}$ in accordance with Eq. (32), is therefore possible, allowing enhancement of the enantiomer ratio for the m_j polarized product.

We have demonstrated the extent of expected control, by considering a model symmetry breaking in the HOH photodissociation in three dimensions, where the two hydrogens are assumed distinguishable [12]. The computation was done using the formulation and computational methodology of Segev et al. [58].

4 Control with Intense Laser Fields

We now consider some extensions of CC to strong laser fields. Parallel work involving other strong field scenarios has been done by Bandrauk et al. [43], Corkum et al. [47], Bardsley et al. [44], Lambropoulos et al. [59] and Guisti-Suzor et al. [45]. Here we concentrate on a strong field control scenario in which the dependence on the relative phase between the two laser beams, and hence on laser coherence, disappears. As a result, coherence plays no role in this scenario (save for being intimately linked with the existence of the narrow-band laser sources needed for its execution). Although the unimportance of coherence means that we lose phase control, the effect still depends on quantum interference phenomena. The scenario is therefore called *interference control.*

To illustrate interference control, we look at the control of the electronic states of Na atoms generated by the photodissociation of Na_2, a process treated in the context of weak field CC in Sect. 2.3. We envision a scenario in which we employ two laser sources: One laser (not necessarily intense) with center frequency ω_1 is used to excite a molecule from an initially populated bound state $|\, E_i \,\rangle$ to a dissociative state $|\, E, m, q^- \,\rangle$. A second laser, with frequency ω_2, is used to couple ("dress") the continuum with some (initially unpopulated) bound states $|\, E_j \,\rangle$. With both lasers on, dissociation to $|\, E, m, q^- \,\rangle$ occurs via one direct $|\, E_i \,\rangle \rightarrow |\, E, m, q^- \,\rangle$, and a multitude of indirect, e.g., $|\, E_i \,\rangle \rightarrow |\, E, m, q^- \,\rangle \rightarrow |\, E_j \,\rangle \rightarrow |\, E, m, q^- \,\rangle$, pathways. The interference between these pathways to form a given channel q at product energy E can be either constructive or destructive. As we show below, varying the frequencies and intensities of the two excitation lasers strongly affects this interference term, providing a means of controlling the photodissociation line shape, and the branching ratio into different products.

With this scenario in mind, we now briefly discuss the methodology of dealing with strong laser fields and the extension of CC ideas to this domain. We consider the photodissociation of a molecule with Hamiltonian H_M in the presence of a radiation field with Hamiltonian H_R whose eigenstates are the Fock states $|n_k\rangle$ with energy $n_k \hbar\omega_k$. (In the case of several frequencies, the repeated index in $n_k\omega_k$ implies the sum over the modes.)

Strong field dynamics is completely embodied [60] in the fully interacting eigenstates of the total Hamiltonian H, $H = H_M + H_R + V$, where V is the light-matter interaction, denoted by $|(E, m, q^-), n_k^-\rangle$:

$$H|(E, m, q^-), n_k^-\rangle = (E + n_k\hbar\omega_k)|(E, m, q^-), n_k^-\rangle \ . \tag{47}$$

The minus superscript on n_k is used in exactly the same way as in the weak field domain: it is a reminder that each $|(E,m,q^-),n_k^-\rangle$ state correlates to a noninteracting $|(E,m,q^-),n_k\rangle \equiv |E,m,q^-\rangle|n_k\rangle$ state when the light-matter interaction V is switched off.

If the system is initially in the $|E_i,n_i\rangle \equiv |E_i\rangle|n_i\rangle$ state and we suddenly switch on V, the photodissociation amplitude to form in the future the product state $|E,m,q^-\rangle|n_k\rangle$ is given simply [60] as the overlap between the initial state and fully interacting state $\langle(E,m,q^-),n_k^-|E_i,n_i\rangle$. This overlap assumes the convenient form

$$\langle(E,m,q^-),n_k^-|E_i,n_i\rangle = \langle(E,m,q^-),n_k|VG(E^+ + n_k\hbar\omega_k)|E_i,n_i\rangle, \qquad (48)$$

by using the Lippmann–Schwinger equation

$$\langle(E,m,q^-),n_k^-| = \langle(E,m,q^-),n_k| + \langle(E,m,q^-),n_k|VG(E^+ + n_k\hbar\omega_k). \quad (49)$$

Here $G(\mathcal{E}) = 1/(\mathcal{E} - H)$ and $E^+ = E + i\delta$, with $\delta \to 0^+$ at the end of the computation. Equation (48) is exact and provides a connection between the photodissociation amplitude and the VG matrix element. It is the latter that we compute exactly using a high field extension of the artificial channel method [61, 62].

Two quantities are of interest: the channel specific line shape,

$$A(E,q,n_k|E_i,n_i) = \int d\hat{\mathbf{k}}\,|\langle(E,\hat{\mathbf{k}},q^-),n_k^-|E_i,n_i\rangle|^2, \qquad (50)$$

and the total dissociation probability to channel q,

$$P(q) = \sum_{n_k}\int dE\,A(E,q,n_k|E_i,n_i). \qquad (51)$$

In Eq. (51), the sum is over photons that excite the molecule above the dissociation threshold. In writing (50), diatomic dissociation is assumed, so that $m = \hat{\mathbf{k}}$.

Consider, for example, the photodissociation of Na_2 from the $|E_i\rangle = |v = 19, {}^3\Pi_u\rangle$ initial state, where v denotes the vibrational quantum number in the ${}^3\Pi_u$ electronic potential [63]. $|E_i\rangle$ is assumed to have been prepared by previous excitation from the ground electronic state. Excitations from $|E_i\rangle$ by ω_1 and mixing of the initially unpopulated $|E_j\rangle$ by ω_2 to the dissociating continua produce Na(3s)+Na(3p) and Na(3s)+Na(4s). Computations were done with ω_1 chosen within the range 15,430 $\mathrm{cm}^{-1} < \omega_1 <$ 15,700 cm^{-1} with intensity $I_1 \sim 10^{10}$ W/cm^2, which is sufficiently energetic to dissociate levels of the ${}^3\Pi_u$ state with $v \geq 19$ to both Na(3s) + Na(3p) and Na(3s) + Na(4s). The second laser has fixed frequency $\omega_2 = 13,964$ cm^{-1} and intensity $I_2 = 3.2\times10^{11}$ W/cm^2 and can dissociate levels with $v \geq 26$ to both products. Under these circumstances the contribution of above threshold dissociation is found to be negligible. However cognizance must be taken of the possibility of dissociation of $|E_i\rangle$ by ω_2 and of

$| E_j \rangle$ by ω_1. These processes do not interfere and cannot be controlled. Hence we must find the range of parameters that minimizes them.

Computational studies [64] show that extensive control over product branching in Na_2 photodissociation is possible. In addition, recent experimental results on this system [38] confirm the theory. Indeed, experimental and theoretical results are in excellent agreement with one another [38, 65].

5 Bimolecular Processes

Until recently, the issue of effectively controlling collisional events was, despite some effort [66], unresolved. The difficulty is that the extension of coherent control scenarios based on the preparation of initial superposition states requires that one optically prepare a coherent superposition of *degenerate* continuum states. This is difficult, but it is required because only states of the same total energy can interfere to alter the reaction products. Quite recently, we demonstrated [67] methods of achieving this goal, as described below.

Consider the collision of a beam of molecules B with a beam of molecules or atoms C, that yields products F and G, i.e.,

$$B + C \to F + G. \tag{52}$$

F and G can be identical to (nonreactive scattering), or different from (reactive scattering) B and C. We call $B + C$ the β arrangement and $F + G$ the γ arrangement. Traditional time-independent scattering theory proceeds by considering (52) with $B + C$ starting in an eigenstate of the free Hamiltonian H^0_β in the β arrangement:

$$H^0_\beta = K_\beta + h_B + h_C, \tag{53}$$

with K_β being the kinetic energy of the $B - C$ relative motion and h_B, h_C denoting the internal Hamiltonians of B and C.

To attain control over the process, consider then the following superposition, which we show below, can be realized experimentally, as the initial asymptotic state

$$| \mathrm{n}, \beta \rangle = | 0, C \rangle \sum_{i=1,2} a_i | i, B \rangle | E^{kin}_\beta(i) \rangle | E_{cm}(i) \rangle, \tag{54}$$

Here

$$E^{kin}_\beta(i) = E - \varepsilon_C(0) - \varepsilon_B(i). \tag{55}$$

where $| i, X \rangle$ with $X = B, C$ are eigenstates, of energy $\varepsilon_X(i)$, of h_B and h_C,

$$[\varepsilon_X(i) - h_X] | i, X \rangle = 0, \quad X = B, C. \tag{56}$$

The $| E^{kin}_\beta(i) \rangle$ and $| E_{cm}(i) \rangle$ states are plane waves describing the free motion of B relative to C and the motion of the $B - C$ center of mass, i.e., $\langle \mathbf{r}_{BC} | E^{kin}_\beta(i) \rangle = \exp(i\mathbf{k}_i \cdot \mathbf{r}_{BC})$, $\langle \mathbf{R}_{BC} | E_{cm}(i) \rangle = \exp(i\mathbf{K}_i \cdot \mathbf{R}_{BC})$, where $|\mathbf{k}_i| = \{2\mu_{BC} E^{kin}_\beta(i)\}^{\frac{1}{2}}/\hbar$, $\mu_{BC} = m_B m_C/(m_B + m_C)$ is the reduced mass of the BC

pair, and $\mathbf{K}_i$ is the BC center of mass momentum. Here $\mathbf{R}_{BC}, \mathbf{r}_{BC}$ are the position of the BC center of mass and the $B-C$ relative vector, respectively.

The superposition state $|\, \mathrm{n}, \beta\,\rangle$ is composed of degenerate eigenstates of H_β^0,

$$[E - H_\beta^0]|\, 0, C\,\rangle|\, i, B\,\rangle|\, E - \varepsilon_C(0) - \varepsilon_B(i)\,\rangle = 0, \qquad (57)$$

and center of mass terms. As a result, we can use standard time-independent scattering theory to calculate the cross-section for scattering from this state to any of the γ-arrangement final states. The latter states, of the form $|\, j, \gamma\,\rangle|\, E^{kin}\,\rangle|\, E_{cm}(i)\,\rangle$, have a component in the center of mass system that satisfies the free Schrödinger equation in the product space,

$$[E - H_\gamma^0]|\, j, \gamma\,\rangle|\, E - \varepsilon_\gamma(j)\,\rangle = 0, \qquad (58)$$

where $|\, j, \gamma\,\rangle$ are the eigenstates of the $F + G$ internal Hamiltonians,

$$[\varepsilon_\gamma(j) - h_F - h_G]|\, j, \gamma\,\rangle = 0, \qquad (59)$$

and $H_\gamma^0 = K_\gamma + h_F + h_G$ is analogous to (53), describing the product in arrangement γ.

The cross-section for forming one of the γ-arrangement final states, having started from the $|\, \mathrm{n}, \beta\,\rangle$ superposition state, is given by

$$\sigma(j, \gamma \leftarrow \mathrm{n}, \beta | E) = |\sum_{l=1,2} \langle\, E_{cm}(l)\,|\langle\, E, j, \gamma^-\,|V_\beta|\, \mathrm{n}, \beta\,\rangle|^2 \qquad (60)$$

Here $V_\beta = H - H_\beta^0$ is the (reactive or nonreactive) interaction potential, with H being the Hamiltonian in the center of mass system. The $|\, E, j, \gamma^-\,\rangle$ are incoming eigenstates of H,

$$[E - H]|\, E, j, \gamma^-\,\rangle = 0, \qquad (61)$$

which go over in the asymptotic limit to a specific free state of the $F + G$ products,

$$\exp(-iEt/\hbar)|\, E, j, \gamma^-\,\rangle \stackrel{t\to\infty}{\to} \exp(-iEt/\hbar)|\, E - \varepsilon_\gamma(j)\,\rangle|\, j, \gamma\,\rangle. \qquad (62)$$

Substituting Eq. (54) in Eq. (60) gives, for the reactive cross-section,

$$\sigma(j, \gamma \leftarrow \mathrm{n}, \beta | E) =$$

$$|\sum_{l=1,2} \langle\, E_{cm}(l)\,|\sum_{i=1,2} a_i \langle\, E, j, \gamma^-\,|V_\beta|\, 0, C\,\rangle|\, i, B\,\rangle|\, E_\beta^{kin}(i)\,\rangle|\, E_{cm}(i)\,\rangle|^2 =$$

$$|a_1|^2 \sigma_{11}^R(j) + |a_2|^2 \sigma_{22}^R(j) + 2\mathrm{R_e} a_1^* a_2 \sigma_{12}^R(j), \quad \gamma \neq \beta \qquad (63)$$

where

$$\sigma_{ii}^R(j) = |\langle\, E, j, \gamma^-\,|V_\beta|\, 0, C\,\rangle|\, i, B\,\rangle|\, E_\beta^{kin}(i)\,\rangle|^2, \; i = 1, 2,$$

$$\sigma_{12}^R(j) = \langle\, E_\beta^{kin}(1)\,|\langle\, 1, B\,|\langle\, 0, C\,|V_\beta|\, E, j, \gamma^-\,\rangle\langle\, E, j, \gamma^-\,|V_\beta|\, 0, C\,\rangle|\, 2, B\,\rangle|\, E_\beta^{kin}(2)\,\rangle$$
$$\times\langle E_{cm}(1)|\, E_{cm}(2)\,\rangle, \quad \gamma \neq \beta. \qquad (64)$$

Although the matrix element $\langle E_{cm}(1)|\, E_{cm}(2)\,\rangle = \int d\mathbf{R}_{BC} \exp[i(\mathbf{K}_2 - \mathbf{K}_1)\cdot \mathbf{R}_{BC}]$ integrated over all space is zero, the proper region of integration in this case is the intersection volume of the B and C beams. Hence if $(\mathbf{K}_2 - \mathbf{K}_1)$ is made sufficiently small, then the integral over this region can be made nonzero, and control over the cross-section is possible. An experimental means of achieving this result is discussed below.

Although (64) indicates that one can control detailed cross sections, often we only want to control the total reactive vs. the total nonreactive cross-section. In that case, the reactive to nonreactive branching ratio is given as

$$\frac{\sigma^R}{\sigma^{NR}} = \frac{\sum_j \sigma(j, \gamma \neq \beta \leftarrow \mathrm{n}, \beta|E)}{\sum_j \sigma(j, \beta \leftarrow \mathrm{n}, \beta|E)} =$$

$$\frac{\sigma_{11}^R + x^2\sigma_{22}^R + 2x|\sigma_{12}^R|\cos(\delta_{12}^R + \theta_{12})}{\sigma_{11}^{NR} + x^2\sigma_{22}^{NR} + 2x|\sigma_{12}^{NR}|\cos(\delta_{12}^{NR} + \theta_{12})}\,, \tag{65}$$

where $x = |a_2/a_1|$, $\theta_{12} = arg(a_2/a_1)$, $\sigma_{ik}^R = \sum_j \sigma_{ik}^R(j)$, $i,k = 1,2$, with similar definitions holding for σ_{ik}^{NR}, and $\delta_{12}^R = arg(\sigma_{12}^R)$, $\delta_{12}^{NR} = arg(\sigma_{12}^{NR})$. Thus, the reactive vs. nonreactive cross-section ratio can be controlled by varying the relative magnitude, x, and the relative phase, θ_{12}, of the a_1 and a_2 coefficients.

Control over the a_i can be attained by a number of routes. One approach prepares the $B - C$ superposition by exciting B to a superposition state and colliding the result with C. Specifically, consider preparing $|\,\mathrm{n}, \beta\,\rangle$ by first irradiating $|\,1, B\,\rangle|\, E_B^{kin}(1)\,\rangle$ to produce

$$|\,\mathrm{n}, B\,\rangle = \sum_{i=1,2} a_i|\, i, B\,\rangle|\, E_B^{kin}(1)\,\rangle, \tag{66}$$

where $|\, E_B^{kin}(1)\,\rangle$ describes the motion of the center of mass of B. Passing this superposition through a hexapole field or using laser cooling techniques allows us to alter the velocities of $|\,1, B\,\rangle$ and $|\,2, B\,\rangle$, giving

$$|\,\mathrm{n}, B\,\rangle = \sum_{i=1,2} a_i|\, i, B\,\rangle|\, E_B^{kin}(i)\,\rangle, \tag{67}$$

where $\langle\, \mathbf{r}_B\, |E_B^{kin}(i)\rangle = \exp(i\mathbf{k}_i^B \cdot \mathbf{r}_B)$, $\mathbf{k}_1^B \neq \mathbf{k}_2^B$, and where $\mathbf{r}_B$ is the lab position of B. Colliding the $|\,\mathrm{n}, B\,\rangle$ superposition state with particle C, of momentum $\mathbf{k}^C$, gives the $B - C$ superposition state

$$|\,\psi\,\rangle = |\,0, C\,\rangle \sum_{i=1,2} a_i|\, i, B\,\rangle|\, E_B^{kin}(i)\,\rangle|\, E_C^{kin}\,\rangle|\, E_{cm}(i)\,\rangle, \tag{68}$$

To produce (54), however, requires that the degeneracy condition (55), be satisfied, i.e., that

$$E_\beta^{kin}(1) - E_\beta^{kin}(2) = \varepsilon_B(2) - \varepsilon_B(1). \tag{69}$$

That is, with $\mathbf{K}_i = \mathbf{k}_i^B + \mathbf{k}_C$ and with the $B - C$ relative center of mass momentum given by $\mathbf{k}_i = (m_C\mathbf{k}_i^B - m_B\mathbf{k}_C)/(m_B + m_C)$, (68) becomes

$$(\hbar^2/2\mu_{BC})(k_1^2 - k_2^2) = \varepsilon_B(2) - \varepsilon_B(1), \tag{70}$$

or

$$\{\hbar^2/[2(m_B+m_C)]\}[(m_C/m_B)((k_1^B)^2-(k_2^B)^2)-2\mathbf{k}_C\cdot(\mathbf{k}_1^B-\mathbf{k}_2^B)] = \varepsilon_B(2)-\varepsilon_B(1). \tag{71}$$

Thus, to achieve control requires that $\mathbf{k}_i^B, \mathbf{k}_C$ be chosen to satisfy the degeneracy condition imposed by (71). Further, $\langle E_{cm}(1)|\, E_{cm}(2)\,\rangle$ must be nonzero. Since by virtue of the definition of $\mathbf{K}_i$, $\int d\mathbf{R}_{BC} \exp[i(\mathbf{K}_1-\mathbf{K}_2)\cdot\mathbf{R}_{BC}] = \int d\mathbf{R}_{BC} \exp[i(\mathbf{k}_1^B - \mathbf{k}_2^B)\cdot \mathbf{R}_{BC}]$, if $(\mathbf{k}_1^B - \mathbf{k}_2^B)$ is made sufficiently small then $\langle E_{cm}(1)|\, E_{cm}(2)\,\rangle$ is nonzero when integrated over the volume of intersection of the B and C beams. Under these circumstances, (55) becomes

$$\{\hbar^2/(m_B+m_C)\}[\mathbf{k}_C\cdot(\mathbf{k}_1^B-\mathbf{k}_2^B)] \approx \varepsilon_B(1)-\varepsilon_B(2), \tag{72}$$

so that large $\mathbf{k}_C$ may be required to satisfy this condition. With such a $\mathbf{k}_C$, bimolecular control, regulated by the amplitude and phases of the a_i, is established.

Alternative methods of preparing the superposition in (54) and maintaining a nonzero $\langle E_{cm}(1)|\, E_{cm}(2)\,\rangle$ can be envisioned. The most obvious deals with superposing degenerate states of B. The energy degeneracy requirement is then automatically satisfied, and since $\mathbf{K}_1 = \mathbf{K}_2$, then $\langle E_{cm}(1)|\, E_{cm}(2)\,\rangle$ is trivially nonzero. Examples include collisions such as $H(2s)+D$ in a superposition with $H(2p)+D$, where D is a molecule and where $H(2s), H(2p)$ result from a

rior coherently controlled photolysis of H_2. Similarly, one can envision using elliptically polarized light to prepare a superposition of m_j states, where m_j is the z-projection of the rotational angular momentum of a diatomic B, and then colliding the result with C. Once again, the degeneracy of the states ensures that control is possible and that the center of mass overlap matrix element is nonzero[68].

Finally, note that the above *formalism* can be readily extended to general superposition states of the form

$$|\,\mathrm{n},\beta\,\rangle = \sum_{i,l} a_{il}|\,i,B\,\rangle|\,l,C\,\rangle|\,E_\beta^{kin}(i,l)\,\rangle|\,E_{cm}(i,l)\,\rangle, \tag{73}$$

with $E_\beta^{kin}(i,l) = E-\varepsilon_C(l)-\varepsilon_B(i)$. Here, the $|\,E_\beta^{kin}(i,l)\,\rangle$ states are plane waves describing the free motion of B relative to C, $(\langle\mathbf{R}|\,E_\beta^{kin}(i,l)\,\rangle \equiv \exp(i\mathbf{k}_{il}\cdot\mathbf{R}))$, where $|\mathbf{k}_{il}| = [2\mu_{BC}E_\beta^{kin}(i,l)]^{\frac{1}{2}}/\hbar$. That is, we can show that such a superposition leads to interference if the $\langle E_{cm}(j,k)|\, E_{cm}(i,l)\,\rangle$ are nonzero, and hence to the possibility of control over the reaction cross-sections. We are currently examining possible methods for experimentally realizing such states.

The possibility of successfully applying coherent control techniques to bimolecular processes opens up a vast new area of control research.

6 Conclusions

Our discussion makes clear that the characteristic features that we invoke in order to control chemical reactions are purely quantum in nature. There is, for example, little classical about the time-dependent picture where the ultimate outcome of the deexcitation, i.e., product H + HD or H_2 + D, depends entirely upon the phase and amplitude characteristics of the wave function. Indeed, as repeatedly emphasized above, if, e.g., collisional effects are sufficiently strong so as to randomize the phases, then reaction control is lost. Hence reaction dynamics are intimately linked to the wave function phases, which are controllable through coherent optical phase excitation.

These results must be viewed in light of the history of molecular reaction dynamics over the past two decades. Possibly the most useful result of the reaction dynamics research effort has been the recognition that the vast majority of qualitatively important phenomena in reaction dynamics are well described by classical mechanics. Quantum and semiclassical mechanics were viewed as necessary only insofar as they correct quantitative failures of classical mechanics for unusual circumstances and/or for the dynamics of very light particles. Considering reaction dynamics in traditional chemistry to be essentially classical in character therefore appeared to be essentially correct for the vast majority of naturally occurring molecular processes. Coherence played no role. The coherent control approach makes clear, however, that coherence phenomena have great potential for application. The quantum phase is always present and can be used to our advantage, even though it is irrelevant to traditional chemistry. By calling attention to the extreme importance of coherence phenomena to controlled chemistry we herald the introduction of a new focus in atomic and molecular science, i.e., introducing coherence in controlled environments to modify molecular processes, thus defining the area of coherence chemistry.

Acknowledgments

. We acknowledge support for this research by the U.S. Office of Naval Research.

References

1. P. Brumer and M. Shapiro, Chem. Phys. Lett. **126**, 541 (1986).
2. M. Shapiro and P. Brumer, J. Chem. Phys. **84**, 4103 (1986).
3. P. Brumer and M. Shapiro, Faraday Disc. Chem. Soc. **82**, 177 (1986).
4. M. Shapiro and P. Brumer, J. Chem. Phys. **84**, 4103 (1986).
5. C. Asaro, P. Brumer, and M. Shapiro, Phys. Rev. Lett. **60**, 1634 (1988).
6. M. Shapiro, J. Hepburn, and P. Brumer, Chem. Phys. Lett. **149**, 451 (1988).
7. P. Brumer and M. Shapiro, J. Chem. Phys. **90**, 6179 (1989).
8. G. Kurizki, M. Shapiro, and P. Brumer, Phys. Rev. B, **39**, 3435 (1989).
9. T. Seideman, M. Shapiro, and P. Brumer, J. Chem. Phys. **90**, 7136 (1989).
10. J. Krause, M. Shapiro, and P. Brumer, J. Chem. Phys. **92**, 1126 (1990).
11. I. Levy, M. Shapiro, and P. Brumer, J. Chem. Phys., **93**, 2493 (1990).
12. M. Shapiro and P. Brumer, J. Chem. Phys. **95**, 8658 (1991).
13. P. Brumer and M. Shapiro, Ann. Rev. Phys. Chem. **43**, 257 (1992).
14. M. Shapiro and P. Brumer, J. Chem. Phys. **97**, 6259 (1992).
15. Z. Chen, P. Brumer, and M. Shapiro, Chem. Phys. Lett. **198**, 498 (1992).

16. X.-P. Jiang, P. Brumer, and M. Shapiro, J. Chem. Phys. **104**, 607 (1996).
17. J. Dods, P. Brumer, and M. Shapiro, Can. J. Chem **72**, (Polanyi Honor Issue) 958 (1994).
18. D. J. Tannor and S. A. Rice, J. Chem. Phys. **83**, 5013 (1985); D. J. Tannor, R. Kosloff, and S. A. Rice, J. Chem. Phys. **85**, 5805 (1986).
19. S. A. Rice, D. J. Tannor, and R. Kosloff, J. Chem. Soc. Faraday Trans. **82**, 2423 (1986).
20. D. J. Tannor and S. A. Rice, Adv. Chem. Phys. **70**, 441 (1988).
21. R. Kosloff, S. A. Rice, P. Gaspard, S. Tersigni, and D. J. Tannor, Chem. Phys. **139**, 201 (1989).
22. S. Tersigni, P. Gaspard, and S. A. Rice, J. Chem. Phys. **93**, 1670 (1990).
23. S. Shi, A. Woody, and H. Rabitz, J. Chem. Phys. **88**, 6870 (1988); S. Shi and H. Rabitz, Chem. Phys. **139**, 185 (1989).
24. A. P. Peirce, M. Dahleh, and H. Rabitz, Phys. Rev. A **37**, 4950 (1988).
25. S. Shi and H. Rabitz, J. Chem. Phys. **92**, 364 (1990).
26. J. L. Krause, R. M. Whitnell, K. R. Wilson, Y. Yan, and S. Mukamel, J. Chem. Phys. **99**, 6562 (1993).
27. W. Jakubetz, B. Just, J. Manz, and H.-J. Schreier, J. Phys. Chem. **94**, 2294 (1990).
28. C. Chen, Y.-Y. Yin, and D. S. Elliott, Phys. Rev. Lett. **64**, 507 (1990) ; *ibid* **65**, 1737.
29. S. M. Park, S.-P. Lu, and R. J. Gordon, J. Chem. Phys. **94**, 8622 (1991); S.-P. Liu, S. M. Park, Y. Xie, and R. J. Gordon, J. Chem. Phys. **96**, 6613 (1992).
30. L. Zhu, U. Kleiman, X. Li, S.-P. Lu, K. Trentelman, and R. J. Gordon, Science **270**, 77 (1995).
31. N. F. Scherer, A. J. Ruggiero, M. Du, and G. R. Fleming J. Chem. Phys. **93**, 856 (1990).
32. K. J. Boller, A. Imamoglu, and S. E. Harris, Phys. Rev. Lett. **66**, 2593 (1991).
33. B. A. Baranova, A. N. Chudinov, and B. Ya. Zel'dovitch, Opt. Comm. **79**, 116 (1990).
34. Y.-Y. Yin, C, Chen, D. S. Elliott, and A. V. Smith, Phys. Rev. Lett. **69**, 2353 1992.
35. E. Dupont, P. B. Corkum, H. C. Liu, M. Buchanan, and Z. R. Wasilewski, Phys. Rev. Lett. **74**, 3596 (1995).
36. B. Sheeny, B. Walker, and L. F.Dimauro, Phys. Rev. Lett. **74**, 4799 (1995).
37. Y.-Y. Yin, R. Shehadeh, D. Elliott, and E. Grant, Chem. Phys. Lett. **241**, 591 (1995).
38. A. Shnitman, I. Sofer, I. Golub, A. Yogev, M. Shapiro, Z. Chen, and P. Brumer, Phys. Rev. Lett. **76**, 2886 (1996).
39. For a discussion see, e.g., J. D. Macomber *The Dynamics of Spectroscopic Transitions* (Wiley, N.Y., 1976)
40. This is the asymptotic condition of scattering theory; see J. R. Taylor, *Scattering Theory* (Wiley, N.Y., 1972).
41. Z. Chen, M. Shapiro and P. Brumer, Chem. Phys. Lett. **228**, 289 (1994).
42. Z. Chen, P. Brumer and M. Shapiro, J. Chem. Phys. **98**, 6843 (1993).
43. S. Chelkowski and A. D. Bandrauk, Chem. Phys. Lett. **186**, 284 (1991); A. D. Bandrauk, J. M. Gauthier, and J. F. McCann, Chem. Phys. Lett. **200**, 399 (1992).
44. A. Szöke, K. C. Kulander, and J. N. Bardsley, J. Phys. B **24**, 3165 (1991) ; R. M. Potvliege and P. H. G. Smith, J. Phys. B **25**, 2501 (1992).
45. E. Charron, A. Guisti-Suzor and F. H. Mies, Phys. Rev. Lett. **71**, 692 (1993).
46. R. Blank and M. Shapiro, Phys. Rev. A **52**, 4278 (1995).

47. S. Chelkowski, A. D. Bandrauk, and P. B. Corkum, Phys. Rev. Lett. **65**, 2355 (1990).
48. The use of perturbation theory does not necessarily imply small total yields. Computational results (P. Brumer and M. Shapiro – to be published) indicate that perturbation theory is quantitatively correct for dissociation probabilities as large as 0.20.
49. M. Shapiro and P. Brumer, J. Chem. Phys. **98**, 201 (1993).
50. N. E.Henriksen and B. Amstrup, Chem. Phys. Lett. **213**, 65 (1993) ; *J. Chem. Phys.* **97** 8285
51. D. Abrashkevich, M. Shapiro, and P. Brumer (to be published).
52. I. Levy I and M. Shapiro, J. Chem. Phys. **89**, 2900 (1988).
53. R. Bavli and H. Metiu, Phys. Rev. Lett. **69**, 1986 (1992).
54. M. Yu. Ivanov, P. B. Corkum, and P. Dietrich, Laser Physics **3**, 375 (1993).
55. L. D. Barron, 1982 *Molecular Light Scattering and Optical Activity* (Cambridge Univ. Press, Cambridge, 1982); R. G. Woolley,Adv. Phys. **25**, 27 (1975) ; *Origins of Optical Activity in Nature*, ed., D. C. Walker (Elsevier, Amsterdam, 1979).
56. For a discussion see L. D. Barron, Chem. Soc. Rev. **15**, 189 (1986).
57. J. M. Hollas,*High Resolution Spectroscopy* (Butterworths, London, 1972).
58. E. Segev E and M. Shapiro, J. Chem. Phys. **77**, 5604 (1982).
59. T. Nakajima and P. Lambropoulos, Phys. Rev. Lett. **70**, 1081 (1993).
60. P. Brumer and M. Shapiro,*Adv. Chem. Phys.* **60**, 371, ed. K. P. Lawley (Wiley-Interscience, N.Y., 1986).
61. M. Shapiro and H. Bony, J. Chem. Phys. **83**, 1588 (1985); G. G. Balint-Kurti and M. Shapiro, *Adv. Chem. Phys.* **60**, 403, ed. K. P. Lawley (Wiley-Interscience, N.Y., 1986).
62. A. D. Bandrauk and O. Atabek, *Adv. Chem. Phys.* **73**, 823 (1989).
63. The potential curves and the relevant electronic dipole moments are taken from I. Schmidt, in Ph.D. Thesis, Kaiserslautern University, 1987.
64. Z. Chen, M. Shapiro, and P. Brumer,,J. Chem. Phys. **102**, 5683 (1995) ; Z. Chen, M. Shapiro, and P. Brumer, Phys. Rev. A **52**, 2225 (1995) ; M. Shapiro, Z. Chen, and P. Brumer, Chem. Phys. (in press)
65. A. Shnitman, I. Sofer, I. Golub, A. Yogev, M. Shapiro, Z. Chen, and P. Brumer (manuscript in preparation).
66. J. Krause M. Shapiro, and P. Brumer, J. Chem. Phys. **92**, 1126 (1990).
67. M. Shapiro and P. Brumer, Phys. Rev. Lett. **177**, 2574 (1996)
68. A. Abrashkevich, M. Shapiro, and P. Brumer (to be published)

This article was processed using the LaTeX macro package with LLNCS style

Coherent Control of Unimolecular Reaction Dynamics Based on a Local Optimization Scheme

Y. Fujimura

Department of Chemistry, Graduate School of Science, Tohoku University, Sendai 980-77, Japan, E-mail: fujimura@mcl.chem.tohoku.ac.jp

Abstract. This chapter reviews recent progress in the theoretical study of coherent control of reaction dynamics developed in our laboratory. The coherent control is based on optimization theory of a linear time-invariant complex system. Since the reaction dynamics of interest are not a linear time-invariant system, the time-dependent Schrödinger equation describing the time evolution of the system from the initial time to the final state is divided into short time stages. The optimization procedure is carried out in each short time stage in which the system can be approximated to a linear time-invariant system. Such an optimization carried out by a succession of short time stages leads to a local optimization scheme. The optimized laser pulse shape in every short time stage is expressed by a feedback form. The local optimization procedure is valid not only for weak laser fields but also for strong fields. The coherent control theory is applied to three types of unimolecular reaction dynamics: dissociation of hydrogen fluoride, isomerization of hydrogen cyanide, and a pump – dump pulse control of a reaction via an upper electronic excited state. The results of these applications show that the local optimization procedure is an effective tool for guiding the pulse shaping of unimolecular reactions.

1 Introduction

In recent years, considerable efforts have been made toward the realization of coherent control of chemical reactions using so-called "laser pulse shaping" [1–5]. Experimental progress in areas such as spectral filtering and chirping techniques has enabled us to create pulse sequences, time-dependent amplitudes, and frequencies for optical control [6, 7]. Coherent control consists of manipulating nuclear wave packets to a desired reaction channel by the use of tailored pulses. There are two fundamental optimization theories for guiding tailored pulses: global and local optimization theories [8–16]. Both of these theories involve optimization of a performance index related to the reaction of interest under the constraint that the system follows the time-dependent Schrödinger equation. In global optimization theory, the initial and final states of the reaction are specified, and the so-called "two-point boundary value problem" under the condition of a given level of laser energy is solved to obtain the optimized pulse shape [2, 3]. This is generally a difficult task. Perturbative treatments based on the global optimization theory have been proposed for practical computations [17–20].

Alternative methods based on classical and quantum optimal control theories have also been considered [21–24]. Problems encountered in global optimization theory can be avoided in local optimization theory [4]. In local optimization theory, the external field is expressed in a feedback form.

In this chapter, first we review the theoretical background for coherent control of reaction dynamics based on a local optimization theory [15, 25]. For this purpose, we introduce an optimization theory for a linear time-invariant system. We show how the optimization theory can be applied to quantum systems satisfying the time-dependent Schrödinger equation. The important point is that the Schrödinger equation can be linearized in the short time limit. The optimization procedure is carried out in each short time stage to reach the desired final state. Therefore, the present optimization is not limited to the control of reactions in weak intensity fields but can also be applied to reaction systems in strong intensity fields. We present an application of this theory to some simple unimolecular reactions: infrared multiphoton dissociation of hydrogen fluoride [26], isomerization of hydrogen cyanide HCN $\rightarrow$ HNC in a one-dimensional model [25], and pump – dump pulse control via an upper electronically excited state [27]. We analyze the time-dependent behaviors of the reaction dynamics under the condition of an optimized laser pulse.

2 Local Optimization Theory

2.1 Optimal Feedback Control of a Linear Time-Invariant System

Figure 1 shows the concept of optimal control of wave packet motion using laser pulses [28]. Here, the ordinate denotes the position of the wave packet, and the abscissa denotes its time development. The external nonstationary radiation field $u(t)$ drives the wave packet to the target at $t = t_f$ from an initial time $t = t_0$. There exist many paths specified by performance index J to reach the target position from t_0. Optimal control means that the reaction proceeds along the path specified by performance index J_2 in the case of Fig. 1. It is not easy to construct the optimal field that positions the final wave packet at the exact target position. Instead of finding the exact position, it is usually sufficient to determine the approximately optimized field by allowing the wave packet to permit some deviation F at the target position at the final time. Q is also a measure of the deviation of the wave packet from the exact optimized path during the time development [15, 20].

To describe quantitatively the optimal control shown in Fig. 1, we take a system described by the differential equation [15, 25]

$$\frac{d\boldsymbol{x}(t)}{dt} = \mathbf{A}\boldsymbol{x}(t) + \mathbf{B}\boldsymbol{u}(t) , \tag{1}$$

where $\boldsymbol{x}(t)$ is a complex time-dependent column vector of the system expressed as $\boldsymbol{x}(t) = [x_1(t), x_2(t), \ldots, x_n(t)]^T$, and $\boldsymbol{u}(t)$ is a complex controlling vector that is expressed as $\boldsymbol{u}(t) = [u_1(t), u_2(t), \ldots, u_m(t)]^T$. In (1), $\mathbf{A}$ is an $n \times n$ time-independent complex matrix, and $\mathbf{B}$ is an $n \times m$ time-independent complex matrix.

In order to optimize the system under the constraint of (1) during a time interval $t_0 \le t \le t_f$, we minimize the performance index J, defined as

$$J \equiv J(\boldsymbol{u}(t),\boldsymbol{x}(t)) \equiv \frac{1}{2}\boldsymbol{x}(t_f)^{\dagger}\mathbf{F}\boldsymbol{x}(t_f) \;+\; \frac{1}{2}\int_{t_0}^{t_f} \mathrm{d}t \;\{\boldsymbol{x}^{\dagger}(t)\mathbf{Q}\boldsymbol{x}(t)+\boldsymbol{u}^{\dagger}(t)\mathbf{R}\boldsymbol{u}(t)$$

$$+\;[\,\boldsymbol{p}^{\dagger}(t)\{\mathbf{A}\boldsymbol{x}(t)+\mathbf{B}\boldsymbol{u}(t)-\frac{\mathrm{d}}{\mathrm{d}t}\boldsymbol{x}(t)\} + \text{complex conjugate}]\}, \tag{2}$$

where the dagger denotes the Hermitian conjugate. $\mathbf{F}$ is an $n \times n$ positive definite Hermitian matrix that specifies a state distribution at the final time (see Fig. 1). The first term on the right-hand side of (2) quantifies a deviation from the optimized system vector at the final time t_f; i.e., the controlling vector is permitted a degree of freedom to relax the degree of constraint factor; $\mathbf{Q}$ is an $n{\times}n$ positive definite Hermitian matrix whose diagonal elements represent a weighting factor of the system; n specifies the eigenstates of the system associated with the optimal control; $\mathbf{R}$ is an $m \times m$ positive definite Hermitian matrix whose diagonal elements represent a weighting factor of controlling vectors. In (2), $\boldsymbol{p}$(t) is a time-dependent Lagrange multiplier vector.

The optimal controlling vector $\boldsymbol{u}^{0}(t)$ is given as [15]

$$\boldsymbol{u}^{0}(t) = -\mathbf{R}^{-1}\mathbf{B}^{\dagger}\boldsymbol{p}(t)\;. \tag{3}$$

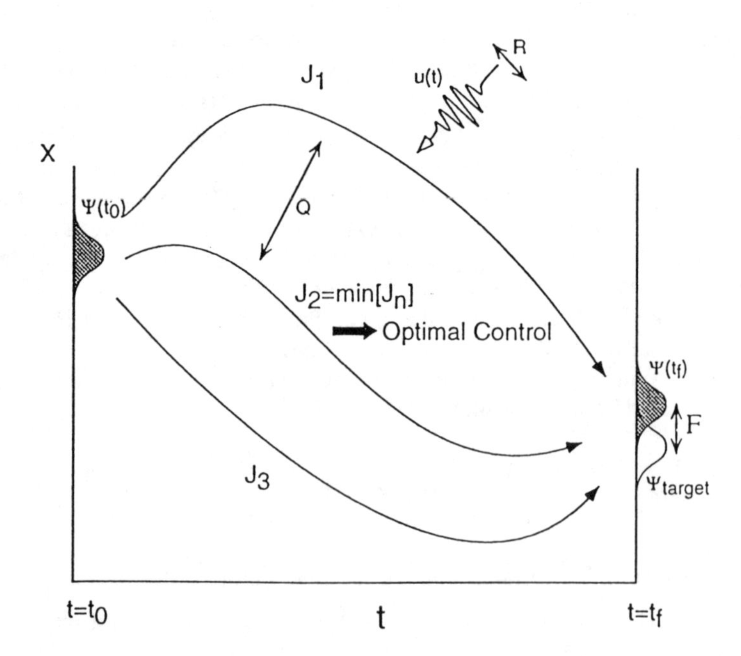

Fig.1. Concept of optimal control wave packet propagation by laser $u(t)$. J denotes the performance index.

By introducing an $n \times n$ matrix $\tilde{\mathbf{P}}(t)$ defined as

$$p(t) = \tilde{\mathbf{P}}(t)x(t), \tag{4}$$

with the boundary condition $p(t_f) = \mathbf{F}x(t_f)$, we obtain an expression for the optimal control vector as

$$u^0(t) = -\mathbf{R}^{-1}\mathbf{B}^{\dagger}\tilde{\mathbf{P}}(t)x(t). \tag{5}$$

The equation for $\tilde{\mathbf{P}}(t)$ is given as

$$\begin{aligned}\frac{\mathrm{d}}{\mathrm{d}t}\tilde{\mathbf{P}}(t) &= -\tilde{\mathbf{P}}(t)\mathbf{A} - \mathbf{A}^{\dagger}\tilde{\mathbf{P}}(t) + \tilde{\mathbf{P}}(t)\mathbf{B}\mathbf{R}^{-1}\mathbf{B}^{\dagger}\tilde{\mathbf{P}}(t) - \mathbf{Q} \\ &\equiv f[\tilde{\mathbf{P}}(t)] .\end{aligned} \tag{6}$$

This equation is called the Riccati equation, which is a nonlinear equation for $\tilde{\mathbf{P}}(t)$. The solution of (6) can be expressed as

$$\tilde{\mathbf{P}}(t) = \mathbf{F} + \int_{t_f}^{t} \mathrm{d}t'\, f[\tilde{\mathbf{P}}(t')]. \tag{7}$$

The optimal external field vector is rewritten as

$$u^0(t) = -\mathbf{K}(t)x(t) , \tag{8}$$

where $\mathbf{K}(t)$, called the feedback gain matrix, is expressed by

$$\mathbf{K}(t) = \mathbf{R}^{-1}\mathbf{B}^{\dagger}\tilde{\mathbf{P}}(t). \tag{9}$$

2.2 Reduction of The Time-Dependent Schrödinger Equation to a Linear Time-Invariant System Equation in a Short Time Regime

The equation of motion of a quantum system in a nonstationary laser field is given by the time-dependent Schrödinger equation within the classical treatment of the laser field as

$$\frac{\partial|\phi(t)\rangle}{\partial t} = -\frac{i}{\hbar}\hat{H}_0|\phi(t)\rangle - \frac{i}{\hbar}\mu \cdot u(t)|\phi(t)\rangle, \tag{10}$$

where $\hat{H}_0$ is the Hamiltonian of the quantum system, $|\phi(t)\rangle$ is the wave function, μ is the dipole moment, and $u(t)$ is a nonstationary laser field that controls the motion of the wave packet. If the wave function $|\phi(t)\rangle$ is expanded in terms of the eigenstates $|i\rangle$ as

$$|\phi(t)\rangle = \sum_{i=0}^{n-1} c_i(t)\,|i\rangle , \tag{11}$$

then (10) can be expressed as

$$\begin{aligned}\frac{\partial\phi(t)}{\partial t} &= -\frac{i}{\hbar}\mathbf{H}_0\phi(t) - \frac{i}{\hbar}\mathbf{D}\phi(t)\boldsymbol{u}(t) \\ &\equiv \mathrm{h}[\boldsymbol{u}(t),\phi(t)] .\end{aligned} \tag{12}$$

Here, the state vector $\phi(t)$ is written as $\phi(t) = [c_0(t), c_1(t), \ldots, c_{n-1}(t)]^T$, and $\mathbf{D}$ is an $n \times m \times n$ dipole moment tensor whose element D_{ijk} represents the magnitude of the optical transition between eigenstates i and k through the nonstationary laser field $u_j(t)$ [15]. The equation of motion, (12) is not the equation of a linear time-invariant system because the second term of (12) has a different structure from that in (1). In order to apply the optimal control theory described in the previous subsection to (12), we divide the time interval $t_f - t_0$ equally into N short stages as shown in Fig. 2. In each stage, the system may be considered as a linear time-invariant system, which means that the optimization is a kind of local optimization. Figure 2 shows the concept of local optimization [28]. The view of an optimization of the ith short stage is magnified. The time interval is set at $\varepsilon \equiv t_f^{(i)} - t_0^{(i)}$ ($i = 1, 2, \ldots, N$; $t_0^{(1)} = t_0$ and $t_f^{(N)} = t_f$). To reduce (12) to the linear time invariant system equation, we use the Taylor expansion of $\mathrm{h}[\boldsymbol{u}^{(i)}(t), \phi^{(i)}(t)]$ around a certain point $(\boldsymbol{u}_0^{(i)}, \phi_0^{(i)})$ at the ith stage. The differences at this point are given as

$$\Delta\phi^{(i)}(t) = \phi^{(i)}(t) - \phi_0^{(i)}, \text{ and } \Delta\boldsymbol{u}^{(i)}(t) = \boldsymbol{u}^{(i)}(t) - \boldsymbol{u}_0^{(i)}, \tag{13}$$

where $t_0^{(i)} \le t \le t_f^{(i)}$.

Equation (12) can then be expressed within the linear approximation as

$$\frac{\partial\phi^{(i)}(t)}{\partial t} = -\frac{i}{\hbar}[\mathbf{H}_0 + \tilde{\mathbf{D}}\boldsymbol{u}_0^{(i)}]\phi^{(i)}(t) - \frac{i}{\hbar}\mathbf{D}\phi_0^{(i)}\Delta\boldsymbol{u}^{(i)}(t). \tag{14}$$

Equation (14) is the equation of motion for a quantum system equivalent to (1).

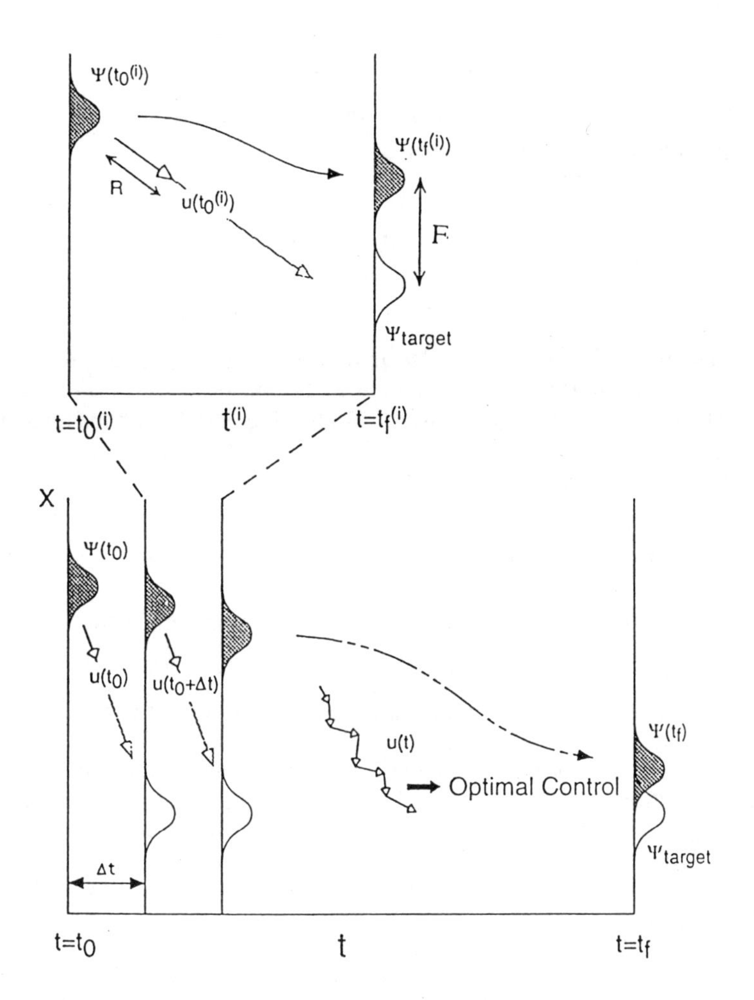

Fig.2. Concept of linearly optimized control of wave packet propagation

2.3 Optimal Feedback Control of Quantum Dynamics

Applying (1) to (14), the optimal nonstationary laser field $\Delta \boldsymbol{u}^{(i)}(t)$ in the ith stage can be expressed as

$$\Delta \boldsymbol{u}^{(i)}(t) = -\frac{i}{\hbar}\mathbf{R}^{-1}[\mathbf{D}\phi(t_0^{(i)})]^{\dagger}\,\tilde{\mathbf{P}}(t)\;\phi^{(i)}(t), \tag{15}$$

where $t_0^{(i)} \le t \le t_f^{(i)}$. Since $\tilde{\mathbf{P}}(t)$ can be replaced by $\mathbf{F}$ from (7) and $\phi(t_f^{(i)})$ by $\phi(t)$ in the limit as $\varepsilon \to 0$, i.e., $t \to t_f^{(i)}$, the optimal control field $\Delta \boldsymbol{u}(t)$ within an infinitesimal time step ε can be expressed as

$$\Delta \boldsymbol{u}(t) = -\frac{\mathrm{i}}{\hbar}\mathbf{R}^{-1}[\mathbf{D}\phi(t)]^{\dagger}\mathbf{F}\phi(t). \tag{16}$$

Here, the superscript i is omitted, since i itself represents the ith stage. Equation (16) shows that the optimized laser pulse is expressed in a feedback form. Once the magnitude of the pulse is given at time t, the pulse at a very short time later is obtained by solving the time-dependent Schrödinger equation with the magnitude of the pulse and substituting the resulting wave function into (16). This procedure is repeated until convergence is obtained.

3 Results and Discussion

3.1 Infrared Multiphoton Dissociation of Hydrogen Fluoride

Hydrogen fluoride (HF) is one of the diatomic molecules whose laser-induced dynamics have been well studied [29–34].

We determine what kinds of pulses can most effectively dissociate HF by using the local optimization procedure described in the preceding section. We also examine the mechanism of the photodissociation dynamics of HF under the condition of an optimized laser field.

The adiabatic potential of HF in the electronic ground state $V(x)$ is approximated by a Morse potential expressed as

$$V(x) = D\{1 - \exp[-\beta(x - x_e)]\}^2 - D, \tag{17}$$

with parameters $D = 46{,}110$ cm^{-1}, $\beta = 2.3$ Å^{-1}, and $x_e = 0.926$ Å [12]. Figure 3 shows the potential curve $V(x)$ together with several eigenstates [26].

The dipole moment is assumed to be expressed as [29]

$$\mu(x) = Bx\exp[-\xi x^4], \tag{18}$$

with $B = 0.24$ Å$\cdot e$ and $\xi = 0.082$ Å^{-4}. An initial guess at the most likely path for the dissociation is a ladder, that is sequential path, $|0\rangle \to |1\rangle \to \dots \to |m-1\rangle \to |m\rangle \to \dots$. The matrix elements of $\mathbf{F}$ are assumed to be satisfied under the condition $F_0 > F_1 > \dots > F_{m-1} > F_m > \dots$. The optimization parameter for the energy cost of the pulse is taken to be $R = 1.07$, which corresponds to an energy fluence of 5.76 J/cm^2.

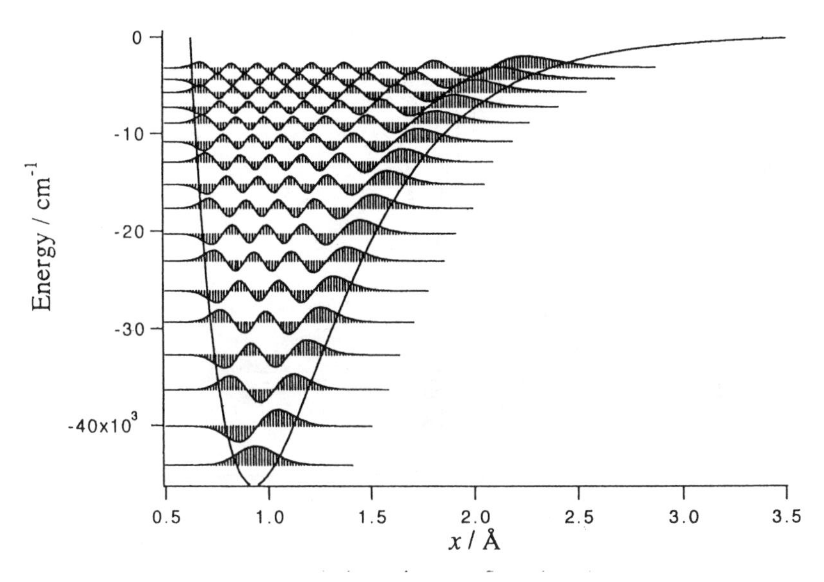

Fig.3. The Morse potential curve of HF in the ground electronic state. Several wave functions for the Morse potential are also shown.

Figure 4 shows the time evolution of the photodissociation dynamics of HF that was calculated by using the local optimization procedure [26]. We can see that the dissociation dynamics are divided into three stages. In the first stage (t = 0 ~ 8 ps), a sequential population transfer starting from the lowest state occurs. This mechanism can be simply explained by using the dipole selection rule $\Delta n = \pm 1$ for optical transitions between vibrational eigenstates in the harmonic approximation, keeping the initial guess for the most likely path. Here Δn refers to the difference in the vibrational quantum numbers between the relevant states. In the second stage (t = 8 ~ 11 ps), transitions other than $\Delta n = \pm 1$ turn out to be favorable transitions. In this stage, the simple dipole transition mechanism breaks down. For example, the optimized field chooses the transition $|7\rangle \rightarrow |9\rangle$ rather than the $|7\rangle \rightarrow |8\rangle$ transition. The same situation can be seen for the $|9\rangle \rightarrow |11\rangle$ transition, which skips the state $|10\rangle$. The third stage starts after about 11 ps. In this stage, the transitions from the bound states to the continuum state occur. In the first and second stages, an almost complete population transfer occurs between the two relevant states, which indicates effective π pulses for the optimal control of the transitions [15].

Figure 5 shows the time variation of the optimized laser field. The optimized laser field consists of a sequence of Gaussian-like envelopes with certain carrier frequencies. The duration of each envelope corresponds to the duration of each transition between the bound states, and the carrier frequency corresponds to the transition frequency. It should be noted that the carrier frequency of each envelope is not given beforehand but is obtained as a result of the optimization procedure.

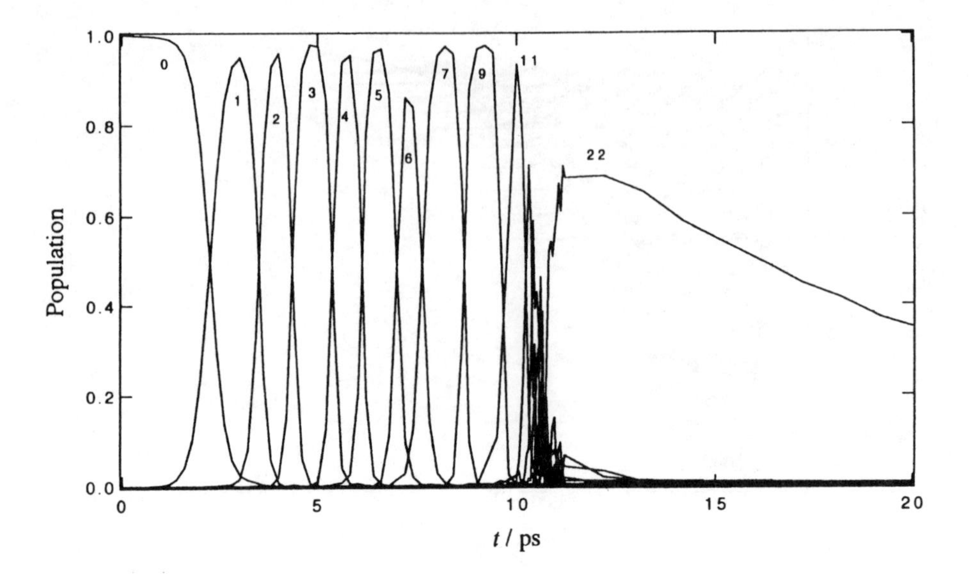

Fig.4. Time variation in the populations of vibrational eigenstates in an optimized laser field

Information on the modulation of these carrier frequencies can be extracted by using a window Fourier transform and by computing the time-resolved power spectrum $S(\omega,t)$ defined as [15]

$$S(\omega,t) \equiv \left| \int_{-\infty}^{\infty} \mathrm{d}\tau \, w(\tau - t, T) u(\tau) \exp(\mathrm{i}\omega\tau) \right|^2 . \tag{19}$$

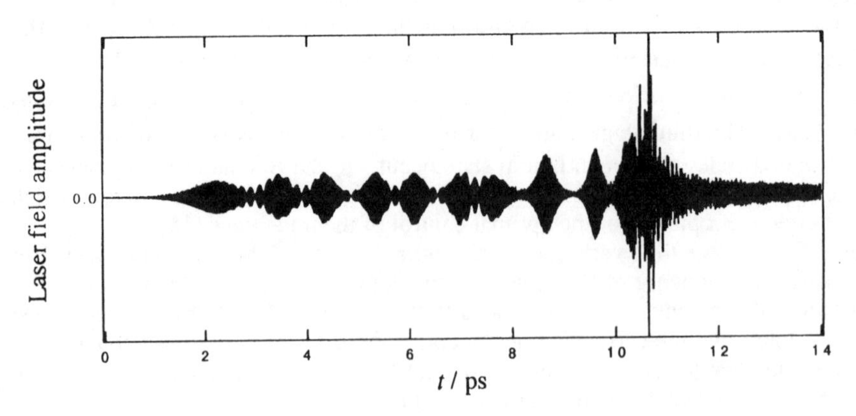

Fig.5. Time variation in the optimized laser field

Here, $w(\tau, T)$ is the Blackman window given as

$$w(\tau,T) = \begin{cases} 0.42 - 0.50\cos\dfrac{2\pi\tau}{T} + 0.08\cos\dfrac{4\pi\tau}{T} & \left(|\tau| \le \dfrac{T}{2}\right), \\ 0 & \left(|\tau| > \dfrac{T}{2}\right). \end{cases} \tag{20}$$

The parameter T, which is a measure of the fineness of the time resolution, is taken to be $T = 0.3$ ps. The frequency resolution $\Delta\omega(= 2\pi / T)$ is 111 cm^{-1}.

Figure 6 shows the contour map of the time-resolved power spectrum $S(\omega,t)$. Initially, the carrier frequency is around 4000 cm^{-1}, i.e., the transition frequency of the $|0\rangle \rightarrow |1\rangle$ quantum transition. During the first stage, the carrier frequency is modulated to become smaller and linear with respect to time. The linear time-dependence of the carrier frequency is called a linear chirping. This feature corresponds to the fact that the transition frequency becomes small as the quantum number increases due to the anharmonicity of the adiabatic potential. This leads to level separation linear with n in the Morse oscillator model. An application of such chirping pulses has been proposed recently for controlling reactions and intramolecular energy redistributions [35–41]. For example, Chelkowski et al. [35] have shown that if pulse frequency $\omega(t)$ decreases at a specific rate, adopted to the molecular anharmonicity of a Morse oscillator, the dissociation probability is many orders of magnitude higher than for a monochromatic pulse of the same intensity. The result of the multiphoton dissociation of HF using a locally optimized laser field supports the important role of the chirp pulse as shown by Chelkowski et al. [35]. Furthermore, the present calculation shows new qualitative results that were not present in previous works. Notice that the frequency jumps from 3000 cm^{-1} to 5200 cm^{-1} around $t = 8$ ps. From $t = 9$ ps to 12 ps in the second to third stages of the multiphoton dissociation

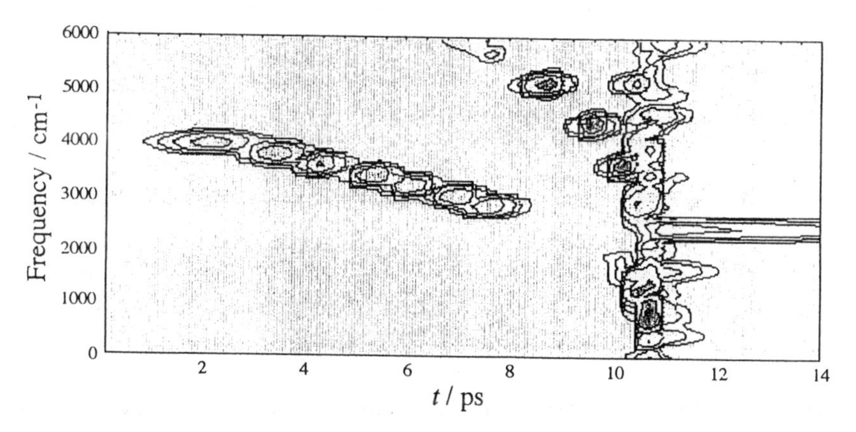

Fig.6. Contour map of the time-resolved power spectrum $S(\omega,t)$

processes, a multistep transition occurs, whose frequencies are approximately expressed by another linear chirping together with other carrier frequency components. Transition moments between higher vibrational states (higher than $n = 8$) are rather large compared to those between lower states ($n = 0, 1,...,7$). Therefore, the transition processes between higher states are faster than those occurring between lower states, which results in a higher chirping rate between 9 and 12 ps than that between 0 and 8 ps. The carrier frequencies other than the linear chirping in the time stage 9 to 12 ps reflect the fact that many eigenstates are involved in the dissociation processes. For a comparison, we calculated the HF multiphoton dynamics induced by a nonoptimized laser field with the same energy fluence as that of an optimized laser field. The result shows that the population transfer occurs only between the two states $|0\rangle$ and $|1\rangle$, and the recurrence feature of this population transfer does not result in any dissociation. In the nonoptimized laser case, dissociation never occurs with the same fluence or energy cost as in the optimized laser case. On the contrary, the locally optimized laser field produces 90% of the dissociation yield at $t = 11$ ps.

3.2 Isomerization of Hydrogen Cyanide

Consider coherent control of hydrogen cyanide (HCN) isomerization, i.e., hydrogen migration of HCN [25, 42].
Assuming that the CN stretching mode is frozen at the equilibrium bond length (1.162 Å) of a linear HCN configuration, we use a 1-dimensional model of isomerization along the minimum energy path as shown in Fig. 7. The isomerization coordinate is denoted by θ. $\theta = 0$ (2π) corresponds to the HCN configuration and $\theta = \pi$ to HNC. The potential energy surface was calculated by the ab initio MCSCF method [43]. Figure 7 also shows the wave functions of the eigenstates selected for the

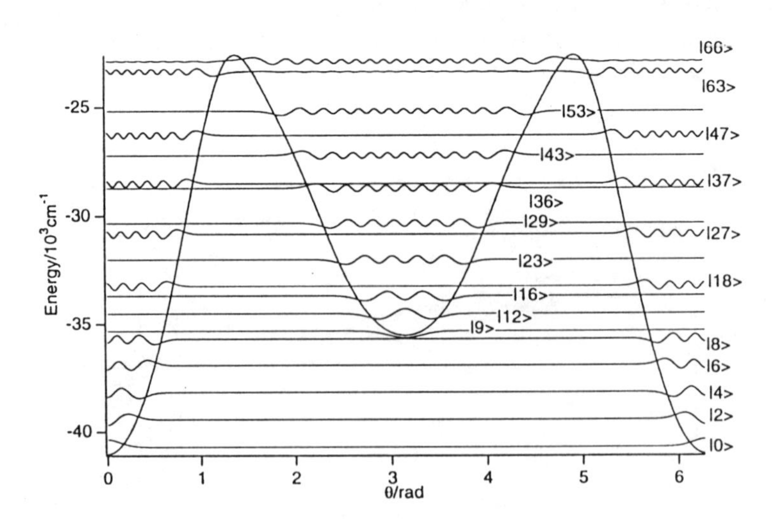

Fig.7. Potential energy curve along the minimum energy path and eigenfunctions selected for isomerization

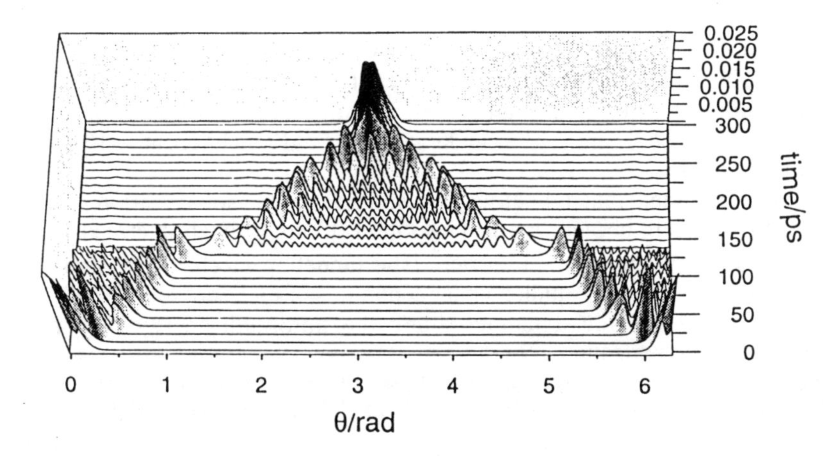

Fig.8. Time evolution of isomerization. The horizontal line refers to the square of the wave functions. The isomerization yield is ~90%.

isomerization. The system has 70 eigenstates below the potential barrier. In the initial state of isomerization, HCN is set to be the lowest state, denoted by $|0\rangle$, and in the final state, HNC is denoted by $|9\rangle$. We take a selective path along which the intermediate states consist of 17 states: $|2\rangle$, $|4\rangle$, $|6\rangle$, $|8\rangle$, $|18\rangle$, $|28\rangle$, $|37\rangle$, $|47\rangle$, $|63\rangle$, $|66\rangle$, $|53\rangle$, $|43\rangle$, $|36\rangle$, $|29\rangle$, $|23\rangle$, $|16\rangle$, and $|12\rangle$. Photon polarization is assumed to be linearly polarized along the linear H–CN axis.

Figure 8 shows the propagation of the nuclear wave packet of isomerization in the presence of the optimized pulses obtained by the local optimization scheme. We can see that the isomerization HCN $\longrightarrow$ HNC occurs with a yield of ~ 0.9.

Figure 9 shows the optimized laser field obtained for the intermediate-state-selective

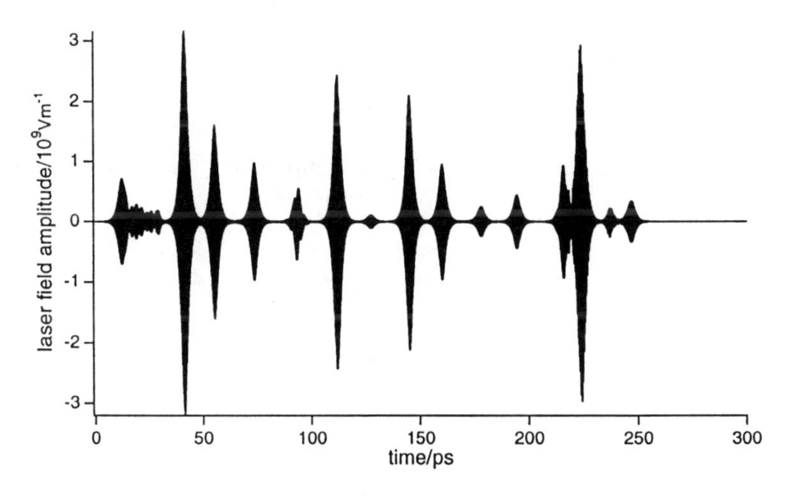

Fig. 9. Time variation in an optimized laser field for isomerization

isomerization. In a previous paper [44], isomerization in an intense IR multiphoton field with ~ 10^{13} W/cm^2 intensity was considered. For such a strong intensity pulse, the isomerization is completed within a few ps. In this review paper, on the other hand, optimization of isomerization under a weak field condition of ~ 10^{10} W/cm^2 intensity is considered.

It should be noted that tunneling is responsible for isomerization along the minimum energy path, because the isomerization occurs through intermediate states $|63\rangle$ and $|66\rangle$ below the reaction barrier. The tunneling mechanisms of HCN $\leftrightarrow$ HNC have been theoretically studied under conditions of no laser field [45–47]. It has been shown that tunneling dynamics along the minimum energy path play an important role in isomerization.

We have considered an intermediate-state-selective isomerization of HCN. Manz and Paramonov [48] and Combariza et al. [49] have also shown that selective excitations by ultrashort pulses are useful for optimal control of chemical reactions. If we consider all the levels below the potential barrier in optimizing the coherent isomerization of HCN, the optimized pulses are expected to consist of two types of chirping pulses: a down chirp pulse, which has a reduced frequency, and an up chirp pulse, of which the frequency increases with time [26, 35, 36].

3.3 Coherent Control by a Pump – Dump Pulse Scheme

In the preceding subsections, we have applied the local optimization procedure to two types of unimolecular reactions that take place in one electronic state. Now we will consider coherent control of a reaction in which wave packets in the initial state are transferred to the final reaction product region via the potential surface of an upper electronic state, as shown in Fig. 10. This type of coherent control scheme, the so-called "pump – dump pulse control" has been proposed by Tannor and Rice [50, 51].

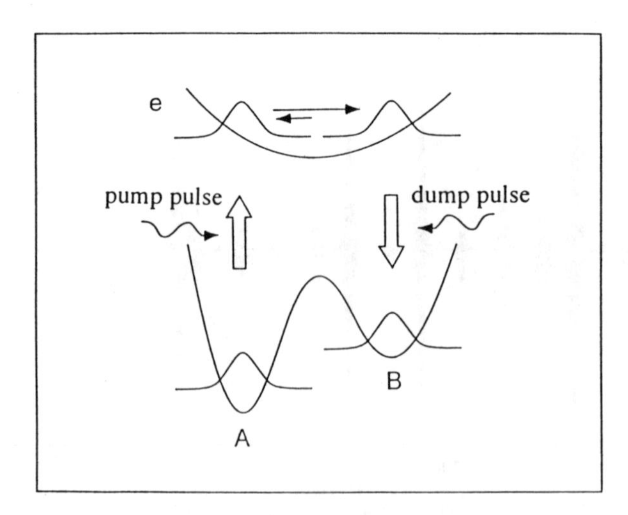

Fig. 10. Model for a pump – dump pulse scheme of coherent control

In this section, we show that the local optimization procedure can also give a promising scenario of the pump – dump pulse control scheme. We can obtain the optimized shapes for both the pump and dump pulses by using the feedback control scheme. Consider the model system for pump – dump pulse control shown in Fig. 10 [27]. Such a model system can be seen in photo-isomerization such as cis–trans isomerization [52, 53]. For simplicity, we use a one-dimensional model in which the potential curve in the ground state has a double minimum given as $V(q) = a\,q^4 - b\,q^2 + c\,q^3$ ($a = 2.9\times10^4$ cm^{-1} Å^{-4}, $b = 9.7\times10^3$ cm^{-1} Å^{-2}, and $c = 2.6\times10^3$ cm^{-1} Å^{-3}) and a harmonic potential for the electronically excited state. The initial state of the reaction is located at A, and the product state is located at B in the ground state. The key point in applying the local optimization procedure to the pump–dump pulse control scheme is that Q is specified by electronic states as well as by vibrational states.

Figure 11 shows the time variation of optimized pulses, population changes, and the

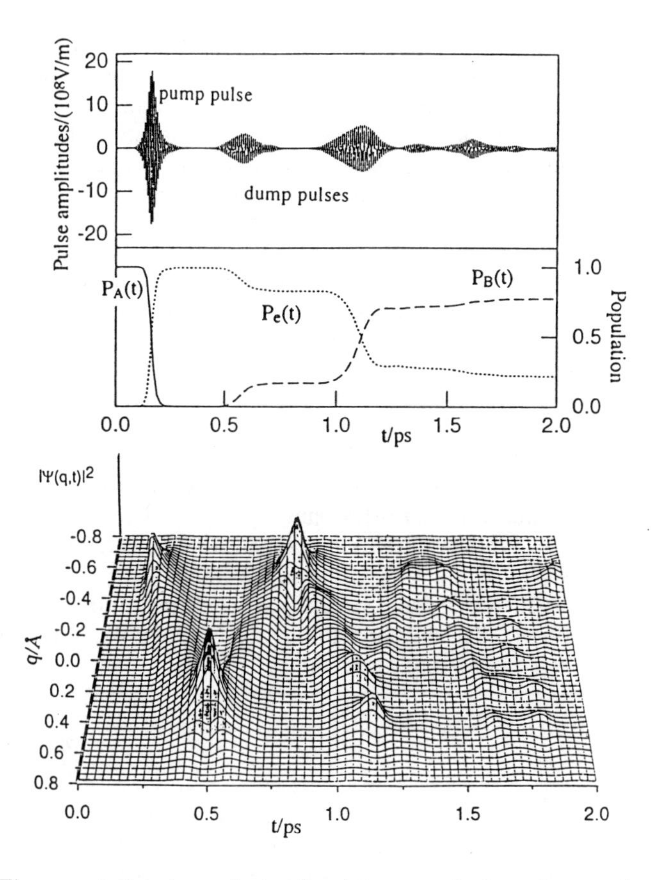

Fig.11. Time variation in optimized pulses, population changes in the initial state (A), upper electronic state (e), and final state (B) and the wave packets on the upper state that are calculated for a model system

excited state wave packet motion calculated under the condition in which the ratio of R between pump and dump pulses is assumed to be 0.3. After almost 100% of thepopulation transfer from the initial ground state A to the excited state e is induced by the pumped pulse, about 80% of the population in the excited state is transferred to the final state B by a sequence of probe pulses within a time scale of 2 ps. The rest of the population remains in the excited state. The optimized dumped pulses are synchronized to the wave packet motion; each dump pulse creates wave packets at the Franck--Condon region in the excited state to connect to the final state B. If we use a dump pulse that has a peak intensity of the same magnitude as that of the pump pulse, as shown in Fig. 11, almost the same yields can be obtained by using a couple of the pump and dump pulses within ~600 fs. Most of the wave packets in the excited state are transferred to the final state B near the opposite site on the excited state potential by its dump pulse at a stroke.

4 Summary

This chapter has presented a theoretical treatment of coherent control of reaction dynamics based on local optimization. In the short time limit, the time-dependent Schrödinger equation determining the motion of a wave packet guided by the pulses can be reduced to the equation of motion of the time-invariant complex system. Three types of model system for coherent control were tested to demonstrate the applicability of the present theory. The results show that the present theory is applicable not only to reaction control in an electronic state, but also to reaction via an electronic excited state. The present optimization procedure can be applied to the control of reaction dynamics under strong pulse conditions as well as that under weak pulse conditions. We have restricted ourselves to the optimal control of unimolecular reactions in isolated systems. The coherent control of reaction dynamics in condensed phases and isolated molecules in the statistical limit is very interesting [16, 54, 55]. Coherent nuclear motions in condensed phases have been recently observed in many femtosecond spectroscopic experiments. Coherent control of reaction dynamics in condensed phases may be possible in the near future.

Acknowledgments: The author wishes to thank the following people, who made an important contribution to this work: Dr. M. Sugawara, Dr. H. Umeda, K.Yamamoto, K. Amano, T. Taneichi, Y. Watanabe, Y. Yawata, Dr. Y. Ohtsuki, and Professor H. Kono. The author gratefully acknowledges the Japan Society for the Promotion of Science for their support of the Japan–Canada joint research project "Control of Reaction Dynamics by Laser." Thanks are also due to Professor A. D. Bandrauk and Dr. S. Chelkowski for their stimulating discussions.

References

1 D. J. Tannor and S. A. Rice, Adv. Chem. Phys. **70**, 441 (1988).
2 P. Brumer and M. Shapiro, Annu. Rev. Phys. Chem. **43**, 257 (1992).
3 D. Neuhauser and H. Rabitz, Acc. Chem. Res. **26**, 496 (1993).
4 B. Kohler, J. L. Krause, F. Raksi, K. R. Wilson, V.V. Yakovlev, R. M.

Whitnell, and Y. J. Yang, Acc. Chem. Res. **28,** 133 (1995).
5 H. Kawashima, M. M. Wefers, and K. A. Nelson, Annu. Rev. Phys. Chem. **46**, 627 (1995).
6 H. Kawashima and K. A. Nelson, J. Chem. Phys. **100**, 6160 (1994).
7 A. P. Heberle, J. J. Baumberg, and K. Köhler, Phys. Rev. Lett. **75**, 2598 (1995).
8 S. Shi, A. Woody, and H. Rabitz, J. Chem. Phys. **88**, 6870 (1988).
9 R. Kosloff, S. A. Rice, P. Gaspard, S. Tersigni, and D. J. Tannor, Chem. Phys. **139,** 201 (1989).
10 W. Jakubetz, J. Manz, and H.-J. Schreiber, Chem. Phys. Lett. **165**, 100 (1990).
11 S. Shi and H. Rabitz, J. Chem. Phys. **92**, 364 (1990).
12 P. Gross, D. Neuhauser, and H. Rabitz, J. Chem. Phys. **96**, 2834 (1992).
13 P. Gross, D. Neuhauser, and H. Rabitz, J. Chem. Phys. **98**, 4557 (1993).
14 L. Shen and H. Rabitz, J. Chem. Phys. **100**, 4811 (1994).
15 M. Sugawara and Y. Fujimura, J. Chem. Phys. **100**, 5646 (1994).
16 M. Sugawara and Y. Fujimura, J. Chem. Phys. **101**, 6586 (1994).
17 J. L. Krause, R. M. Whitnell, K. R . Wilson, Y. Yan, and S. Mukamel, J. Chem. Phys. **99**, 6562 (1993).
18 Y. Yan, R. E. Gillian, R. M. Whitnell, K. R . Wilson, and S. Mukamel, J. Phys. Chem. **97**, 2320 (1993).
19 L. Shen, S. Shi, and H. Rabitz, J. Phys. Chem. **97**, 8874 (1993).
20 L. Shen, S. Shi, and H. Rabitz, J. Phys. Chem. **97**, 12114 (1993).
21 V. Dubov and H. Rabitz, Chem. Phys. Letters **235**, 309 (1995).
22 J. Botina, H. Rabitz, and N. Rahman, **102**, 226 (1995).
23 V. Dubov and H. Rabitz, J. Chem. Phys. **103**, 8412 (1995).
24 J. Botina, H. Rabitz, and N. Rahman, **104**, 4031 (1996).
25 Y. Watanabe, H. Umeda, Y. Ohtsuki, H. Kono, and Y. Fujimura, Chem. Phys., **217**, 317 (1997).
26 M. Sugawara and Y. Fujimura, Chem. Phys. **196**, 113 (1995).
27 Y. Ohtsuki, Y. Yawata, H. Kono, and Y. Fujimura (in preparation).
28 H. Umeda, D. Sc. thesis, Tohoku University (1997).
29 J. R. Stine and D. W. Noid, Opt. Commun. **31**, 161 (1979).
30 D. W. Noid and J. R. Stine, Chem. Phys. Lett. **65**, 153 (1979).
31 K. M. Christoffel and J. M. Bowman, J. Phys. Chem. **85**, 259 (1981) .
32 R. B. Walker and R. K. Preston, J. Chem. Phys. **67**, 2017 (1977) .
33 C. Leforestier and R. E. Wyatt, J. Chem. Phys. **78**, 2334 (1983) .
34 A. Guldberg and G. D. Billing, Chem. Phys. Lett. **186**, 229 (1991).
35 S. Chelkowski, A. D. Bandrauk, and P. B. Corcum, Phys. Rev. Lett. **65**, 2355 (1990) .
36 S. Chelkowski and A. D. Bandrauk, Chem. Phys. Lett. **186**, 264 (1991).
37 J. E. Combariza, S. Görtler, B. Just, and J. Manz, Chem. Phys. Lett. **195**, 393 (1992) .
38 B. Just, J. Manz, and G. K. Paramonov, Chem. Phys. Lett.**193**, 429 (1992).
39 S. Chelkowski and A. D. Bandrauk, J. Chem. Phys. **99**, 4279 (1993) .
40 J. Jansky, P. Adam, An, V. Vinogradov, and T. Kobayashi, Chem. Phys. Lett. **213**, 368 (1993).
41 T. Taneichi, T. Kobayashi, Y. Ohtsuki, and Y. Fujimura, Chem. Phys. Lett. **231**, 50 (1994).
42 H. Umeda and Y. Fujimura, J. Chin. Chem. Soc. **42**, 353 (1995).

43 H. Umeda, M. Sugawara, Y. Fujimura, and S. Koseki, Chem. Phys. Lett. **229**, 233 (1994).
44 C. Dion, S. Chelkowski, A. D. Bandrauk, H. Umeda, and Y. Fujimura, J. Chem. Phys. **105**, 9083 (1996).
45 S. K. Gray, W. H. Miller, Y. Yamaguchi, and H. F. Schaefer III, J. Chem. Phys. **73**, 2733 (1980).
46 B. A. Waite, J. Phys. Chem. **88**, 5067 (1984).
47 Z. Bacic, R. B. Gerber, and M. A. Ratner, J. Phys. Chem. **90**, 3606 (1988).
48 J. Manz and G. K. Paramonov, J. Phys. Chem. **97**, 12625 (1993).
49 J. E. Combariza, S. Görtler, B. Just, and J. Manz, Chem. Phys. Lett, **195**, 393 (1992).
50 D. J. Tannor and S. A. Rice, J. Chem. Phys. **83**, 5013 (1985).
51 D. J. Tannor, R. Kosloff, and S. A. Rice, J. Chem. Phys. **85**, 5805 (1985). For pump-dump studies from an alternative point of view, see T. Seideman, M. Shapiro and P. Brumer, J. Chem. Phys. **90**, 7132 (1989). X. Jiang, M. Shapiro, and P. Brumer, J. Chem. Phys. **104**, 607 (1996).
52 W. T. Pollard, C. H. B. Cruz, C. V. Shank, and R. A. Mathiees, J. Chem. Phys. **90**, 199 (1989).
53 G. Cerullo, C. J. Bardeen, Q. Wang, and C.V. Shank, Chem. Phys. Lett. **262**, 362 (1996).
54 M. Messina, K. R. Wilson and J. L. Krause, J. Chem. Phys. **104**, 173 (1996).
55 H. Tang, R. Kosloff and S. A. Rice, J. Chem. Phys. **104**, 5457 (1996).

Dynamics

Photodissociation Dynamics of Chlorinated Benzene Derivatives

Teijiro Ichimura

Department of Chemistry, Tokyo Institute of Technology, Meguro, Tokyo 152-8551, Japan,
E-mail: tichimur@chem.titech.ac.jp

Abstract. Dynamics of photoexcited chrorinated benzene derivatives have been described. 193 nm photolyses of chlorobenzene, pentafluorochlorobenzene, dichlorobenzenes and chlorotoluenes, and a 248 nm photolysis of chlorobenzene have been investigated by photofragment spectroscopy, and the resonance-enhanced two-photon ionization and laser-induced fluorescence spectra were measured for the S_1 chlorotoluenes.

1 Introduction

A prototype aromatic molecule, benzene, has been extensively studied with various techniques [1]. Of particular interest is the so-called third channel process, which starts suddenly at ca. 2800 cm^{-1} above the origin of the S_1 state. In contrast, benzene derivatives containing heavy atoms such as chlorine or bromine may have rapid photodissociation channels, since optically excited levels can interact with dissociative states.

Photodecomposition mechanisms of simple molecules and also fairly large molecules have been successfully investigated by photofragment spectroscopy. Statistical distributions of the photofragment translational energy for the molecules clarify the dissociation mechanism. Photofragment spectroscopy for small polyatomic molecules of alkyl chlorides such as CH_3Cl [2], C_2H_5Cl [3] and *iso*-C_3H_7Cl [3] have revealed that a dominant primary process for the photodecomposition is the photodissociation due to a direct excitation of molecules into the repulsive (σ^*, n) or (σ^*, π) state, whose photofragments show a distinct anisotropic distribution. For chlorinated ethylene derivatives such as vinylchloride [4] and dichloroethylene [4] fast photodecomposition is suggested to occur through both (σ^*, n) and (σ^*, π) states crossed and internally converted from the optically excited (π^*, π) state at 193 nm. In photolyses of acetylene derivatives such as CH_3CCCl [3] and CH_3CCBr [3] the C - X (X = Cl or Br) bond dissociation also has been suggested to occur from both $\sigma^* \leftarrow \sigma$ and $\pi^* \leftarrow \pi$ transitions. These results obtained by photofragment spectroscopy, indicate that as the molecular size of chlorinated polyatomic increases, predissociation becomes more dominant over direct dissociation. C_6H_5Cl also seems to have a fast photodissociation channel as well as slow decomposition pathways from vibrationally excited levels of both the low lying triplet and singlet states. Pioneering work on the photodecomposition of aryl halides at 193 and 249 nm by Freedman et al. [5] reported that the photodecomposition of C_6H_5Cl at 193 nm occurs from the lowest excited singlet state via a quick internal conversion of the S_3 state brought about by 193 nm light, though the presence of the fast dissociation channel was not clarified.

We have further explored the photodissociation mechanism of chlorinated benzene derivatives, such as chlorobenzene [6], pentafluorochlorobenzene [6], dichlorobenzenes

[7], and chlorotoluenes [7]. Pentafluorines, a methyl group, or an additional chlorine atom is introduced into the chlorobenzene molecule to investigate their substituent effects by analyses of time-of-flight (TOF) distributions of the photofragment, Cl atoms. Angular distribution measurement of photofragments also have been carried out, since it is essential to prove a fast photodissociation.

Photofragment spectroscopy is a very powerful method to investigate the dissociation process if the concentration of photofragment is high enough to be detected. This means that the S/N ratio is quite low for the small absorption cross section. Therefore, the S_1 states of chlorotoluene vapor [8 - 11] have been studied under photostationary conditions, which revealed that the excited molecules undergo intersystem crossing within a few nanoseconds to triplet states, where decomposition and vibrational relaxation to the stable triplet state competitively occur. Absorption spectra of chlorotoluene vapor are congested with vibrationally hot and rotational bands. Accordingly, the laser-induced fluorescence technique is applied to clarify photochemical dynamics in the S_1 state of jet-cooled chlorotoluene molecules. Observed fluorescence excitation spectra exhibit a feature characteristic of the methyl internal rotation and the C - Cl vibrational modes [12, 13]. The comparison of these spectra with the resonance-enhanced two-photon ionization (RE2PI) spectra reveals the mode selectivity of the nonradiative process (presumably intersystem crossing) in each level of the S_1 state [14].

2 UV and VUV Absorption Spectra

A supersonic free jet apparatus set on the BL2A beam line of UVSOR, Okazaki, was used to measure the absorption spectrum of jet-cooled chromophore. The sample vapor, at an appropriate temperature, in He or Ar carrier gas at a total stagnation pressure of several hundred torr was expanded into a free jet chamber through a diverging conical nozzle (entrance diameter 0.38 mm, exit diameter 1.16 mm, and channel length 3 mm) attached to a fuel injector. The fuel injector was operated at a 10 Hz repetition rate with an open duration of 40 msec. The nozzle temperature was usually set at higher temperatures than that of the sample tube in order to avoid forming sample clusters. The free jet chamber was evacuated through a liquid N_2 baffle by an oil diffusion pump backed by a mechanical booster pump, and the pressure in the chamber was $\approx 5 \times 10^{-4}$ torr during a run. The vacuum chamber was isolated from the chambers of the monochromator and the last focusing mirror by a LiF window. UV and VUV light monochromatized by a 1 m Seya - Namioka type monochromator was focused on and crossed the free jet. The cross sectional profile of the light beam was 5 mm long along the center line of the jet stream and 1 mm wide. The distance between the nozzle tip and the center of the focused light was 3.5 mm. The transmitted light intensity was monitored by a combination of sodium salicylate coating on the exit window and a photomultiplier tube.

Signals from the transmitted light were fed to two counting scalers, each enabled by each timing signal (gate width 30 ms), one corresponding to the free jet on and the other to the free jet off. After accumulating for 10 on - off signals at each wavelength,

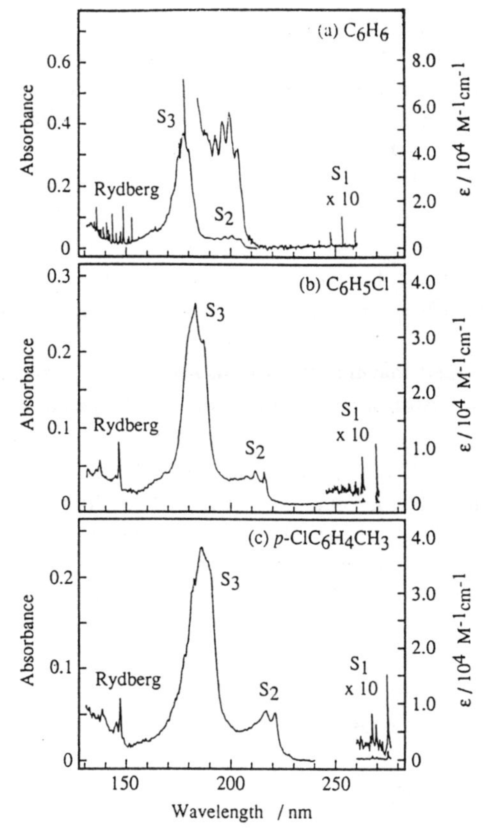

Fig. 1. Absorption spectra of benzene (**a**), chlorobenzene (**b**), and *p*-chlorotoluene (**c**) in a supersonic jet: (**a**), benzene (92 torr) seeded in He (308 torr) with nozzle temperature 323 K and spectral resolution of 0.065 nm, (**b**), chlorobenzene (120 torr) in Ar (290 torr) with nozzle temperature 367 K and spectral resolution of 0.2 nm, (**c**) *p*-chlorotoluene (100 torr) in Ar (250 torr) with nozzle temperature 373 K and spectral resolution of 0.2 nm.

the transmitted photon counts corresponding to the free jet on, I_{on}, and that to the free jet off, I_{off}, were obtained. The absorbance of a sample in a free jet at the wavelength was determined as $\log(I_{off} / I_{on})$, regardless of any fluctuations in the incident light intensity, and the absorption spectrum was obtained by scanning the wavelength of the monochromatized synchrotron radiation by an appropriate step width.

Figure 1 shows direct absorption spectra of jet-cooled benzene (a) [15], chlorobenzene (b) [16], and *p*-chlorotoluene (c) [17]. Here only the absorption spectrum (b) of chlorobenzene is described. The observed wavelength region includes the S_1 (240 - 271 nm), S_2 (225 - 195 nm), S_3 (195 - 160 nm), and Rydberg (170 - 130 nm) states. The ordinates of the figure are the absorbance, $\log(I_{off} / I_{on})$, and the molar extinction coefficient, ε. The absolute values of ε at a given wavelength λ_{exc} are determined by normalizing the observed absorbance, $\log(I_{off} / I_{on})$ to that for the 215.0 nm band of the S_2 state, where the absolute ε value is measured in the room-temperature vapor ($\varepsilon = 6.07 \times 10^3\ M^{-1}\cdot cm^{-1}$), assuming that (i) the ε values at 215.0 nm are identical in both vapor and jet conditions, and (ii) the effective product, *cl*, of the concentration *c* and the path length *l* determined at the reference wavelength (215.0 nm) can be used at other wavelengths. The former assumption is justified because of the broadness of the band even in a jet. Thus the absolute ε values of some S_1 vibronic bands have been determined: The ε value for 0 - 0 transition is found to be 730 $M^{-1}\cdot cm^{-1}$ for a spectral resolution of 0.2 nm. The minimum value of 0 - 0 transition at the S_2 state is determined as 150 $M^{-1}\cdot cm^{-1}$ corresponding to 0.3% standard deviation in the $(1 - I_{off} / I_{on})$ value. That value is the minimum observable absorption in this study due to the low concentration of the sample in the jet. For the strong absorption at the broad maximum of the S_3 state (185.0 nm), saturation effects should be taken into account under the present conditions.

It should be emphasized that the absorption spectrum measured under the present conditions is free from absorption due to chlorobenzene clusters $(C_6H_5Cl)_n$ ($n \geq 2$). In contrast, when the Ar carrier gas pressure is increased to 760 torr, broad absorption bands are found to appear around 200 nm in addition to the bands of isolated chlorobenzene. This new feature may be attributed to chlorobenzene clusters rather than $C_6H_5Cl\cdot Ar_n$ ($n \geq 1$) clusters, since the absorption spectra of jet-cooled benzene at a stagnation pressure of 760 torr with different carrier gases were essentially the same, indicating that the major clusters formed are benzene clusters rather than $C_6H_6\cdot Rg_n$ (Rg = He, Ar, Xe).

3 Photofragment Spectroscopy

3.1 Experiment and Analysis

TOF Spectrum Measurement. The apparatus for the photofragment spectrometer has been described in detail elsewhere [4]. A supersonic molecular beam seeded in 1.5 atm Ar is irradiated by 193 nm ArF excimer laser pulses (≈30000 shots) or 248 nm KrF excimer laser pulses (150000 shots) at a frequency of ≈5 Hz. The focused laser light was crossed at right angles with both the pulsed molecular beam and a quadrupole mass (Q-mass) filter axis. For an angular distribution measurement, a pile-of-plates polarizer consisting of ten suprasil quartz plates was used to linearly polarize the excimer laser light. The plane of polarization was rotated with respect to the detector axis. The time-of-flight (TOF) spectrum is generated by synchronously gating a multichannel scaler with the laser firing pulse and accumulating photofragment signals as a function of time after the laser pulse through a Q-mass filter. The Q-mass, operating at unit mass resolution, is separated from the reaction chamber and pumped differentially by a turbomolecular pump and an ion pump. The TOF spectrum is then converted by using a suitable Jacobian factor [18] to a center-of-mass total translational energy distribution $[P(E_T)]$ after subtracting the drift time in the Q-mass tube. The flight path was 16 cm and the time resolution was 1 μs. The detector is situated perpendicular to the molecular beam axis. This configuration probably reduces the detection probability of the photofragments with kinetic energies below 3 kcal/mol.

Samples of chlorobenzene, *o*-, *m*-, and *p*-chlorotoluene, and *o*-, *m*-, and *p*-dichlorobenzene (stated purity, 99%) were obtained from Tokyo Kasei and used without further purification. A sample of pentafluorochlorobenzene (stated purity, 99%) obtained from PCR Research Chemicals Inc. was also used without further purification. The sample was heated to ≈400 K before injection into the beam chamber.

Translational Energy Distributions. TOF spectra of the Cl fragments, $P(t)$, were converted to center-of-mass translational energy distributions, $P(E_T)$, by using the following equations. For the slowest fragments a Maxwell - Boltzmann distribution is assumed as in (1).

$$P(E_T) \propto E_T^{1/2} \exp(-E_T/kT). \tag{1}$$

For the faster fragments, the following Gaussian function [18] distributed around t_0 for a given vibrational state 0 is assumed:

$$P(t) \propto t \exp\{-[(t - t_0)/\Delta t]^2\}, \tag{2}$$

where Δt indicates the width. Here, not only the apparatus shape factor but also the width originating in the narrow vibrational state distribution around the central state 0 is taken into account. This equation can lead to the following equation for $P(t)$ [19]:

$$P(t) \propto t^3 (l^2/t^2 + c^2)^{1/2} \exp[-(t-t_0)^2/A], \tag{3}$$

where l and c are the flight length (16 cm) and the mean velocity of the molecule in a molecular beam, respectively, and the correction of the velocity dependence of the detector ionization efficiency is included. The constant A is used as an adjusting parameter. Since the fragments are detected in a perpendicular direction to the molecular beam axis, the total translational energy E_T is expressed as follows [4]:

$$E_T = (m_1/m_2)(m_1 + m_2)\{l^2/(t - d)^2 + c^2\}, \tag{4}$$

where m stands for the mass number and the subscripts 1 and 2 refer to the detected fragment and the counterpart, respectively, d is the drift time in the Q-mass tube. In order to correct for the drift time, the conversion calculation using (1) and (3) was carried out with the value of the corrected time.

The probability of each dissociation process can be estimated from the numerical integration of each simulated distribution function, using (1) and (3), which is listed in Table 1. The total dissociation probability is assumed to be unity. The value of the dissociation energy (D_0) is determined from the literature [20]. The available energy (E_{avl}) for photofragments is calculated from the following relation:

$$E_{avl} = h\nu - D_0 + 2RT. \tag{5}$$

The value of the average translational energy (E_T) is taken from the average of each simulated distribution.

Angular Distribution. The laboratory angular distribution $W(\Theta)$ of fragments is analyzed using the following relation [21]:

$$W(\Theta) = A\{1 + 2BP_2[\cos(\Theta - \Theta_0)]\}, \tag{6}$$

where A is a normalization factor, B the shape factor, P_2 the second-degree Legendre polynomial, and Θ_0 a phase shift caused by the transformation from center-of-mass to laboratory coordinates. The smooth curves in Fig. 10 are the least-square fittings to (6). Uncertainties in the values of B and Θ_0 are within the standard deviation of the least-square fitting.

The shape factor B can be expressed by the following relation,

$$B = P_2(\chi)P_2(\beta), \tag{7}$$

where χ is the angle through which the molecule rotates during the photodissociation and β is the angle between the transition dipole moment and the initial direction of fragment recoil. The angle χ can be calculated from the classical equation as shown by Zare and Hershbach [22]. Axial recoil ($\chi = 0°$) is approached when the transition from the initially pumped state to the dissociative is expected in a time scale less than a rotational period. In the present experiment a supersonic molecular beam is used to generate rotationally cooled molecules. Accordingly, the effect of rotation for the high-energy component can be negligible. If one assumes $\chi \approx 0$ for a rotationally frozen limit, the term B is expressed by the following simple equation:

$$B = P_2(\beta). \tag{8}$$

The observed angular distributions of the chlorine TOF signals were analyzed using (8).

3.2 TOF Spectra and Translational Energy Distributions

Chlorobenzene. Figure 2 left (L) (a) shows the TOF signal of Cl (*m/e* = 35) for the photolysis of chlorobenzene (C_6H_5Cl) molecular beams at 193 nm. It appears that the TOF spectrum in Fig. 2L(a) consists of two components: the fast and narrow distribution and the slow and broad one. The fast-moving component has a peak at around 100 μs after the laser pulse, and the slow one at 120 μs is as in Fig. 2L(a) (uncorrected for the drift time in the mass filter). Reduction of the intensity of laser pulse from 100 to less than 30 mJ/pulse did not give any change of the TOF profile. Thus two photon absorption is negligible under the present experimental conditions.

Two kinetic energy components clearly appear in the translational energy distribution $P(E_T)$ of photofragments in Fig.2 right(R) (a). The low-energy one corresponding to the slow-moving fragment in Fig. 2(a) sharply drops at around 5 kcal/mol, and the higher-energy one is distributed from ≈7 up to 50 kcal/mol and gradually decreases for higher E_T values. The energy difference of the two distributions is approximately 20 kcal/mol, which is far beyond the spin-orbit splitting (≈2.5 kcal/mol [23]) between $Cl(^2P_{3/2})$ and $Cl^*(^2P_{1/2})$. Thus, these two distributions must arise from different excited species. The distribution in Fig. 3(a) was simulated using (1) and (3). First, the low-energy component is expressed by (1), and then the residual higher-energy component after subtraction of the low energy one from the original distribution is expressed by (3). The higher-energy distribution, however, cannot be expressed by a single Gaussian function, but by the summation of two functions. The simulated peak value of E_T for the higher-energy distribution is approximately 30 kcal/mol and that for the middle energy one is ≈13 kcal/mol, as shown by the broken lines in Fig. 2R(a). Accordingly, the whole $P(E_T)$ distribution in Fig. 2R(a) can be expressed by the superposition of three different energy distributions.

Pentafluorochlorobenzene and Chlorobenzene at 248 nm Photolysis. Figure 2L(b) shows the TOF signal of Cl for the photolysis of C_6F_5Cl molecular beams at 193 nm. The TOF profile exhibits a prompt rise at early time, indicating the presence of a fast component, and a gradual decrease at long delay times, corresponding to a slow component. The total translational energy distribution, $P(E_T)$ of photofragments in Fig. 2R(b) has a peak at 3 kcal/mol and a small hump at 13 kcal/mol, which is simulated by the sum of one Boltzmann and one Gaussian function. The simulated distributions in Fig. 2R(b) were well reconverted to the TOF spectrum in a previous

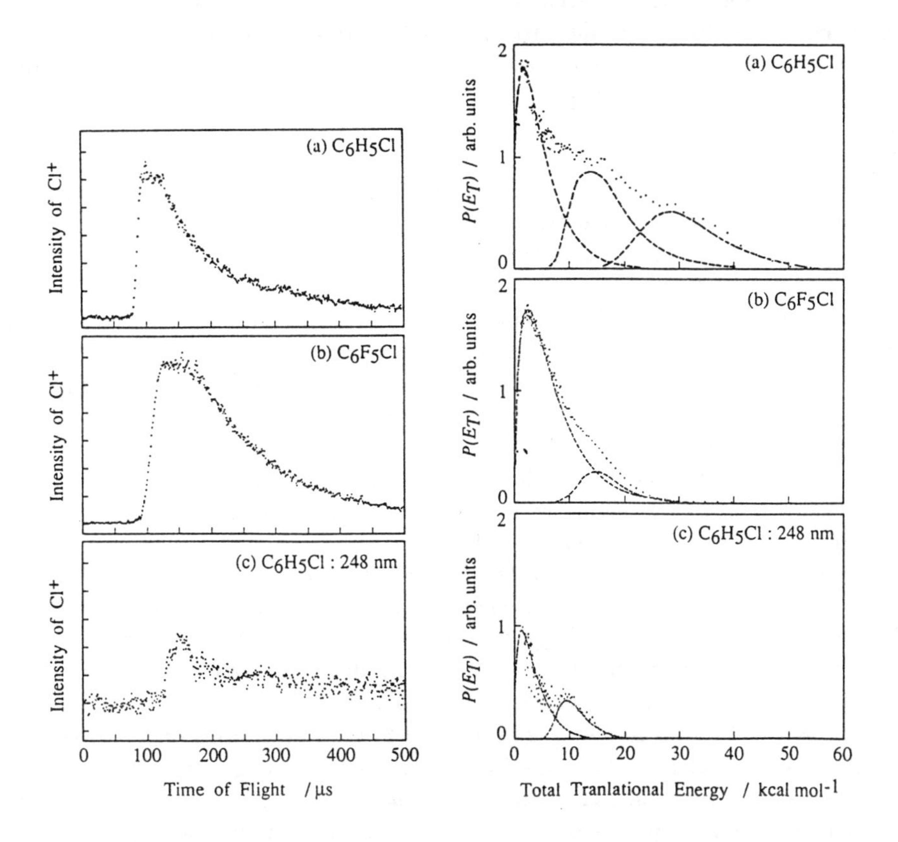

Fig. 2. *Left*: Time-of-flight distributions for Cl^+(m/e = 35) signals in the photolyses of chlorobenzene (**a**), and pentafluorochlorobenzene (**b**) at 193 nm, and chlorobenzene at 248 nm (**c**). *Right*: Total center-of-mass translational energy distribution $P(E_T)$ obtained from Cl photofragments in the photolyses of chlorobenzene (**a**), and pentafluorochlorobenzene (**b**) at 193 nm, and chlorobenzene at 248 nm (**c**). Broken lines show the best-fitting line assuming two Gaussian functions (3) for higher-energy components in (**a**), and one Gaussian function in (**b**) and (**c**) for the higher-energy component, with one Maxwell - Boltzmann function (1) for the low energy.

paper [24], confirming the superposition of the two components. The $P(E_T)$ distribution at translational energies below 3 kcal/mol may be slightly distorted. The major kinetic energy peak around 3 kcal/mol, however, suggests the presence of a dissociation channel that leads to the low-kinetic-energy photofragments. This feature is in contrast with that (Fig. 2(a)) seen in the energy distributions of chlorobenzene photolysis at the same excitation wavelength (193 nm) as this case.

In Fig. 2L(c) is shown the TOF spectrum of Cl photofragments obtained in the photolysis of chlorobenzene at 248 nm. The absorption coefficient of chlorobenzene vapor at 248 nm is only 200 $M^{-1}\cdot cm^{-1}$, which is about one-fifteenth of the value at 193 nm. Thus the S/N ratio in Fig. 2 (c) is not very good, although 150,000 laser shots were fired and the signal was accumulated to improve the S/N ratio. Two peaks also appear in the spectrum, and the fast Cl photofragment also exhibits a narrow time distribution, which is well separated from the slow one. The translational energy distribution $P(E_T)$ shown in Fig. 2R(c) is deduced from the TOF spectrum in Fig. 2L(c). Two apparent peaks in $P(E_T)$ with peak E_T values at 3 and 9 kcal/mol are observed, and the profile of $P(E_T)$ can be well simulated by the summation of one Boltzmann function (1) and one Gaussian function (3). In analogy to the analytical treatment of chlorobenzene photolysis at 193 nm, the partitioning of available energy and the probability of the two dissociation channels were obtained.

Chlorotoluenes and Dichlorobenzenes. TOF signals of Cl (m/e = 35) for 193 nm photolyses of chlorotoluenes and dichlorobenzenes are shown in Fig. 3. In Fig. 3L the sharp rise observed at around 100 μs corresponds to the fast component, and the second peak, centered at around 140 μs indicates the presence of slow fragments. The gradual decrease at long delay times indeed shows the existence of the slowest component, though the peak is not clearly seen in the figure due to the fairly weak and broad distribution of the component. The relative abundance of each component seems to be different among three isomers of chlorotoluene.

For dichlorobenzenes as shown in Fig. 3R, the TOF profiles are similar to those in Fig. 3 L. The slowest component, however, is abundant in comparison with that for chlorotoluene, since long delayed signals are observed up to 500 μs for dichlorobenzene, but only 300 μs for chlorotoluene (Fig. 4L). These differences are clearly seen when TOF spectra are converted to the kinetic energy distribution.

TOF spectra observed in Fig. 3 are converted to the total translational energy (E_T) by using (4). The obtained $P(E_T)$ distributions are simulated by superposition of three components whose distribution functions are (1) and (3). These results for chlorotoluenes and dichlorobenzenes are shown in Fig. 4. Each component for the best simulated distribution is numerically integrated to estimate the contribution of each dissociation process. Under the assumption of unit quantum yield for the 193 nm photodissociation, the probability of each process can be determined. These values for chlorotoluenes and dichlorobenzenes are summarized in Table 1, together with those for chlorobenzene etc., for the purpose of comparison.

The kinetic energy distributions in the chlorotoluene photodissociation (Fig. 4L) differ from those of chlorobenzene (Fig. 2R(a)) and also dichlorobenzenes (Fig. 4R). The medium energy fragment emerges, and the lowest-energy fragment submerges. This is most characteristic of *p*-chlorotoluene. The value of E_T is higher by 3 - 4 kcal/mol than those of the *m*- and *o*-isomers. On the other hand, the E_T value at the peak of the highest energy distribution of the *p*-isomer is slightly small, and its probability is also lower than those of the *m*- and *o*-isomers (Table 1). This trend is also seen for *p*-dichlorobenzene.

The E_T distributions and probabilities for dichlorobenzenes in Table 1 are generally in agreement with those for chlorobenzene. Some features observed for dichlorobenzenes are: (1) small E_T and, accordingly, small f_T values for the highest-energy fragment; (2) the probability for the medium-energy fragment is slightly larger than that of chlorobenzene; (3) the ortho-isomer gives a slightly lower E_T value for the medium-energy fragment and a substantially lower E_T value for the highest-energy fragment, with higher probability than the *m*- and *p*-isomers; (4) the *m*-isomer dissociates in a similar fashion to chlorobenzene; and (5) the substantially smaller E_T and probability for the highest-energy fragment of the *p*-isomer than the *m*-isomer.

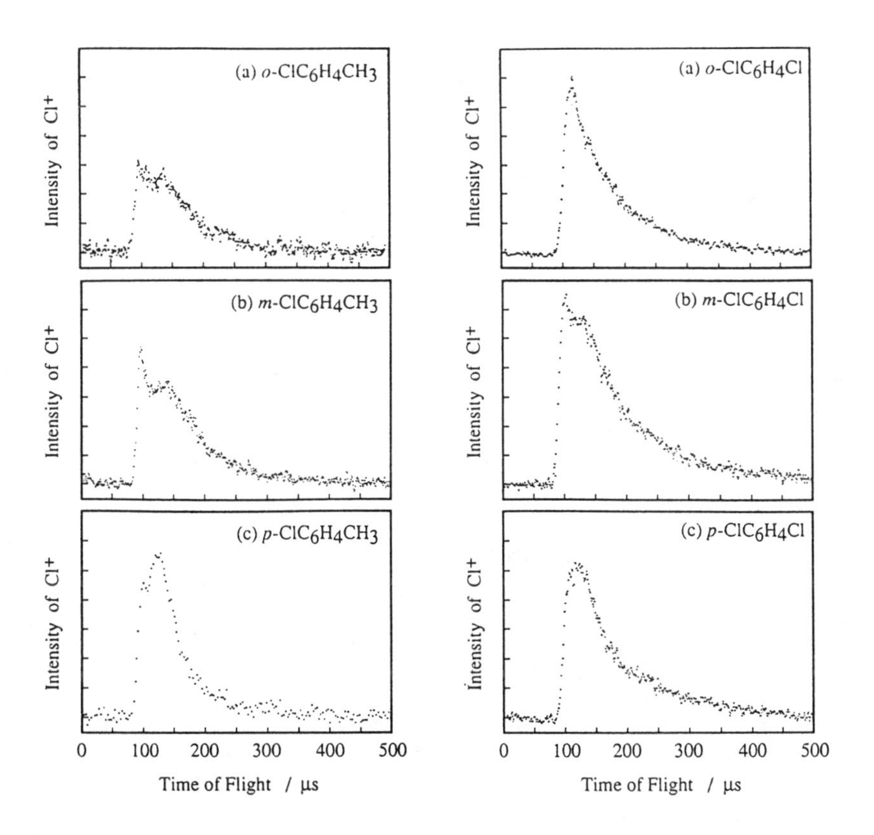

Fig. 3. Time-of-flight distributions for Cl^+(m/e = 35) signals for the 193 nm photolyses of chlorotoluenes (*left*) and dichlorobenzenes (*right*). Signals were accumulated for ≈50,000 laser shots.

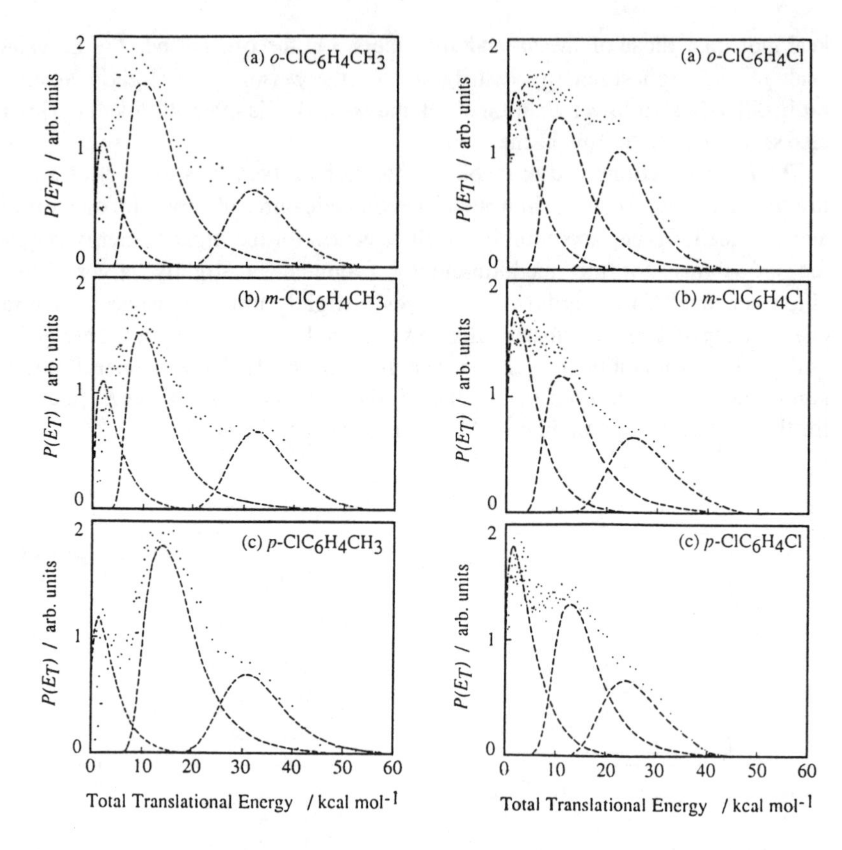

Fig. 4. Total center-of-mass translational energy distribution $P(E_T)$ for chlorotoluenes (*Left*) obtained from Cl photofragments in the 193 nm photolyses of chlorotoluenes (Fig. 3L) and for dichlorobenzenes (*Right*) obtained from Fig. 3R. Broken lines show the best-fitting line assuming two Gaussian functions ((3)) for higher-energy components, and one Maxwell - Boltzmann function ((1)) for the low energy.

3.3 Angular Distributions of the Cl Fragments

When the dissociation mechanism involves several different channels of different delay times, the anisotropy measurement of each photofragment is necessary to clarify the dissociation mechanism. Thus fragment energy distributions have to be measured at various laboratory angles against the electric vector direction of the linearly polarized light. Qualitative experiments have been carried out on parallel and perpendicular excitation of *p*-dichlorobenzene [25]. The TOF spectra for *p*-chlorotoluene and *p*-dichlorobenzene were observed at various laboratory angles in order to get quantitative information on the dissociation mechanism, consisting of three different dissociation channels.

Figure 5L shows the TOF distribution of Cl at m/e = 35 for the photolysis of *p*-chlorotoluene ($ClC_6H_4CH_3$) molecular beams with the linearly polarized light at 193

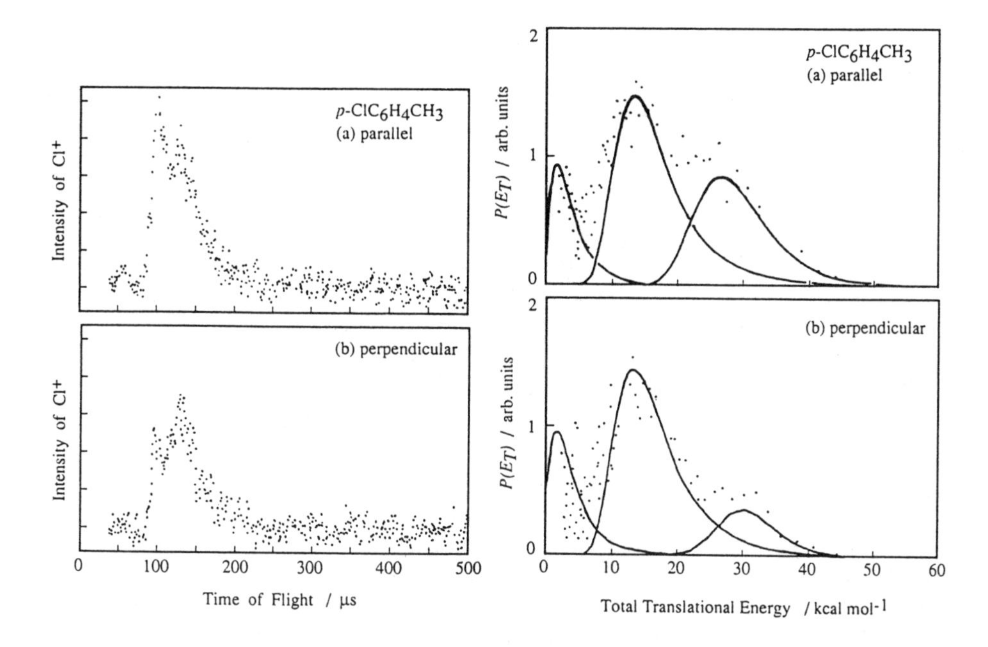

Fig. 5. *Left*:Two typical examples of time-of-flight distributions for Cl^+ (m/e = 35) signals resulting from the linearly polarized 193 nm excitation of a *p*-chlorotoluene molecular beam: **(a)** the electric vector of the polarized light is parallel to the detector axis ($\Theta = 0°$ excitation), and **(b)** the electric vector is perpendicular to the detector axis ($\Theta = 90°$ excitation). Signals for each laboratory angle were accumulated for about 30,000 laser shots. *Right*: Total center-of-mass translational energy distributions $P(E_T)$ obtained from Cl photofragments formed by the linearly polarized 193 nm excitation of a *p*-chlorotoluene molecular beam with 0° excitation (**a**) and 90° excitation (**b**) shown in the left-hand figure. $P(E_T)$ is simulated by the superposition of three components using a Maxwell-Boltzmann function and two Gaussian functions. The best-fitted simulated functions are drawn by solid lines. The highest energy distribution clearly demonstrates an anisotropy, whose integrated intensity is plotted in Fig. 6(a). See text for details.

nm. Only two typical examples are shown for parallel (0°) and perpendicular (90°) excitation in the figure. The polarized plane of the excitation laser beam was set (a) perpendicular to the molecular beam axis and (b) parallel to the molecular beam axis in Fig. 5L. Since the detector axis is perpendicular to the molecular beam axis, the electric vector of the light in Fig. 5L(a) is parallel to the laboratory recoiling axis of the detector. The intensity of the fast component observed at around 100 μs delay for parallel excitation (Fig. 5L(a)) is much stronger than that for perpendicular excitation (Fig. 5L(b)). The TOF profile observed with the perpendicular excitation in Fig. 5L(b) seems to be very similar to that observed for the nonpolarized laser beam excitation (Fig. 3L(c)).

The total translational energy distribution, $P(E_T)$, of fragments derived from TOF spectra in Fig. 5L is shown in Fig. 5R. The simulated distribution with E_T = 28

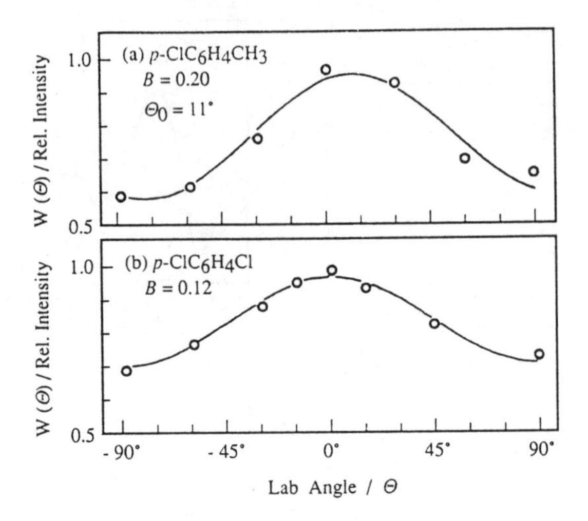

Fig. 6a, b. Angular distributions of the Cl fragments (the highest-energy component in Fig. 5R). The circles are the data points for the time-of-flight signals from 193 nm photolyses of *p*-chlorotoluene (**a**) and *p*-dichlorobenzene (**b**). Examples of $P(E_T)$ distributions observed for the laboratory angle $\Theta = 0°$ and $\Theta = 90°$ in (**a**) are shown in Fig. 5R. The solid lines show the best fitted simulation curves using (6). See the section on angular distribution in the text for the parameter definition.

kcal/mol corresponds to the fast fragment in Fig. 5L. Integrated areas for the lower energy distributions with $E_T = 3$ and 13 kcal/mol in Fig. 5R were almost constant for both perpendicular and parallel excitations. These results indicate that these slow fragments observed in Fig. 5L are isotropic for variation of the excitation electric vector of the polarized light.

The fast fragment, however, is anisotropic, and the integrated intensity of the fast component is plotted against the laboratory angle in Fig. 6. The angular distribution of the fast fragment for *p*-dichlorobenzene is also shown in Fig. 6(b). Slow fragments for *p*-dichlorobenzene also resulted in isotropic distributions. Anisotropic angular distributions in Fig. 10 were simulated by the least-square fitting method using (6). The best-simulated curves are drawn by the solid line. The angular distribution of signal intensity has a phase shift (Θ_0) caused by the transformation from center-of-mass to laboratory coordinates. Because of the rather large mass difference of the two fragments (Cl and C_6H_4Cl), this phase shift was negligibly small in the present analysis of *p*-dichlorobenzene. In the case of *p*-chlorotoluene the mass difference between Cl and $C_6H_4CH_3$ is relatively small, and the phase shift was observed to be 11°.

Best-fitted curves have led to the value of the shape factor (B) in (6). The B value (0.20) for *p*-chlorotoluene is larger than that (0.12) for *p*-dichlorobenzene. Putting these B values into (8) led to the estimate of β values under the assumption of a rotationally frozen limit. The derived β values for *p*-chlorotoluene and *p*-dichlorobenzene were 47.0° and 50.0°, respectively.

3.4 Dissociation Mechanism

Dissociation processes of chlorobenzene, chlorotoluenes, and dichlorobenzenes under supersonic beam conditions can be clarified from the present results mentioned above. The quantum yield, however, can't be directly determined, since photon flux absorbed by a molecule and its concentration in the beam are hard to estimate. The dissociation mechanism in the vapor phase may be similar to that in the supersonic beam. Thus the quantum yield in the gas phase is cited here. The assumption of unit probability for the dissociation of chlorobenzene is likely to be feasible, since our previous stationary photolysis measurements revealed that the photodecomposition quantum yields of chlorobenzene [26] in the presence of ethane at 184.9 nm and 206.2 nm are 1.0 and 0.7, respectively. These values might have included nominal experimental errors of 20%. The presence of a foreign gas (ethane) has given some ambiguity to estimating the quantum yield at zero pressure, but even the pressure effect should play a converse role to lead to a lower value for collision-free conditions. Therefore, the approximation of a unit quantum yield for a 193 nm photolysis of a molecular chlorobenzene beam seems to be valid. Our previous results of vapor-phase photolysis experiments [9, 27] also give evidence of a unit yield for chlorotoluenes and dichlorobenzenes. The value of the probability in Table 1, therefore, can be regarded as photodecomposition quantum yield (Φ).

Fast Photodissociation (First Channel). The highest E_T value of 24 - 33 kcal/mol obtained in Table 1 may be explained by a similar mechanism to that of alkyl and alkene halides. The rather fast and narrow photofragment energy distribution may be compared to those of photodissociation of the C - Cl bond in alkyl chlorides [2, 3]. In the photolyses of alkyl chlorides at 193 nm, the molecules are excited into the continuum of a repulsive state by the (σ^*, np) transition [2] and immediately undergo dissociation to fragments within a time scale of molecular vibration. Peak energies in the E_T distribution of photofragments obtained in the photolyses of CH_3Cl, C_2H_5Cl and i-C_3H_7Cl were 57, 36, and 35 kcal/mol, respectively [3]. The narrow nonstatistical distribution of E_T values was successfully explained by a model calculation of a Poisson distribution [3]. The narrow E_T distribution with peak energy at 30 kcal/mol observed for chlorobenzene photolysis at 193 nm may be explained by a similar mechanism.

The first absorption bands of alkyl chlorides are observed at ≈200 nm, which corresponds to the transition $\sigma^* \leftarrow$ np within the valence shell, where np is the outermost lone-pair p orbital of the chlorine atom, and σ^* is the antibonding C - Cl molecular orbital [28]. The extinction coefficient of the $\sigma^* \leftarrow$ n transition of CH_3Cl, however, is much smaller than that of the $\pi^* \leftarrow \pi$ transition of chlorobenzene at 193 nm by two orders of magnitude. Although the extinction coefficient of the $\sigma^* \leftarrow$ n transition in chlorobenzene is not known, the irradiation of chlorobenzene at 193 nm should result predominantly in the (π^*, π) state, and hence the transition to the repulsive (σ^*, n) state is minor. However, this cannot eliminate the contribution of the photodissociation via the repulsive (σ^*, n) state for the high E_T component. Therefore, the high translational energy peak observed for chlorobenzene is attributed to the dissociation through an excitation to a repulsive state and/or through a fast predissociation competing with the intramolecular vibrational relaxation or internal conversion from the initially excited state. The fast primary process (the first channel) for chlorobenzene, therefore, may be represented as follows:

$$C_6H_5Cl + h\nu \rightarrow C_6H_5Cl(\pi^*, \pi) \rightarrow C_6H_5Cl(\sigma^*, \pi) \rightarrow C_6H_5 + Cl(Cl^*).$$
$$\rightarrow C_6H_5Cl(\sigma^*, n) \rightarrow C_6H_5 + Cl(Cl^*).$$

This is very similar to the photodecomposition of vinylchloride and dichloroethylene at 193 nm [4]. The rapid photodissociation occurs through the (σ^*, n) dissociative state due to level crossing from the optically excited (π^*, π) state and also the internal conversion to the low-lying (σ^*, π) state competitively takes place. According to an *ab initio* calculation, the two (σ^*, π) states are localized in the low-energy region, and the lowest (σ^*, π) state is low enough to be the lowest excited state at the ground-state geometry. The internal conversion from the optically excited (π^*, π) state to the (σ^*, π) state can efficiently induce quick predissociation and/or the formation of highly vibrationally excited levels of the ground state.

This intramolecular relaxation mechanism is also operative for chlorobenzene photolysis. The triplet (σ^*, π) state of chlorobenzene is located energetically close to the triplet (π^*, π) state, which is estimated from an *ab initio* calculation by Nagaoka et al. [29]. Thus the singlet (σ^*, π) state should be lower than the singlet (π^*, π) state, since the energy splitting between the singlet and the triplet (σ^*, π) states is in general smaller than that for (π^*, π) states [30]. Therefore, chlorobenzene molecules excited to the S_3 (π^*, π) state by 193 nm irradiation can be quickly converted to the (σ^*, π) state, giving rise to fast predissociation.

The formation of $Cl^*(^2P_{1/2})$ seems to be minor. The $Cl^*(^2P_{1/2})/Cl(^2P_{3/2})$ ratio at 212.6 nm photolysis of chlorobenzene was determined to be of order 0.1 by S. Satyapal et al. [31]. They measured the ratio by VUV laser-induced fluorescence and found that the Cl*/Cl ratio is small when the Cl atom is bound directly to an aromatic ring. Chlorobenzene is just the case, though the excitation wavelength of 193 nm is slightly shorter than 212.6 nm.

Triplet Route (Second Channel). The broad distribution has the second peak at the E_T value of 13 kcal/mol in Fig. 2R(a). The $P(E_T)$ of slower photofragments is not explained by a simple statistical model. This fact indicates that the excess energy of chlorobenzene is still not completely distributed over whole vibrational modes. The nonstatistical distribution in slower photofragments suggests that a part of the dissociation process should be faster than complete thermalization. Based on studies of other chlorobenzene derivatives [11], the most probable exit state for this nonstatistical distribution may be the vibrationally excited triplet state, presumably the (π^*, π) state, generated through fast intersystem crossing from the initial optically excited singlet (π^*, π) state.

At the translational energy of about 13 kcal/mol, a small and narrow lump of the photofragment distribution also can be seen overlapping with the main distribution in Fig. 2R(b). This corresponds to the fairly sharp rise of the TOF distribution at the short flight time in Fig. 2L(b). The distribution with peak E_T value of 13 kcal/mol in Fig. 2R(b) cannot be represented by a Maxwell - Boltzmann distribution function, but

Table 1. Average partitioning of available energy and probability of dissociation channels in the photolyses of chlorobenzene, pentafluorochlorobenzene, chlorotoluenes, and dichlorobenzenes at 193 nm or 248 nm.

Parent molecule	D_0(C - Cl)[a] (kcal/mol)	E_{avl}[b] (kcal/mol)	E_T[c] (kcal/mol)	f_T[d]	Exit state	Probability (Φ)
ClC_6H_5	97	52	5.0	0.096	S_0*	0.39
			16	0.31	T*	0.34
			30	0.58	S(σ*, π)	0.27
ClC_6F_5	97	52	6.5	0.13	S_0*	0.86
			15	0.29	T*	0.14
o-$ClC_6H_4CH_3$	97	52	4.0	0.077	S_0*	0.17
			13	0.25	T*	0.57
			32	0.62	S(σ*, π)	0.26
m-$ClC_6H_4CH_3$	97	52	4.0	0.077	S_0*	0.16
			12	0.23	T*	0.57
			32	0.62	S(σ*, π)	0.27
p-$ClC_6H_4CH_3$	97	52	4.0	0.077	S_0*	0.15 (0.13)[f]
			16	0.31	T*	0.62 (0.53)[f]
			31	0.60	S(σ*, π)	0.23 (0.34)[f]
o-ClC_6H_4Cl	97	52	5.0	0.096	S_0*	0.27
			12	0.23	T*	0.40
			24	0.46	S(σ*, π)	0.33
m-ClC_6H_4Cl	97	52	5.0	0.096	S_0*	0.32
			13	0.25	T*	0.40
			28	0.54	S(σ*, π)	0.28
p-ClC_6H_4Cl	97	52	5.0	0.096	S_0*	0.33 (0.31)[f]
			15	0.29	T*	0.41 (0.39)[f]
			27	0.52	S(σ*, π)	0.26 (0.30)[f]
ClC_6H_5[g]	97	19	3.8	0.20	S_0*	0.64
			11	0.58	T*	0.36

[a] Dissociation energies of these molecules are assumed to be the same with chlorobenzene [20].
[b] $E_{avl} = h\nu - D_0 + 2RT$.
[c] Average translational energy.
[d] Fraction of E_T in E_{avl}.
[e] Total probability is assumed to be 1.0. Probability values can be regarded as photodecomposition quantum yields(Φ).
[f] These values in parentheses were obtained by parallel excitation (Θ = 0°) of a linearly polarized laser beam to the detector.
[g] Photolysis at 248 nm.

by a Gaussian function. All of the excess energy of pentafluorochlorobenzene seems not to be distributed over all vibrational modes. Thus the most probable exit state to give the faster photofragments is attributed to the vibrationally excited triplet state.

The probability of the triplet route in 193 nm dissociation of pentafluorochlorobenzene is much smaller than that for chlorobenzene. The low probability of the dissociation through triplet levels indicates an aspect of the fluorination effect, namely deformation of the benzene ring. The ring deformation enhances the rate of internal conversion. Thus intersystem crossing from the optically excited singlet level must compete with quick internal conversion, resulting in a minor contribution of the dissociation through the triplet route. If we assume a unit quantum yield for the photodecomposition of pentafluorochlorobenzene at 193 nm, the probability (0.14) for the dissociation through triplet levels in Table 1 may be attributed to the effect of the substituent chlorine atom, since unit quantum yield of internal conversion was obtained in a 193 nm photolysis of hexafluorobenzene [32].

Hot Molecule Mechanism (Third Channel). The smallest E_T distribution with the peak value of around 3 kcal/mol in Fig. 2R may be represented by a Maxwell - Boltzmann distribution. The photodecomposition process to give the distribution should be slower than thermal randomization of the excess energy. At the present stage a fast internal conversion from S_3 to S_1 prior to dissociation is not clear, but is a probable pathway. The measurement of fluorescence decay of chlorobenzene[1] revealed that molecules in S_1 decay within one nanosecond. Dissociation slower than thermal randomization seems to occur from the vibrationally excited S_0 state (S_0*) to give vibrationally hot phenyl radicals. The formation of vibrationally hot molecules was qualitatively confirmed by another shape fitting of the distribution in Fig. 2R(a) using the expression $P(E_T) \propto [(E_{avl} - E_T)/E_{avl}]^{s-1}$, where the number ($s$) of effective vibrational modes in RRK formulation [34] is derived to be 23. The high s number reveals that most of the excitation energy is distributed to vibrational modes. This hot molecule mechanism was also postulated in a 193 nm photolysis of benzyl chloride [5].

Intensities in energies lower than 3 kcal/mol are not accurate, since the detector axis is set at a right angle to the molecular beam axis. The detection probability is relatively smaller for photofragments with kinetic energies less than 3 kcal/mol. The distribution of the lowest E_T value due to the third channel in Table 1 should be understood to give a lower limit of the photodecomposition quantum yield.

The formation of a vibrationally hot molecule has been observed in a 193 nm photolysis of benzene vapor [35]. Yokoyama et al. [36] have studied a photodissociation of a molecular benzene beam at 193 and 248 nm, and they suggested that photodissociation occurs following internal conversion to the vibrationally excited ground state. The formation yield of hot benzene in a 193 nm photolysis is likely to be unity. The probability of this hot mechanism in a 193 nm photolysis of chlorobenzene is approximately 0.4 (Table 1). The difference must be caused by the heavy atom effect of the substituent chlorine in the benzene ring. The intersystem crossing induced by one chlorine atom is competitive with internal conversion from S_3 to S_0*. About 60% of initially excited S_3 molecules give rise to dissociation via first and second channels.

---

[1] The fluorescence lifetime of chlorobenzene vapor (0.9 torr) was determined to be 0.96 ns at a 266 nm picosecond laser excitation [33].

Excitation Wavelength and Fluorination Effects. The difference between TOF spectra of 193 nm and 248 nm photolyses of chlorobenzene is the absence of the fast photofragment in the case of 248 nm excitation. This result should indicate that the chlorobenzene molecules excited by a 248 nm photon cannot reach the energy level giving a fast predissociation or a direct photodissociation. The absorption peak of the (σ^*, n) transition localized in the C - Cl bond of alkyl chloride appears at a shorter wavelength than 200 nm, and its transition probability is negligibly small at 248 nm. Likewise, the chlorobenzene molecule is not excited to the repulsive state, but to the S_1 state, which is prohibited from directly crossing over to the repulsive potential. The absence of such a fast dissociation was also confirmed by the results of stationary photolysis measurements [26]. The photolysis of chlorobenzene vapor at 253.7 nm revealed that the excited molecule in the S_1 state decomposes from the triplet state through the intersystem crossing with a quantum yield of 0.4, and the residual hot molecules through internal conversion can be relaxed to the ground state by collisions with the third body (ethane).

The $P(E_T)$ distributions of the photofragments in a 193 nm photolysis of C_6F_5Cl is well reproduced by a Maxwell - Boltzmann distribution. This Maxwellian translational energy distribution can imply a statistical vibrational energy distribution among vibrational modes in hot molecules. The dissociation process is slower than thermal randomization of the excess energy. Therefore, the state to produce the low-energy photofragments has a long lifetime.

The average E_T value (6.5 kcal/mol in Table 1) of the Boltzmann distribution for C_6F_5Cl is slightly higher than the C_6H_5Cl value (5.0 kcal/mol in Table 1). The C_6F_5Cl should be cooler than the C_6H_5Cl because it has a higher specific heat. This discrepancy may be caused by the different dissociation mechanisms. Excited C_6F_5Cl molecules dissociate via two channels without the fast predissociation, while the C_6H_5Cl dissociates via three channels including the fast predissociation. Eventually, the effective E_{avl} value of the dissociation from the S_0* of C_6F_5Cl must be higher than that of C_6H_5Cl. Another possibility is simply the difference of the bond strength. If the C - Cl bond in C_6F_5Cl is weaker than that in C_6H_5Cl, more energy is available for translation.

The fluorescence lifetime of C_6F_6 and C_6F_5H vapor has been found to be very short [37], indicating the very quick internal conversion to the ground electronic state. The unit formation yield of vibrationally highly excited ground-state molecules has been observed in the 193 nm photolyses of C_6H_6 [35] and C_6F_6 [38]. The contribution of a distorted benzene ring conformation postulated for C_6F_5I and C_6F_5Br [5] is probable for the present pathway. The benzene ring deformation should enhance the internal conversion from S_3 to S_0. The excitation energy in the molecule after a quick internal conversion can be distributed over various low vibrational frequency modes introduced by C - F bonds. Thus the formation of a large amount of vibrationally excited ground state molecules results in dissociation to give vibrationally hot pentafluorophenyl radicals and hence the low kinetic energy distributions.

The lack of direct photodissociation from the (σ^*, n) state in C_6F_5Cl is probably attributed to the very low absorption coefficient of the transition due to the shift by fluorination. The $\sigma^* \leftarrow$ n transition probability of CHF_2Cl at 193 nm is indeed lower than that of CH_3Cl [39]. The singlet (σ^*, π) state of C_6F_5Cl is also localized in the higher-energy region due to fluorination.

The Methyl Substituent Effect. The photodecomposition quantum yields (Φ) of the second channel for chlorotoluene becomes much larger than that of chlorobenzene (Table 1). This seems to be independent of the molecular symmetry, since all three isomers give similar results. The increased amount in the Φ value of the second channel is equal to the decrease of the third channel. Thus among the two competitive channels, the second channel is found to be enhanced substantially by the methyl group substitution. The first channel seems to be unchanged, although the *p*-isomer has a higher molecular symmetry than the *o*- and *m*-isomers.

The methyl substitution effect on the photodecomposition mechanism seems to be attributable to the enhanced intersystem crossing due to the methyl group. The methyl internal rotation in S_1 fluorotoluene has been studied for the first time by Okuyama et al. [40] and more recently by Moss et al. [41, 42]. They postulated that the methyl torsional mode accelerates the intramolecular vibrational redistribution (IVR) due to the increased level density situated at a lower energy level than the excited single vibronic level at > 400 cm^{-1} above the S_1 origin. On the other hand, the mode selectivity of the methyl torsion in intersystem crossing of acetophenone has been studied by Kamei et al. [43]. They observed a sensitized phosphorescence excitation spectrum of acetophenone in a supersonic jet, and their results strongly suggest that the internal rotation of the methyl group plays an important role in inducing the intersystem crossing. The characteristic feature of an S_3 chlorotoluene molecule may be different from that of the above-mentioned molecules. The intersystem crossing of the excited singlet chlorotoluene is induced by the spin - orbit interaction, namely the L - S coupling mainly taking place on the heavy chlorine atom, and the vibronic coupling, in analogy to that of the S_1 chlorobenzene [26]. The process should go through the intermediate state, $^1(\sigma, \pi^*)$ or $^3(\sigma, \pi^*)$, that should be coupled with the $^3(\sigma, \sigma^*)$ state to eventually give rise to the dissociation. The presence of the methyl group in chlorotoluene, therefore, must provide higher level density in these intermediate states that can be coupled with the initially excited $^1(\pi, \pi^*)$ state.

Our previous photostationary experiments on dichlorobenzenes [9] and chlorotoluenes [27] optically excited with ultraviolet radiation have revealed that these excited molecules undergo C - Cl bond dissociation via vibrationally excited triplet levels through intersystem crossing. The predominant formation of triplet molecules from the S_1 levels has been demonstrated by the observation of biacetyl phosphorescence sensitized by dichlorobenzenes [44, 45] and chlorotoluenes [10, 11]. The formation quantum yield of stable triplet molecules relaxed by the collision with the foreign gas molecules was higher for chlorotoluene than for dichlorobenzene at the same pressure of the third body. The results were then explained by the longer lifetime for the triplet chlorotoluene molecules than the triplet dichlorobenzene. From the present results the initial triplet formation yield from the S_3 state of chlorotoluene can be regarded as higher than that of dichlorobenzene. Namely, the relative rate of intersystem crossing from the S_3 state of chlorotoluene must be faster than that of dichlorobenzene. Accordingly, the effect of the methyl group on the intersystem crossing rate is larger than that of the additional chlorine atom.

Additional Chlorine Atom and Molecular Symmetry Effects. The photodecomposition mechanism of dichlorobenzene is in general similar to that of chlorobenzene, though some differences are observed in dissociation quantum yields (Table 1). The similarity declares that additional chlorine substitution in chlorobenzene does not cause drastic change in the photodecomposition mechanism of the S_3 state excited at 193 nm. The Φ values for the second channel overwhelm the third channel in photodissociation of three isomers of dichlorobenzene, while the value for the third channel in chlorobenzene is the biggest among the three channels. This difference is due to the increased spin - orbit interaction by the additional chlorine atom. The electron affinity of the chlorine atom may have some effect on the electron density of the *o*- and *p*-positions. The dissociation mechanism characteristic in *o*-dichlorobenzene seems to be attributed to the close location (ortho conformation) of the two chlorine atoms. The electron densities of these two atoms must affect each other in the course of energy delocalization of the π-orbital to the σ-orbital of the C - Cl bond, giving rise to the bond dissociation.

Among three isomers of dichlorobenzene, the vibrationally hot triplet molecules of *p*-dichlorobenzene formed by intersystem crossing from the S_1 state live rather long, and the decomposition is enhanced by collisions with foreign gas molecules [27a]. The formation quantum yield of stable triplet molecules also indicated a longer lifetime of vibrationally hot triplet *p*-dichlorobenzene [27b]. Likewise, the para-isomer of chlorotoluene indicated a relatively longer lifetime among three isomers [10, 11]. The higher symmetry molecule of disubstituted benzene derivatives generally seems to have a longer lifetime in the liquid phase [46]. The longer lifetime should reflect on the kinetic energy of fragments observed in this study. The second E_T value for *p*-chlorotoluene, however, is slightly higher than those of the *m*- and *o*-isomers, and the dissociation yield of the *p*-isomer is also higher than those of *m*- and *o*-isomers (Table 1). The f_T value for the fragment may be sensitive to the energy possessed in the excited molecule, and indeed the higher values for the *p*-isomer are seen in both chlorotoluene and dichlorobenzene. These results are therefore indicative that the excited triplet *p*-isomer to give the fragment should have a longer lifetime than *m*- and *o*-isomers.

Anisotropy of the First Channel. The fast photodissociation channel (first channel) of *p*-dichlorobenzene and *p*-chlorotoluene molecules is clearly demonstrated by the anisotropic parameters (Fig. 6). From the molecular symmetry of C_{2v} (*p*-chlorotoluene) or D_{2h} (*p*-dichlorobenzene) the electronic transition dipole of the S_3 molecule initially induced by a 193 nm photon is directed parallel or perpendicular to the long axis of the molecular plane ($Cl - C_6H_4 - CH_3$ or $Cl - C_6H_4 - Cl$). Taking into account the symmetrical characters of their molecular orbitals and the experimental results that the derived B values are positive, the initial transition dipole must be oriented along the long axis of the molecule. And then the fast Cl fragment is likely to recoil with the angle of β (47.0° and 50.0° for *p*-chlorotoluene and *p*-dichlorobenzene, respectively) against the molecular long axis, though the assumption of $\chi = 0$ suggests that the values of β would have to be either zero or 90°. The values obtained based on the assumption are far from the expected ones. This may suggest that the dissociation occurs in a time scale of the rotational motion. The lifetime of the excited molecule to give the Cl fragment would reflect the time for the molecular rotation, which should have order a picosecond or less.

The fast dissociation of the C - Cl bond in *p*-dichlorobenzene and *p*-chlorotoluene must involve a C - Cl stretching motion along the reaction coordinate. From the energetics of a 193 nm photon and the C - Cl bond, two C - Cl bonds in *p*-dichlorobenzene cannot be dissociated simultaneously. Accordingly, a symmetric stretching motion of the C - Cl bonds does not give rise to the bonds dissociation. The dissociation counterpart, the chlorophenyl radical, may be vibrationally (presumably C - Cl stretching) excited and then is quickly followed by fast intramolecular vibrational redistribution. This situation is similar to *o*- and *m*-dichlorobenzenes, and the simple dissociation of ozone [47]. Once the molecule of ozone is dissociated into a O atom and O_2 molecule, available energy for fragment molecules is distributed to stretching vibration of the fragment O_2.

4 RE2PI Spectra of S_1 Chlorotoluenes

The experimental apparatus for RE2PI spectrum measurement is already described elsewhere [48]. A chlorotoluene molecular beam was prepared in a pulsed supersonic free jet expansion of chlorotoluene vapor (ca. 10 torr) seeded in Ar (ca. 900 torr). The molecule in the beam was selectively photoionized by RE2PI through its S_1 state with the second harmonic of a dye laser pumped by an excimer laser. The ions produced are analyzed by an angular-type reflectron TOF mass spectrometer.

4.1 Isotope Separation

Figure 7(a) shows the mass selected RE2PI excitation spectra of $^{13}C^{12}C_6H_7{}^{35}Cl$ (*m*/*e* = 127) and $^{12}C_7H_7{}^{35}Cl$ (*m*/*e* = 126) for the 0 - 0 band region of *o*-chlorotoluene. The peak wavelengths of the two excitation spectra were separated by approximately 3.0 cm^{-1}. This separation is attributed to the difference in the 0 - 0 transition energy of the two isotopes. The "heavier" molecule containing ^{13}C is found to be excited to the S_1 state with a wavelength shorter (higher energy) by ≈0.02 nm than that of the "lighter" molecule. Thus using this character of absorption wavelength separation it is possible to perform the isotope enrichment. Typical TOF mass spectra are also illustrated in Fig. 7(a)R. The mass spectra were obtained by RE2PI excitation at the two peak wavelengths in corresponding excitation spectra followed by TOF mass analysis. The ^{13}C isotope ions with mass numbers 127 and 129 are exclusively obtained by excitation at 271.183 nm (36875.5 cm^{-1}). On the other hand, 271.206 nm (36872.3 cm^{-1}) excitation leads to the production of almost 100% pure ^{12}C molecule, since the naturally abundant ^{13}C isotope of about 7% is totally discriminated for this excitation. The enrichment factor(*f*) of the ^{13}C isotope containing molecule (^{13}CClT) in naturally the abundant ^{12}C molecule (^{12}CClT) can be calculated by the following equation [49],

$$f(^{13}\mathrm{CClT}/^{12}\mathrm{CClT}) = \{N(^{13}\mathrm{CClT})/N(^{12}\mathrm{CClT})\}_{ex}/\{N(^{13}\mathrm{CClT})/N(^{12}\mathrm{CClT})\}_0. \quad (9)$$

Here *N* stands for the number density of the molecule, presumably ion peak intensity in the present study, and ex and 0 indicate after enrichment and before enrichment

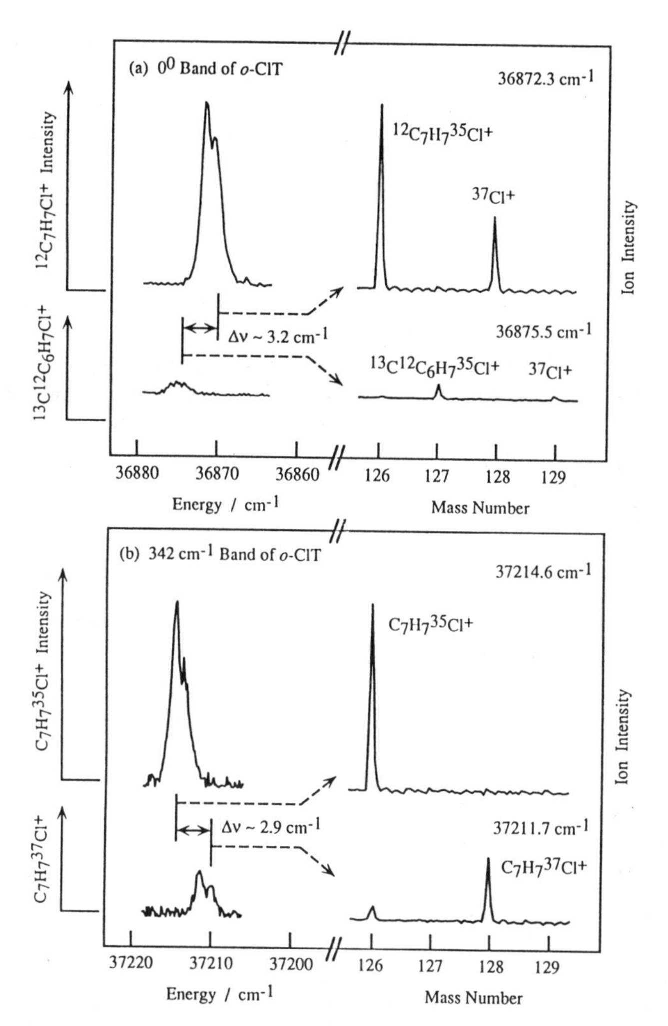

Fig. 7. (a) ^{13}C isotope shift. *(Left)* RE2PI excitation spectra of mass-selected ions for the 0 - 0 band region of *o*-chlorotoluene. The step width of the scan was 0.27 cm^{-1}. The laser power was low enough to avoid fragmentation of $^{12}C_7H_7{}^{37}Cl^+$ into $^{12}C_7H_6{}^{37}Cl^+$, which might overlap with the ^{13}C signal. *(Right)* TOF mass spectra obtained at the peak wavelengths of the two isotopes. **(b)** ^{37}Cl isotope shift. *(Left)* RE2PI excitation spectra of mass-selected ions for a vibronic band (vibrational frequency: 342.3 cm^{-1}) region of *o*-chlorotoluene. *(Right)* TOF mass spectra obtained at the peak wavelengths of the two isotopes.

(namely the natural abundance), respectively. The calculation of the *f* value from the data in Fig. 7 is ≈120. This value is much higher than those of other reports [50, 51, 52]. This result is due to the high selectivity of the present apparatus, since the enrichment efficiency is mainly dependent upon the molecular beam conditions, wavelength resolution of light source, and resolution of the mass spectrometer.

Figure 7(b) shows the mass selected RE2PI excitation spectra of $^{12}C_7H_7{}^{35}Cl$ (*m*/*e* = 126) and $^{12}C_7H_7{}^{37}Cl$ (*m*/*e* = 128), and mass spectra measured at the peak wavelengths in the excitation spectra for the vibronic band wavelength region of the *o*-chlorotoluene molecule. The ^{37}Cl isotope shift is specific to vibrational mode, as reported in other papers [53, 54, 55]. The isotope shift has been observed for a couple of vibrational bands, but only one example for the vibrational band at 0+342.3 cm^{-1} is shown in Fig. 7(b). The absorption peak of $C_7H_7{}^{37}Cl$ is shifted ≈2.9 cm^{-1} to the red compared with that of $C_7H_7{}^{35}Cl$. The f value calculated from the relation similar to (9) was ≈15, which is similar to that (≈16) obtained for $^{35}Cl^{37}Cl$ over $^{35}Cl^{35}Cl$ in dichlorotoluene [53]. It should be noted here that the absorption wavelength of the vibrational mode for the ^{37}Cl-containing molecule is shifted to the red, and the vibronic band consists of some rotational contours, as is clearly seen to the red of the Q band head. Thus at the peak absorption wavelength for the ^{37}Cl-isotope-containing molecule, a small amount of the ^{35}Cl molecule is also ionized. So the excitation at a slightly red-shifted wavelength results in much higher discrimination of the ^{35}Cl isotope, although the absolute ion intensity of $C_7H_7{}^{37}Cl^+$ itself decreases.

4.2 Excitation Spectra

A mass-tuned (m/e = 126) excitation spectrum has been observed for the excitation wavelengths scanned from 257 nm to 276 nm with a laser intensity of a few μJ. It should be noted here that the fragmentation of $C_7H_7Cl^+$ is very efficient, but it can be negligibly small, with laser intensities less than 10 μJ/pulse. Under these low laser powers, the observed $C_7H_7Cl^+$ ion intensity should reflect the population density of the resonant S_1 state. The excitation spectra are shown in Fig. 8L, where the ion intensity is normalized by the laser power measured.

This result suggests that those intensities of the RE2PI spectrum should reflect the absorption cross-section of the $S_1 \leftarrow S_0$ transition, because the ionization occurs from the S_1 state within one pulse duration of the pump laser, which is fast enough to be almost independent of the nonradiative decay process. However, there is only one problem that the ionization cross-sections of all the transitions are different. In this case, this difference would be negligibly small, since the ionization cross-section would not drastically change within some different energy regions. Under this assumption, therefore, the RE2PI excitation spectrum can be regarded as the absorption spectrum.

5 LIF Spectra of S_1 Chlorotoluenes

5.1 Vibrational bands

The fluorescence excitation spectra of *o*-, *m*-, and *p*-chlorotoluene (ClT) were measured in a supersonic free jet for the wave number region of the S_1 state, which are shown in Fig. 8R, and the strongest bands observed in the spectra were assigned as the origins, whose wave numbers are 36,863, 36,602, and 36,281 cm^{-1}, respectively. Prominent bands including the origin observed in the spectrum of *m*-ClT are doublets. The lower-frequency band of the doublet is shifted by 4.9 cm^{-1}, which turned out to be due to the methyl internal rotation, and that is described in the following section. In the

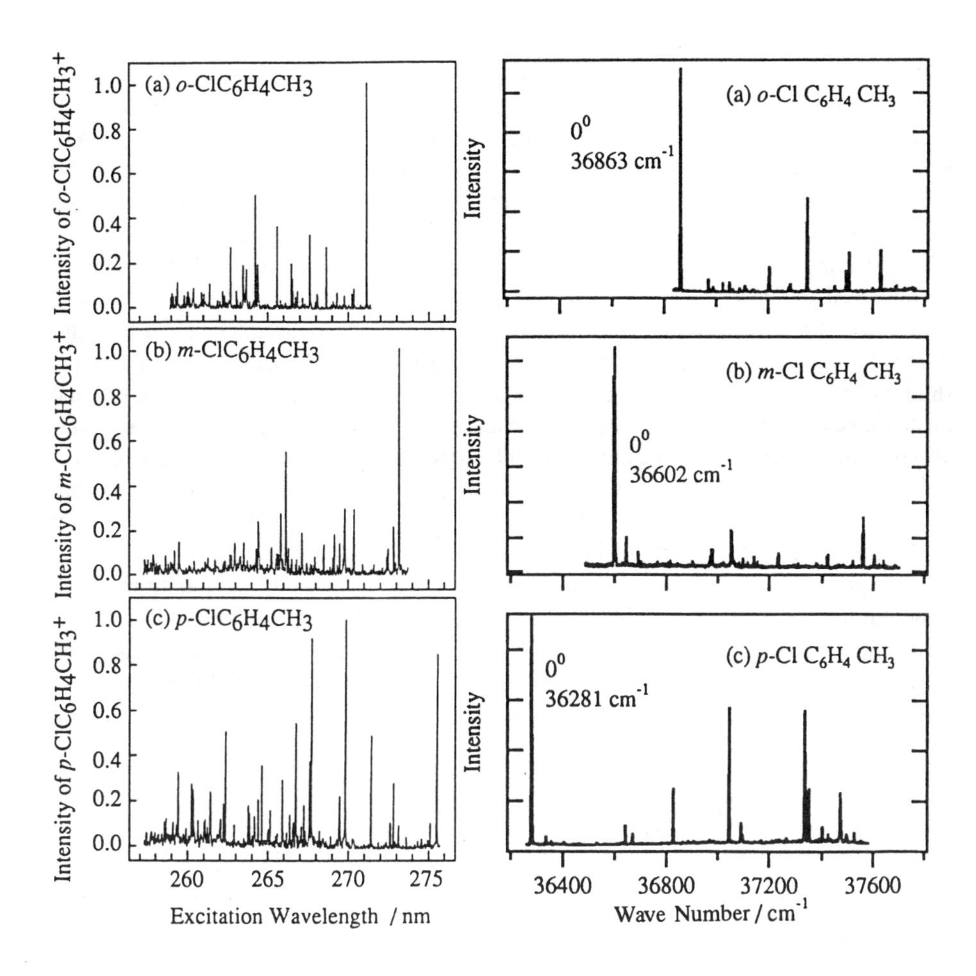

Fig. 8. *Left*: RE2PI excitation spectra of mass-selected (m/e=126) ions for the S_1 region of *o*-, *m*-, and *p*-chlorotoluene. The abscissa wavelength is not corrected. The laser power was low enough to avoid fragmentation of $C_7H_7Cl^+$. *Right*: LIF excitation spectra of *o*-, *m*-, and *p*-chlorotoluene. The abscissa indicates corrected wave number. Assignments are given in Tables 2 and 3.

wave number region from the origin to 200 cm^{-1} a couple of weak bands are observed for *p*-ClT, and these are also due to S_0 - S_1 transitions of the methyl internal rotation.

Some vibrational structures appear in the region from 300 to 1500 cm^{-1} above the origin. In order to make assignments of these bands, dispersed fluorescence spectra were taken, with pumping of each vibrational band observed in the excitation spectrum. A fluorescence spectrum observed in excitation of the 360 cm^{-1} band for *p*-ClT demonstrated that the strongest peak appears at 368 cm^{-1} above the origin [12], which corresponds to the vibrational frequency of the equivalent mode in the S_0 state. Values of vibrational frequencies for *o*-, *m*-, and *p*-ClT molecules in S_0 states have been given in the literature [56]. Accordingly, the comparison of the observed vibrational frequency in the dispersed fluorescence spectrum with the literature value enable us to assign the vibrational mode of the observed band in the excitation spectrum. The 368 cm^{-1} band is in accord with the vibrational energy of the 7a mode in the ground state. Here the vibrational mode is designated by the Wilson notation. The correspondence between the observed frequency in the S_0 state and the literature frequency is likely to be good within experimental error. Table 2 summarizes the vibrational frequencies in both the S_1 and S_0 states obtained from the above analyses together with ^{35}Cl - ^{37}Cl isotope shifts obtained from the RE2PI measurement mentioned above. Among the vibrational modes appearing in the fluorescence spectrum, the C - Cl stretching modes for *o*-, *m*-, and *p*-chlorotoluene molecules are determined to have vibrational energies of 341, 378, and 360 cm^{-1} in S_1 states, respectively. The stretching 7a mode of heavy chlorine in the molecule, however, does not have the lowest frequency, but the 15 mode for *p*-ClT does by a frequency about 110 cm^{-1} lower than the 7a mode.

In comparison with reported vibrational frequencies of the C - Cl stretching mode of *p*-dichlorobenzene (301 cm^{-1}) [54, 55] and 2,6-dichlorotoluene (342.2 cm^{-1}) [53], these values for *m*- and *p*-chlorotoluenes are slightly higher. The value for the 7a mode of *o*-ClT is 341 cm^{-1}, which is in excellent agreement with that of 2, 6-dichlorotoluene, revealing that the additional chlorine substitution in *o*-ClT does not affect the C - Cl

Table 2. Vibrational frequencies (cm^{-1}) of *o*-, *m*-, and *p*-chlorotoluene in S_1 and S_0 states, ^{35}Cl - ^{37}Cl isotope shifts (cm^{-1}) in S_1 states, and vibrational band assignments.

o-$ClC_6H_4CH_3$				*m*-$ClC_6H_4CH_3$				*p*-$ClC_6H_4CH_3$			
S_1	S_0	shift	Assig.	S_1	S_0	shift	Assig.	S_1	S_0	shift	Assig.
0	0	---	0-0	0	0	---	0-0	0	0	---	0-0
109	246	0	$10a^1$	374	387	1.6	15^1	249	252	1.7	15^1
123	---	---	---	378	416	3.6	$7b^1$	360	368	0.3	$7a^1$
188	---	---	---	454	522	0	$6a^1$	386	---	---	$(16a^1)^a$
269	---	---	---	498	---	---	---	547	639	0	12^1
341	445	3.2	$7a^1$	542	---	---	---	765	801	0	$6a^1$
422	437	1.4	$16b^1$	635	683	1.3	1^1	812	---	0	$(10a^1)^a$
483	553	0	$6b^1$	823	858	0.4	$6b^1$	817	---	0	$(10a^1)^a$
643	678	0.5	$6a^1$	964	1002	0	$(12^1)^a$	1056	1102	0.9	1^1
655	678	0	$6a^1$	1008	---	---	---	1072	1156	0	$9a^1$
768	808	0.3	12^1					1194	---	0	$(13^1)^a$

[a] Tentative assignment.

vibrational frequency. The C - Cl stretching mode is very sensitive to the isotope of the chlorine atom. The largest isotope shift has been observed among various vibrational modes for *p*-dichlorobenzene [54]. The 7b mode defined for *m*-ClT has shown the largest isotope shift (3.7 cm^{-1}), while the 7a mode for *p*-ClT has given rise to a small isotope shift (0.27 cm^{-1}) in Table 2. Instead of the 7a mode, the isotope shift of the 15 mode is 1.71 cm^{-1}, which is largest among all the vibrational modes. This discrepancy may become clear when the *ab initio* calculation of the vibrational frequencies and normal modes is carried out. Qualitatively, displacements for the 7a mode vibration in *p*-ClT seem to be smaller than for the 15 mode vibration.

All the vibrational frequency values for S_1 states are definitely smaller than those for S_0 states, as shown in Table 2. This tendency was also seen for fluorotoluene molecules [40]. This result suggests that the S_1 potential surface becomes slightly shallower than the S_0 potential.

5.2 Methyl Internal Rotation

The prominent bands including the origin observed in Fig. 8R(b) for *m*-ClT are doublets whose lower frequency bands are shifted by 4 cm^{-1}. A couple of low frequency weak bands that appeared in Fig. 8R(c) for *p*-ClT cannot be ascribed to the vibrational modes in the S_1 state, since such low-frequency modes are not expected. Hence these bands represent the methyl internal rotational levels in the S_1 state.

Internal rotational energy levels are denoted by a combination of the rotational quantum number *m* of a one-dimensional free rotor and the symmetry species of the permutation inversion group. *o*- and *m*-ClT belong to the G_6 permutation inversion group, and *p*-ClT belongs to the G_{12} group [57, 58]. The selection rule in an electronic transition for *o*- and *m*-ClT is that $a_1 \leftrightarrow a_1$, $a_2 \leftrightarrow a_2$, and $e \leftrightarrow e$ transitions are allowed.

Doublets of prominent vibronic bands for *m*-ClT are attributed to two origins of the internal rotation, which are $0a_1(S_1) \leftarrow 0a_1(S_0)$ and $1e(S_1) \leftarrow 1e(S_0)$ transitions. Accordingly, the internal rotational bands were assigned as transitions originating from these two initial levels, $0a_1$ and $1e$, in the S_0 state with the selection rule. These assignments were made in comparison with those of *m*-fluorotoluene [40], because the band structures are quite similar to each other. Some assignments have been described in a previous paper [12], where the dispersed fluorescence spectra of each internal rotational band in the S_1 state were measured. Assignments and observed energies are listed in Table 3. Any band due to the $a_1 \leftrightarrow a_2$ transition was not observed owing to a forbidden transition from the selection rule.

In order to confirm the assignments and determine the potential for the internal rotation in the S_0 and S_1 states of *m*-ClT, their level energies were calculated by reproducing observed energies with *B* of the reduced rotational constant, V_3 and V_6 of the potential barriers as fitting parameters. The best values were in consequence of B=5.32 cm^{-1}, V_3=1 cm^{-1} and V_6= - 8 cm^{-1} for the S_0 state, and B=4.15 cm^{-1}, V_3=129 cm^{-1} and V_6= - 39 cm^{-1} for the S_1 state. Table 3 also lists those calculated level

Table 3. Internal rotational levels (cm^{-1}) of *m*- and *p*-$ClC_6H_4CH_3$ in S_0 and S_1 states.

m-$ClC_6H_4CH_3$					*p*-$ClC_6H_4CH_3$				
	S_0		S_1			S_0		S_1	
species	obsd.	calcd.[a]	obsd.	calcd.[b]	species	obsd.	calcd.[c]	obsd.	calcd.[d]
$0a_1$	0	0	0	0	$0a_1'$	0	0	0	0
$1e$	0	0	-4.4	-4.4	$1e''$	0	0	0	0
$2e$	15	15.9	48.2	48.3	$2e'$		16.1		14.1
$3a_2$		42.6		51.6	$3a_2''$		45.0		41.3
$3a_1$	50	49.9	89.1	89.1	$3a_1''$	53	53.0	54.7	54.4
$4e$	80	79.9	105.9	105.8	$4e'$	80	81.4	75.3	76.1
$5e$	126	127.8		144.4	$5e''$	114	129.9		119.5
$6a_2$		186.2		183.3	$6a_2'$		189.0		171.9
$6a_1$	193	191.6		185.0	$6a_1'$	195	194.7		178.6

[a] V_3=1 cm^{-1}; V_6=−8 cm^{-1}; B=5.32 cm^{-1}, [b] V_3=129 cm^{-1}; V_6=−39 cm^{-1}; B=4.15 cm^{-1},
[c] V_6=−17 cm^{-1}; B=5.40 cm^{-1}, [d] V_6=−38 cm^{-1}; B=4.91 cm^{-1}

energies whose values are in good agreement with observed ones.

The low-frequency bands for *p*-ClT also can be ascribed to the internal rotation. The origin band provides a single peak, not a doublet, suggesting that the energy difference between the two transitions, $0a_1'(S_1)\leftarrow 0a_1'(S_0)$ and $1e''(S_1)\leftarrow 1e''(S_0)$, should be less than 1 cm^{-1}, which is the present experimental resolution. Therefore, internal rotational bands were assigned as transitions originating from the initial levels, $0a_1'$ and $1e''$, in the S_0 state. The low-frequency band structures in *p*-ClT also resemble those in *p*-fluorotoluene [40], so that the former assignment followed the latter. Table 3 also contains the observed energies from the origin, where the energy of the $1e''$ species is assumed to be 0. Observed peak intensities for $3a_1''(S_1)\leftarrow 0a_1'(S_0)$ and $4e'(S_1)\leftarrow 1e''(S_0)$ transitions are extremely weak, due simply to forbidden transitions, and they become slightly allowed owing to the coupling between the internal rotation and the rotation of the whole molecule.

The internal rotational level energies for *p*-ClT were calculated similar to those for *m*-ClT. Since *p*-ClT has a higher symmetry, only two parameters, V_6 and B, were fitted to all the observed level energies. The best-fitting calculated level energies are B=5.40 cm^{-1} and V_6= - 17 cm^{-1} for the S_0 state, and B=4.91 cm^{-1} and V_6= - 38 cm^{-1} for the S_1 state. These calculated level energies also are listed in Table 3.

5.3 Vibrational Mode Selectivity in Nonradiative Decay Process

A comparison between relative intensities, which is taken relative to the origin band intensity, of the LIF excitation spectrum (black bars) and the RE2PI excitation spectrum (white bars) observed in Fig. 8 is illustrated in Fig. 9. In the three figures each band assignment (Wilson's notation) is indicated above each bar. Relative intensities of the LIF spectrum of all the vibronic transitions are much less than those of the RE2PI spectrum of the equivalent transitions, including the transitions due to the internal rotation. The decreased amount of the LIF intensity indicates that the excitation of the methyl internal rotation and/or the vibrational motion in the S_1 state should enhance the nonradiative decay process.

Methyl Internal Rotation. The relative intensities of the LIF spectrum for the transitions above the 2*e* level are much less than those of the RE2PI spectrum of the equivalent transitions. The fluorescence decrease of the internal rotational bands indicates that

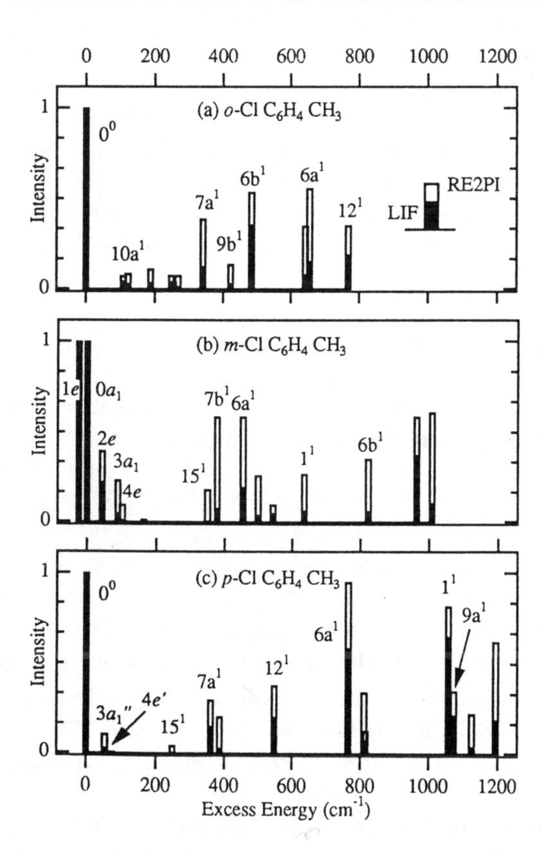

Fig. 9. Comparisons between relative intensities of the LIF spectra and the RE2PI spectra observed for *o*-, *m*-, and *p*-chlorotoluene (Fig. 8). Relative intensities are normalized by the origin band. White bars are values observed by the RE2PI spectrum and black bars those observed by the LIF spectrum.

the excitation of the internal rotation may enhance the nonradiative decay process, and those intensities in the LIF spectrum of each internal rotational band may reflect the fluorescence quantum yield of each internal rotational level.

Possible candidates for the nonradiative decay process enhanced by the internal rotation are internal conversion (IC) and intersystem crossing (ISC). These processes take place competing with each other from the internal rotational level. It may not be possible to decide only from the present results which one of the processes is responsible for the fluorescence decrease. The IC process is not so efficient for the neighbor levels of the S_1 origin because the IC rate from the S_1 state is generally slow under the collision free condition in comparison with the ISC process. So the ISC would be dominant. It cannot be denied that the intramolecular vibrational redistribution (IVR) process occurs before the ISC process does. Generally speaking, the IVR process does not take place so quickly at around the internal rotational levels of ≈ 100 cm^{-1} above the S_1 origin, since the IVR process becomes prominent for ≈ 1000 cm^{-1} above the S_1 origin. This is caused by the fact of the low-level density near the zero-point vibration in the S_1 state, though the level mixing of the internal rotation with other vibronic levels was postulated by Moss et al. [42]. One possibility is the methyl internal rotation coupled with other motions, such as the rotation of the whole molecule, probably due to the Coriolis interaction. Kamei et al. concluded that the levels of the *e* species for the internal rotation of the methyl group in the methylglyoxal have a larger ISC efficiency than those of a_1 species from the sensitized phosphorescence spectrum measurement [43]. The faster decay process may be the ISC similar to the methylglyoxal, though the electronic excited $^1(\pi^*, \pi)$ state of the chlorotoluene is different from the $^1(n, \pi^*)$ state of the methylglyoxal.

C - Cl Vibration. An important feature observed in Fig. 9 is that the LIF intensity of the vibronic band whose isotope shift is observable is much less than the RE2PI intensity. Such a vibronic band involves the C - Cl vibration, since the ^{35}Cl - ^{37}Cl isotope shift is observed. Take the 7b mode, which is the C - Cl stretching mode, and the 6a mode, which is the ring C - C vibrational mode, for example. Though the vibrational energy difference between the two modes is very small, the fraction of the LIF intensity of the former mode is much less than that of the latter. Accordingly, the C - Cl vibration should enhance a nonradiative decay process.

S_1 chlorotoluene molecules in the gas phase have a very small fluorescence quantum yield, on the order of 10^{-2} [26] and hence a large formation yield for a triplet state to eventually decompose [10, 11]. The facts strongly suggest that the S_1 chlorotoluene molecules undergo the ISC process through the spin – orbit interaction and the vibronic coupling to produce the $^3(\sigma^*, \sigma)$ state. The contribution of the Coriolis interaction to the spin - orbit interaction and/or the vibronic interaction is not known at present.

The enhanced nonradiative decay process by the C - Cl vibration should be the ISC process. There exists the ISC, taking place through the spin - orbit interaction of the heavy chlorine atom. Moss et al. [42] indicated the direct contribution of the methyl internal rotation in *p*-fluorotoluene to vibrational level mixing at a vibrational energy of only ≈ 400 cm^{-1}. Hence the IVR process may be enhanced from the 7b mode of *m*-ClT. So the ISC process must be competitive with the IVR. The relaxation process from the S_1 (π^*, π) state is eventually the dissociation of the C - Cl bond, whose quantum yield is approximately unity under the collision-free condition. The dissociation would occur from the repulsive $^3(\sigma^*, \sigma)$. However, the direct transition from $^1(\pi^*, \pi)$ to $^3(\sigma^*, \sigma)$ by spin - orbit interaction is forbidden, since two electrons must change their orbitals [26]. Hence the transition from the $^1(\pi^*, \pi)$ state to the $^3(\sigma^*, \sigma)$ state occurs via an intermediate $^3(\pi^*, \sigma)$ or $^3(\sigma^*, \pi)$ state. The $^3(\pi^*, \sigma)$ state is situated at much higher energy than the $^1(\pi^*, \pi)$ state. Thus the probable intermediate state is $^3(\sigma^*, \pi)$. In consequence the ISC process from $^1(\pi^*, \pi)$ to $^3(\sigma^*, \pi)$ is enhanced by the spin - orbit interaction mainly taking place on the Cl atom of the C - Cl vibration.

References

1 For example, T. Ichimura, H. Shinohara, K. Ohashi, and N. Nishi, Bull. Chem. Soc. Japan, **65**, 234 (1992), and related references therein.
2 H. Okabe, in *Photochemistry of Small Molecules* (Wiley-Interscience, New York, 1978).
3 M. Kawasaki, K. Kasatani, H. Sato, H. Shinohara, and N. Nishi, Chem. Phys. **88**, 135 (1984).
4 M. Umemoto, K. Seki, H. Shinohara, N. Nishi, M. Kinoshita, and R. Shimada, J. Chem. Phys. **83**, 1657 (1985).
5 A. Freedman, S. C. Yang, M. Kawasaki, and R. Bersohn, J. Chem. Phys. **72**, 1028 (1980).
6 T. Ichimura, Y. Mori, H. Shinohara, and N. Nishi, Chem. Phys. **189**, 117 (1994).
7 T. Ichimura, Y. Mori, H. Shinohara, and N. Nishi, J. Chem. Phys. **107**, 835 (1997)
8 A. Shimoda, T. Hikida, T. Ichimura, and Y. Mori, Chem. Lett. 265 (1979).
9 T. Ichimura, Y. Kohso, T. Hikida, and Y. Mori, J. Photochem. **26**, 17 (1984) .
10 T. Ichimura, N. Nahara, Y. Mori, M. Sumitani, and K. Yoshihara, Chem. Phys. **95**, 9 (1985).
11 T. Ichimura, N. Nahara, R. Motoshige, and Y. Mori, Chem. Phys. **96**, 453 (1985).
12 T. Ichimura, A. Kawana, T. Suzuki, T. Ebata, and N. Mikami, J. Photochem. Photobiol. A.Chem. **80**, 145 (1994) .
13 H. Kojima, T. Suzuki, T. Ichimura, A. Fujii, T. Ebata, and N. Mikami, J. Photochem. Photobiol. A.Chem. **92**, 1 (1995).
14 H. Kojima, T. Suzuki, and T. Ichimura, submitted to J. Phys. Chem.
15 A. Hiraya and K. Shobatake, J. Chem. Phys. **94**, 7700 (1991).
16 T. Ichimura, A. Hiraya, and K. Shobatake, UVSOR Activity Report 1990, 20 (1990).
17 T. Ichimura, A. Hiraya, and K. Shobatake, unpublished results.
18 G. S. Ondrey, S. Kanfer, and R. Bersohn, J. Chem. Phys. **79**, 179 (1983).
19 T. Ichimura, Y. Mori, H. Shinohara, and N. Nishi, Chem. Phys. Lett. **122**, 55 (1985).
20 S. W. Benson, F. R. Cruickshank, D. M. Golden, G. R. Haugen, H. E. O'Neal, A. S. Rodgers, R. Shaw, and R. Walsh, Chem. Rev. **69**, 279 (1969).
21 G. E. Busch and K. R. Wilson, J. Chem. Phys. **56**, 3626, 3638, 3655 (1972).
22 R. N. Zare and D. R. Herschbach, Proc. IEEE **51**, 173 (1963).
23 C. E. Moore, *Atomic Energy Levels*, Natl. Stand. Ref. Data Ser. Natl. Bur. Stand. (U.S.) **35**, Vol. I,195 (1949).
24 T. Ichimura, Y. Mori, H. Shinohara, and N. Nishi, Chem. Phys. Lett., **125**, 263 (1986).
25 T. Ichimura, Y. Mori, H. Shinohara, and N. Nishi, J. Spectrosc. Soc. Jp. **38**, 55 (1985).

26 T. Ichimura and Y. Mori, J. Chem. Phys. **58,** 288 (1973).
27 (a) A. Shimoda, Y. Kohso, T. Hikida, T. Ichimura,, and Y. Mori, Chem. Phys. Lett. **64,** 348 (1979). (b) T. Ichimura, A. Shimoda, K. Kikuchi, Y. Kohso, T. Hikida, and Y. Mori, J. Photochem. **31,** 157 (1985).
28 M. B. Robin, in *Higher Excited States of Polyatomic Molecules* (Academic Press, New York, 1974).
29 S. Nagaoka, T. Takemura, H. Baba, N. Koga, and K. Morokuma, J. Phys. Chem. **90,** 759 (1986).
30 S. P. McGlynn, F. J. Smith, and G. Cilento, Photochem. Photobio., **3,** 269 (1964).
31 S. Satyapal, S. Tasaki, and R. Bersohn, Chem. Phys. Lett. **203,** 349 (1993).
32 T. Ichimura, Y. Mori, N. Nakashima, and K. Yoshihara, J. Chem. Phys. **83,** 117 (1985).
33 T. Ichimura and Y. Mori, unpublished results.
34 R. A. Marcus, J. Chem. Phys. **62,** 1372 (1975).
35 N. Nakashima and K. Yoshihara, J. Chem. Phys. **79,** 2727 (1983).
36 A. Yokoyama, X. Zhao, E. J. Hintsa, R. E. Continetti, and Y. T. Lee, J. Chem. Phys. **92,** 4222 (1990).
37 D. V. O'Connor, M. Sumitani, J. M. Morris, and K. Yoshihara, Chem. Phys. Lett. **93,** 350 (1982).
38 T. Ichimura, Y. Mori, N. Nakashima, and K. Yoshihara, J. Chem. Phys. **83,** 117 (1985).
39 C. Hubrich, C. Zetzsch, and F. Stuhl, Ber. Bunsenges. Physik. Chem. **81,** 437 (1977).
40 K. Okuyama, N. Mikami, and M. Ito, J. Phys. Chem. **89,** 5617 (1985).
41 C. S. Parmenter, and B. M. Stone, J. Chem. Phys. **84,** 4710 (1986).
42 D. B. Moss and C. S. Parmenter, J. Chem. Phys. **98,** 6897 (1993).
43 S. Kamei, K. Okuyama, H. Abe, N. Mikami, and M. Ito, J. Phys. Chem. **90,** 93 (1986).
44 T. Ichimura, K. Nahara, and Y. Mori, J. Photochem. **33,** 49 (1986).
45 T. Ichimura, K. Nahara, Y. Mori, M. Sumitani, and K. Yoshihara, J. Photochem. **33,** 173 (1986).
46 K. W. Holtzclaw and M. D. Schuh, Chem. Phys. **56,** 219 (1981).
47 P. J. Hay, R. T. Pack, R. B. Walker, and E. J. Heller, J. Phys. Chem. **86,** 862 (1982), and D. Imre, J. L. Kinsey, A. Sinha, and J. Krenos, J. Phys. Chem. **88,** 3956 (1984).
48 T. Ichimura, H. Shinohara, N. Nishi, Chem. Phys. Lett. **146,** 83 (1988).
49 V. S. Letokhov, in *Chemical and Biological Applications of Lasers,* Vol. 3 (C. B. Moore, Academic Press, New York, 1977).
50 U. Boesl, H. J. Neusser, and E. W. Schlag, J. Am. Chem. Soc. **103,** 5058 (1981).
51 S. Leutwyler and U. Even, Chem. Phys. Lett. **81,** 578 (1981).
52 O. Dimopoulou-Rademann, K. Rademann, B. Brutschy, and H. Baumgartel, Chem. Phys. Lett. **101,** 485 (1983).
53 D. M. Lubman, R. Tembreull, and C. H. Sin, Anal. Chem. **57,** 1084 (1985).
54 E. A. Rohlfing, and C. M. Rohlfing, J. Phys. Chem. **93,** 94 (1989).
55 W. D. Sands and R. Moore, J. Phys. Chem. **93,** 101 (1989).
56 G. Varsanyi, in *Assignments for Vibrational Spectra of Seven Hundred Benzene Derivatives,* Vol. 1 (Adam Hilger, London, 1974).
57 H. C. Longuet-Higgins, Mol. Phys., **6,** 445 (1963).
58 P. Bunker, in *Molecular Symmetry and Spectroscopy* (Academic Press, London, 1979).

Ultrafast Relaxation and Nonlinear Localized Excitations in Conjugated Polymers

Takayoshi Kobayashi

Department of Physics, University of Tokyo, 7-3-1 Hongo, Bunkyo-ku, Tokyo 113, Japan

Abstract. In order to discuss the relaxation dynamics in conjugated polymers they are classified into two groups, one with a degenerate ground state and the other with a nondegenerate ground state. The characteristic features in the excitations in the two groups are summarized. The ultrafast dynamic was studied for three different poly(phenylacetylene)s (PPAs) with weakly-nondegenerate ground state to clarify the transition of the nonlinear properties between the two limitting cases.

The transient absorption spectra over a wide spectral range from visible to near-infrared were measured for thin films of three substituted poly(phenylacetylene)s, which are considered to have weakly nondegenerate ground 1^1A_g state and the excited 2^1A_g state below the 1^1B_u state. The broad photoinduced absorption (PIA) was observed in the band-gap. Three PIA peaks were commonly found at early delay-times, and the PIA spectral shape changed within 1 ps. The middle PIA peak with about 100 fs decay constant is assigned as a hot self-trapped exciton (STE). The single-peak transition energy of the hot STEs (1.4 - 1.5 eV) was explained as a transition from the hot STE either to an electron-hole threshold or to a biexciton. The higher- and lower-energy peaks show the same dynamics and hence both are due to the same species. They are attributed to an oppositely-charged, overall neutral, spatially-confined soliton-antisoliton pair decay via a geminate recombination following a power-law.

Using the gap-states splitting obtained from the femtosecond data and gap energy determined from stationary absorption spectra the electronic correlation energy, a conjugation length (λ_C) and a soliton-antisoliton distance (d) were evaluated. The distance was found to be nearly proportional to the conjugation length with a ratio of $d/\lambda_C = 0.40$. The ratio of full width of the soliton pair ($2\xi + d$) and the conjugation length λ_C ranges between 0.80 and 0.93 for the three PPPAs studied. This result indicates that an overall size of the soliton-antisoliton pair is strongly confined within a segmented conjugation chain.

The formation process of a soliton-antisoliton pair was discussed using a potential surface in a configuration space. The time-dependent spectra observed particularly within 1 ps confirm that the soliton-antisoliton pair is formed by way of a hot STE in a degenerate systems. The formation time of the soliton-antisoliton pair was evaluated as 55 - 98 fs consistent with the results of the theoretical estimation. As we move to more strongly nondegenerate systems the branching to the soliton-pair formation is hindered because of an inherent confinement of photoexcitations, and the relaxation and thermalization of the hot STE follow.

1. Introduction

1.1 Conjugated Polymers with a Degenerate or Nondegenerate Ground State

Conjugated polymers are attracting scientists' interest because of their possibility of being used as optical device materials utilizing their large and often ultrafast optical nonlinearity. In order to clarify the mechanism of the ultrafast optical nonlinear properties common to the π-conjugated polymers and further to design such polymers with even larger and faster nonlinearity extensive studies have been made. However, until our recent works [1, 2] had been performed there had been two groups of studies especially experimental ones between which there seemed not to be active arguments. The first group discusses the polymers with a degenerate ground state, represented by *trans*-polyacetylene, in which most contributing excited species to the resonant ultrafast optical nonlinearity just after excitation are solitons. The second group studies the polymers with a nondegenerate ground state such as polydiacetylenes, in which excitons are contributing to the ultrafast resonant optical nonlinearities of the systems.

The large optical nonlinearity of both conjugated polymers with degenerate and nondegenerate ground states is commonly induced by their large polarizability and strong vibronic coupling or in other words electron-lattice interaction. The lattice here indicates the main-chain vibration. A large spectroscopic change is induced by the nonlinear excitations such as solitons and polarons, both of which are generated by the geometrical relaxation due to the interaction between electronic excitation and lattice vibration. Photoexcitation triggers chain distortions resulting in solitons between the so called A- and B-phases in polyacetylene with a degenerate ground state or in polarons between the benzenoid- and quinoid-forms in polythiophene with a nondegenerate ground state.

In *trans*-polyacetylene the localized excitations such as charged solitons propagate along the chain to contribute both dark- and photo-conductivities [3]. The charge-transport properties are hence quite different from those of inorganic materials. The localized excitations such as solitons and polarons can be created either by chemical doping or by photoexcitation, and they induce changes in the complex refractive index namely absorption, reflection, and transmission constants.

1.2 Summary of Theoretical Works of the Localized Excitations

In this subsection the most simple model describing the electronic state coupled with lattice (main chain distortion) is given by the following the SSH hamiltonian [4].

$$H = -\sum_{n,s}\left(t_{n+1,n}c^{+}_{n+1,s}c_{n,s} + \text{h.c.}\right) + \sum_{n}\frac{1}{2}K\left(u_{n+1} - u_{n}\right)^2 + \sum_{n}\frac{1}{2}M\dot{u}_n^2 \tag{1}$$

$$t_{n+1,n} = t_0 - \alpha(u_{n+1} - u_n) \tag{2}$$

Here $c_{n, s}$ and $c_{n, s}^+$ are the annihilation and creation operator, respectively, U_n is the displacement of n-th CH from its equilibrium, M, the mass of CH, K, the spring constant of the lattice, $t_{n+1, n}$, the transfer integral of π-electrons between the site n and $n+1$, and α, the coupling constant of electron-lattice interaction, and h.c., Hermite conjugate.

It was found that the soliton formation induces the lattice distortion over the several sites [4] and hence the continuum-model approximation can be made [5]. In this approximation, the following gap parameter $\Delta(x)$ is used as an order parameter, which is a function of the position $x = na$ along the main chain direction. In case if the system has a perfect bond alternation the gap parameter $\Delta(x)$ is constant namely either Δ_0 or $-\Delta_0$. The former and the latter correspond to the A- and B-phases, respectively. The one-electron energy is then given by

$$\varepsilon(k) = \pm\sqrt{(kv_F)^2 + \Delta_0^2}, \tag{3}$$

where k is the wave number of electron and v_F is the Fermi velocity.

A soliton is the electronic excitation associated with the lattice distortion between the two phases of $\Delta(x) = \Delta_0$ and $\Delta(x) = -\Delta_0$. The distortion is given by

$$\Delta_s(x) = \Delta_0 \tanh\left(\frac{x}{\xi}\right), \tag{4}$$

where ξ is the half-width at half-maximum. The soliton excitation energy E_s, obtained by using the above trial function is given by a function of $r \equiv \xi/\xi_0 \equiv \xi\Delta_0/n_F$, which has the highest value of $E_s = 2\Delta_0/\pi$ when r is equal to unity. The electronic excitation is formed by using one state per spin from both valence and conduction bands.

When the gap levels are occupied by 2, 1, and 0 electrons, the localized excitations are called negatively charge soliton (S^-), neutral soliton (S^0), and positively charge soliton (S^+), respectively. The neutral soliton has a spin 1/2 and the charged solitons have no spin.

1.3 Confinement of the Gap States of a Polymer with a Nondegenerate Ground State

In many one-dimensional conjugated polymers with nondegenerate ground state such as *cis*-polyacetylene, polydiacetylene, polythiophene, and polythienylenevinylene, the photoexcitations are confined in space along the main chain because of unequal energy between the phases such as benzenoid- and quinoid-forms [6].

In such systems, polaron (P), bipolaron (BP), and self-trapped exciton (STE) are the localized excitations [7-9]. The absorption spectra of several polydiacetylenes exhibit a characteristic exciton peak together with a vibrational structure. Ultrafast relaxation process in polydiacetylene after photoexcitations by excitonic transition has been studied in detail and explained in terms of the formation of self-trapped exciton (STE) associated with the geometrical relaxation from the acetylene type configuration to that of burtatriene in the main chain [10-15]. Polythiophenes have been studied from the same viewpoint of the self-trapping process of excitons [16]. They have been studied also in terms of different excitations, namely some kind of soliton pairs, because the polythiophenes can be considered as slightly modified systems from *cis*-polyacetylene since the sulfur atom in the polymer does not change much the electronic state of the C-C chain. The nondegeneracy in the ground state is not as strong as in polydiacetylenes [17, 18]. On the other hand the existence of exciton absorption in *trans*-polyacetylene was also discussed recently [19]. The spectroscopic properties of the reflection spectrum of *trans*-polyacetylene is highly dependent on the degree of isomerization from *cis*-polyacetylene to *trans* form, namely the contents of both isomers on the isomerization procedure and also on the concentrations of impurities. The structure becomes obscured when the isomerization degree from cis- to *trans*-polyacetylene increases. This gives some doubt about the existence of the exciton in an intrinsic *trans*-polyacetylene. Therefore, except the above paper [19], it is broadly believed that the exciton does not exist in the intrinsic *trans*-form polyacetylene.

In such a situation it is of interest and importance to study the ultrafast relaxation in an intermediate case between the degenerate ground-state systems represented by *trans*-polyacetylene and strongly nondegenerate ground-state systems represented by polydiacetylenes. Localized excitations in the former and the latter are solitons and the self-trapped excitons (STEs) (sometimes called exciton polarons or neutral bipolarons), respectively, the difference between them is dependent on the separation between electron and hole and hence depends on the confinement strength.

Since the strength of the confinement is dependent on the nondegenerate energy in the ground state, distortions, and defects in the main chain, the species of the localized excitations and their relaxation processes are dependent on the main chain configuration and the conjugation length. It is, therefore, necessary to perform systematic studies of the time-resolved spectroscopy in the spectral region corresponding to such localized excitations as solitons and STEs for many polymers with various main chains and side groups. The spectral region is extending from the near-ultraviolet to near-infrared and even to mid-infrared regions.

1.4 Difference Absorption Spectrum due to the Creation of the Localized Excitations

Typical band-gap energies in various conjugated polymers are between 2 and 4 eV. Peaks of the induced absorption spectrum due to the generation of the gap states are considered to appear at the corresponding transition energies from the valence-band top to the gap states, from the gap states to the bottom of the conduction band, and between the gap states because of the maxima of the density of states in one-dimensional systems. Hence the induced absorption spectra are expected to be in the near-infrared region. Therefore, in order to clarify the ultrafast dynamic behavior of the localized excitations it is necessary to measure the time-resolved spectra in the spectral range. The experimental time-resolved study of polymers in such spectral region was first performed by Rothberg et al. [20, 21] They observed the induced absorption due to the charged solitons in the spectral range of 0.45-0.50 eV with the time resolution of about 500 fs. However, since the number of spectral data points in their works is only five and the signal to moise ratio is low, it is difficult to discuss the detailed spectral feature and its time dependence. It is expected that the transient spectrum in the conjugated polymers changes dramatically its shape very rapidly because of the fast energy exchange between the electronic system (electron-hole pair or exciton) and lattice (main chain) system which are strongly coupled with each other [22]. Therefore it is very important to measure ultrafast time-resolved *continuous* spectrum in the spectral range for the clarification of the relaxation mechanism of the localized excitations which are the origins of the resonant optical nonlinearity.

2. Experimental

2.1 Samples

Three poly(phenylacetylene)s in the form of a film were used as samples. They are poly[(*o*-trimethylsilyl)phenylacetylene] (PMSPA), poly[(4-*tert*-butyl-2,6-dimethylphenyl)acetylene (PMBPA), and poly[(*o*-isopropylphenyl)acetylene] (PPPA), of which chemical structures are shown in Fig.1. It is known that these polymers contain much of *trans*-form from the Raman spectra [23], especially in samples prepared by using WCl_6 as a catalyst [23]. These polymers are chemically stable even in air at slightly elevated temperatures than room temperature but below 200°C.

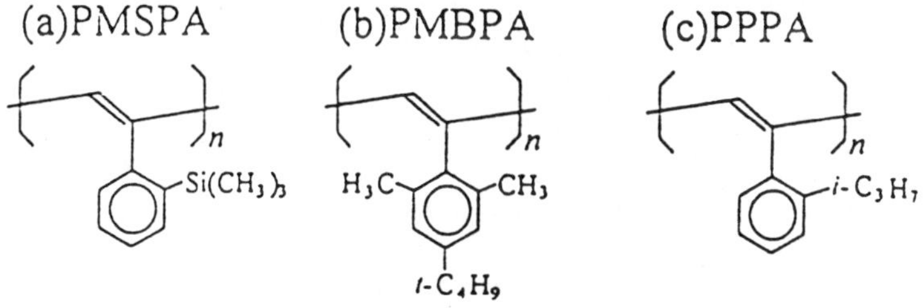

(a)PMSPA (b)PMBPA (c)PPPA

$Si(CH_3)_3$ H_3C CH_3 t-C_4H_9 i-C_3H_7

Fig. 1. The chemical structures of the three polymers investigated.

2.2 Apparatus

Femtosecond absorption spectroscopy apparatus are based on a regeneratively amplified Kerr-lens mode-locked Ti:saphire laser [1] and that based on a four-stage dye amplified colliding-pulse mode-locked (CPM) dye laser [2, 10] described elsewhere.

3. Stationary Absorption Spectra

PMSPA, PMBPA, and PPPA have solid fibril structured of dark violet, purple, and orange colors, respectively. Because of the bulky side groups they are amorphous and soluble in solvents such as toluene and trichloromethane. Homogeneous thin films are formed by evaporating solvent completely from the cast films of the polymers. For the near-infrared measurement, CaF_2 substrates were used. The absorption spectra of the samples of three polymers are shown in Fig.2 together with the calculated spectra by the convolution of the shape function of the absorption spectrum of one-dimensional electronic system given below [6] and the inhomogeneously broadened gap energy E_g with the Gaussian distribution. The fitted curves to the half-band of the lower energy side of the absorption spectra are obtained by changing the value of the inhomogeneous width as a fitting parameter.

$$\alpha(\omega) = A \cdot \frac{E_g / \omega}{\sqrt{\omega^2 - E_g^2}} \qquad (\omega > E_g) \quad . \tag{5}$$

Here A is an amplitude coefficient and ω is the photon energy. The mean values of E_g thus determined are listed in Table 1 together with several other parameters determined from femtosecond experiments.

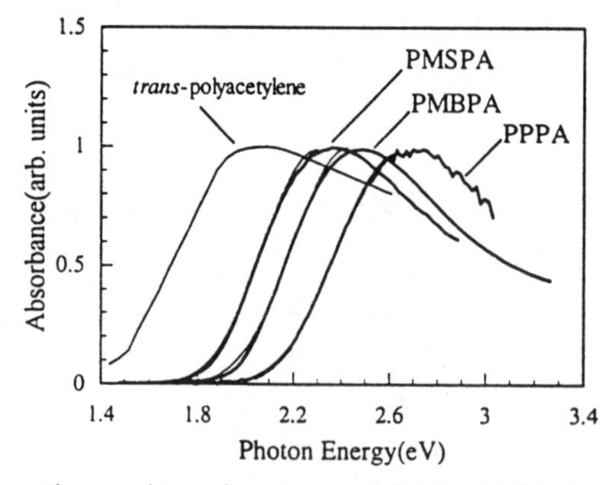

Fig. 2. The stationary absorption spectra of PMSPA, PMBPA, PPPA, and *trans*-polyacetylene normalized a the peak absorbance. The dashed curves are obtained by the eq.(5) convoluted with the Gaussian distribution of the gap energy E_g.

Table 1 . The parameters determined for the power-law components in the three different polyphenylacetylenes, PMSPA, PMBPA, and PPPA. The band-gap energy (E_g), power (n), delay time of the power-law component (t_d), high-energy peak (E_H), low-energy peak (E_L).

sample	PPPA	PMBPA	PMSPA
E_g (eV)	2.51 ± 0.02	2.33 ± 0.01	2.20 ± 0.01
n	0.86 ± 0.07	0.78 ± 0.07	0.65 ± 0.05
t_d (fs)	55 ± 30	65 ± 30	98 ± 8
E_H (eV)	2.09	1.98	1.80
E_L (eV)	0.860	0.870	0.820

It was found that these polymers exhibit the doping effect on the absorption spectra. When $FeCl_3$ is doped in a PMSPA film, depletion of the fundamental absorption is observed and a new absorption band appears around 1.1 μm near the center of the band gap, and it was attributed to charged solitons from the high content of *trans*-form in the polymer [24].

4. Femtosecond Absorption Spectra

4.1 Observed Transient Absorption Spectra

Femtosecond transient spectra of PMSPA in two spectral regions are shown in Figs.3 and 4. The former and the latter are measured with the fundamental of the CPM laser at 630 nm [1] and the second harmonic at 400 nm of the Ti:sapphire laser [2, 10], respectively, as pump sources, respectively, using the self-phase modulation (SPM) continuum and parametric signal of the SPM and the fundamental as probes, respectively. In Fig.3, the transient spectrum of PMSPA 150 fs after the excitation with the second harmonic at 315 nm of the CPM laser is also shown by a thin curve, but there is not much difference between 315 nm excitation and 630 nm excitation at all the delay times observed.

Transient spectra of PMBPA and PPPA are also shown in Figs.5 and 6. In all the observed spectra there are several common features. Very broad absorption appears below the stationary absorption and bleaching in the latter spectral region. There are at least two decaying components, one of the two is very rapidly decaying and the other decays much more slowly. There seems to be one more component which does not decay in the observed time range up to 100 ps. In the spectrum shown in Fig.3, there is a negative dip at 1.82 eV at delay times of -0.05 and 0.00 ps. This is due to the Raman gain spectrum corresponding to the Stokes shift of 1440 cm^{-1}, which is the difference between the pump energy of 2.00 eV and the signal photon energy at 1.82 eV [25].

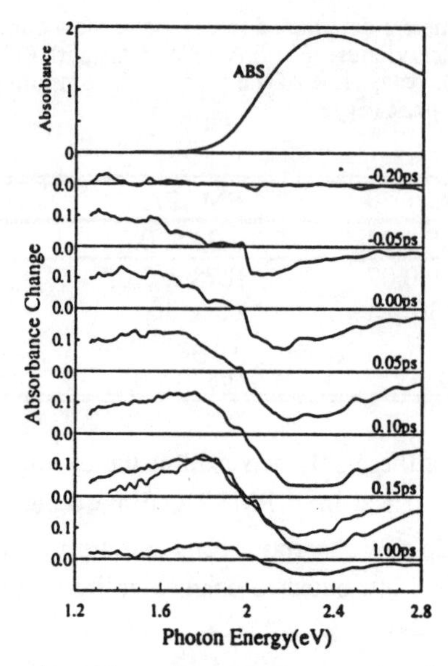

Fig. 3. Femtosecond transient absorption spectrum of a thin film of PMSPA in the visible range at delay times between -0.2 ps and 1.0 ps. The pump source is the amplified CPM dye laser with 100 fs pulse width oscillating at 620 nm corresponding to 2.0 eV. The excitation density is about 2×10^{16} photons/cm^2. The polarizations of pump and probe light pulses are parallel to each other. ABS is the stationary absorption spectrum of the sample. A thin solid line at 0.15 ps corresponds to the induced absorption excited at 310 nm (4.0 eV).

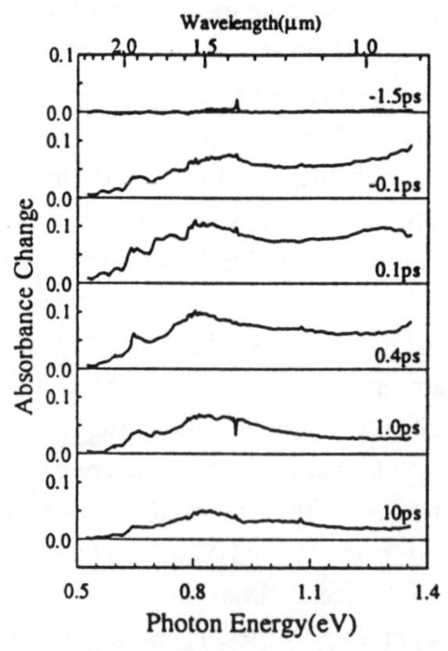

Fig. 4. Transient absorption spectrum of a thin film of PMSPA in the near infrared region. The pump laser is the second harmonic of the regeneratively amplified Ti:sapphire laser at about 2×10^{16} photons/cm^2. The excitation photon energy is 3.1 eV corresponding to 400 nm. The polarizations of pump and probe light pulses are parallel to each other.

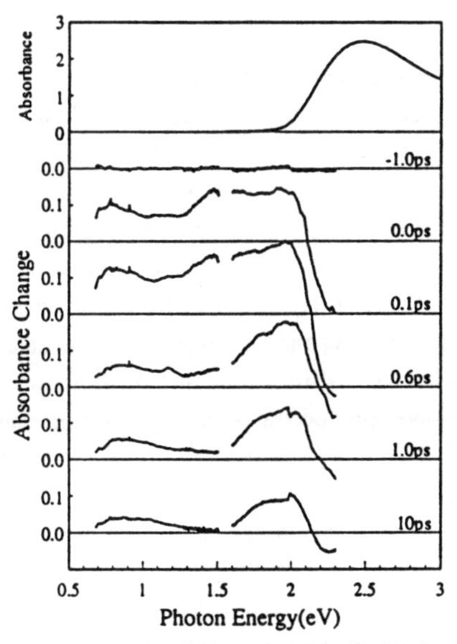

Fig. 5. Time-resolved spectrum of the sample of PMBPA thin film. Excitation pulse width, photon energy and photon density are 100 fs, 3.1 eV, and 2×10^{16} photons/cm^2, respectively. The polarizations of pump and probe beams are parallel to each other. ABS is the stationary absorption spectrum of the PMBPA thin film sample.

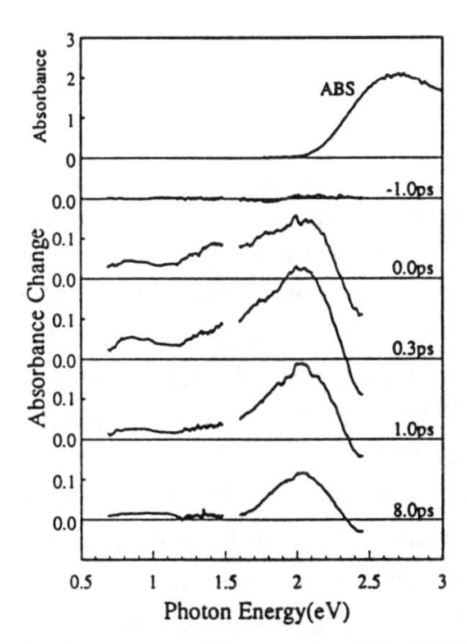

Fig. 6. Time-resolved spectrum of the sample of the PPPA thin film sample. Excitation pulse width, photon energy, and photon density are 100 fs, 3.1 eV, and 2×10^{16} photons/cm^2, respectively. The polarizations of pump and probe beams are parallel to each other. ABS is the stationary absorption spectrum of the PPPA thin film sample.

From the experimental results shown in Figs.3 and 4, the features of the transient spectrum of PMSPA are summarized as follows. (1) Induced absorption grows in the 1.2-1.4 eV region just after excitation. (2) There is a continuous change in the spectrum for 200 fs after the process (1). (3) Finally the spectrum has two peaks at 0.8-0.9 eV and 1.8 eV about 1 ps after excitation. Several spiky structures in the near-infrared (NIR) region are artifacts due to some fractures in the detector. The spectral feature lower than 0.8 eV is due to the fluctuation of the continuum spectrum in this photon energy region. In the later experiment, this was avoided by improving the laser power, lasing mode pattern, and data analysis system. The data analysis was performed by accumulating the data by the real-time rejection of those which show much different probe spectrum from the average spectral shape induced probably by the laser power fluctuation.

4.2 Time Dependence of the Absorbance Change and the Separated Spectra

The time dependence of the induced absorption/bleaching is shown in Fig.7. The decay curve can be separated into the following three contributions composed of exponential decay, power-law decay, and constant terms. The last term represents a component which does not decay even 100 ps after excitation at all probe photon energies studied.

$$\Delta A(t,\nu) = A(\nu)\cdot erf\left[\{\sigma(t-t_{\rm d})\}^{-n}\right] + B(\nu)\cdot \exp(-t/\tau) + C(\nu) \quad . \qquad (6)$$

The average of the time constant, τ, of the exponential decay fitted by using the above functions between 1.3 and 1.5 eV is determined to be 135 fs in average and it is within ± 20 fs in the proved photon energy range. While the absorbance change peaked at 1.8 eV has a finite delay time $t_{\rm d}$, and it decays following a power law. The power n determined in the probe-photon energy region of 1.75-1.9 eV is 0.65 in average and it varies within ± 0.05 in the region. By analyzing the time-resolved spectral data, the spectra $A(\nu)$, $B(\nu)$, and $C(\nu)$ are obtained, the results are shown in Fig.8 together with the fundamental stationary absorption spectrum.

The power-law-decay component $A(\nu)$ has a low-energy peak at $E_{\rm L} = 0.82$ eV and a high-energy peak at $E_{\rm H} = 1.80$ eV with an inbetween valley around 1.1-1.2 eV. The exponential component $B(\nu)$ has a single peak at $E_{\rm ex} = 1.40$ eV with a full-width at half-maximum of $\Gamma = 0.24$ eV. The long-lived component $C(\nu)$ has two peaks at 0.8 and 1.7-1.8 eV closely resembling those of $A(\nu)$.

The time dependence of absorbance change and the decomposed spectra $A(\nu)$, $B(\nu)$, and $C(\nu)$of PMBPA are shown in Figs.9 and 10, respectively. The time dependence of absorbance and the decomposed spectra of PPPA are shown in Figs. 11 and

12, respectively. They have similar features to those of PMSPA except a difference in the positions of the peaks. The parameters determined from the experimental data are listed in Table 1.

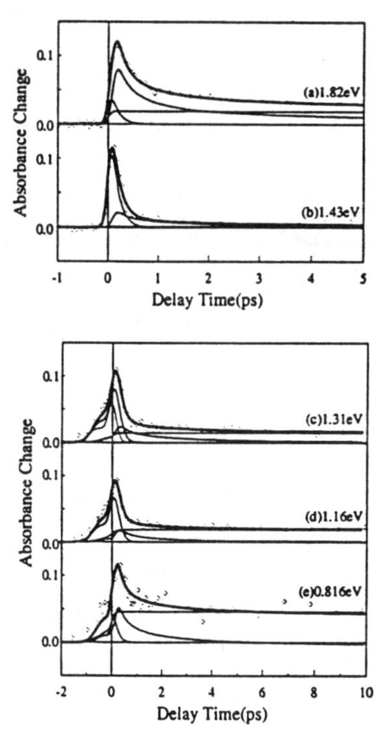

Fig. 7. Time dependence of the absorbance change in the thin sample film of PMSPA probed at (a) 1.82 eV and (b) 1.43 eV excited at 2.0 eV of CPM dye laser, and probed at (c) 1.31 eV, (d) 1.16 eV, and (e) 0.816 eV excited at 3.1 eV of the second harmonic of Ti:sapphire laser.

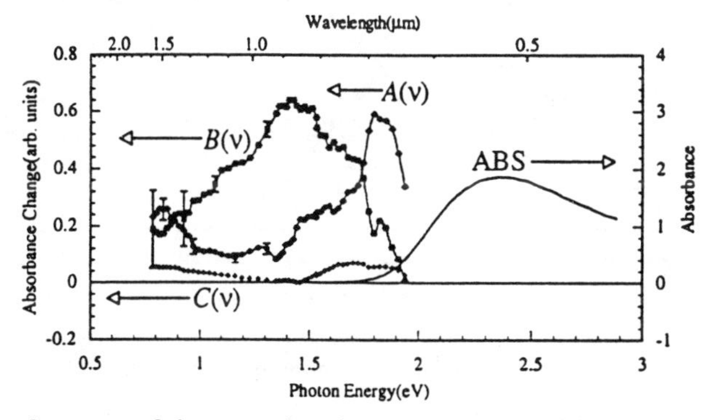

Fig. 8. Spectrum of the power-law decay component $A(\nu)$, exponential decay component $B(\nu)$, and the very long-life component $C(\nu)$, obtained from the time-resolved spectrum of the PMSPA thin film sample.

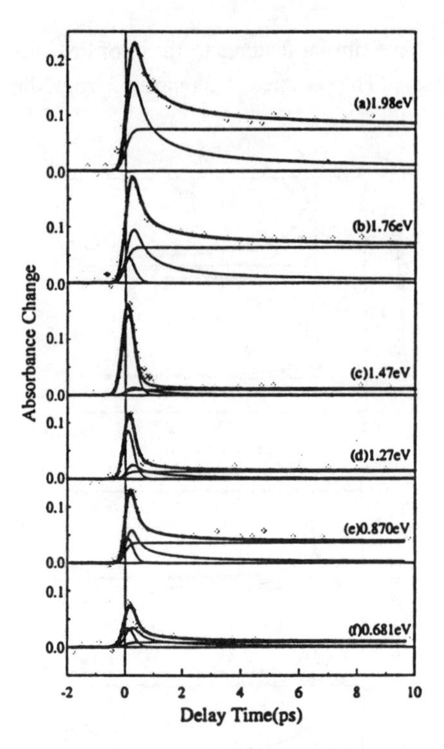

Fig. 9. Time dependence of absorbance change probed at several probe photon energies. The circles are experimental data points and the thick curves are fitting using eq.(6) to the experimental results. Thin curves are the time dependence corresponding to the decays of $A(\nu)$, $B(\nu)$, and $C(\nu)$ components.

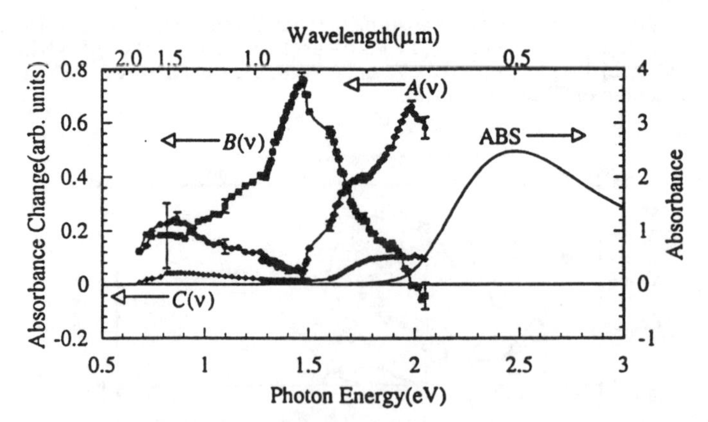

Fig. 10. Spectrum of the power-law decay component $A(\nu)$, exponential decay component $B(\nu)$, and the very long-life component $C(\nu)$, obtained from the time-resolved spectrum of the PMBPA thin film sample.

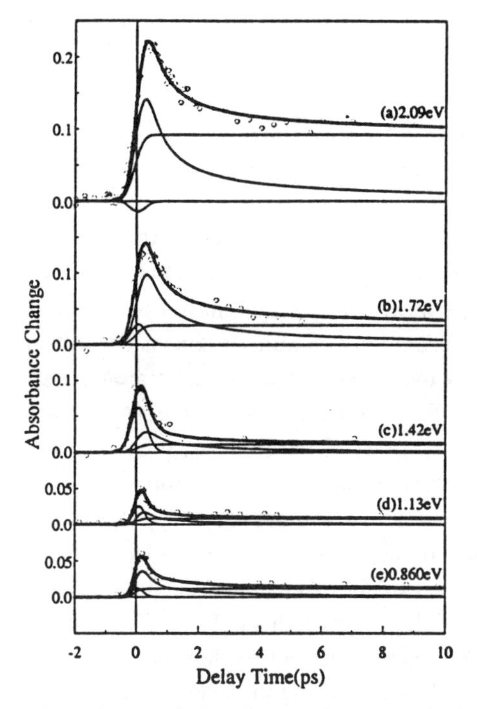

Fig. 11. Time dependence of absorbance change probed at several probe photon energies. The circles are experimental data points and the thick curves are fitting using eq.(6) to the experimental results. Thin curves are the time dependence corresponding to the decays of $A(\nu)$, $B(\nu)$, and $C(\nu)$ components.

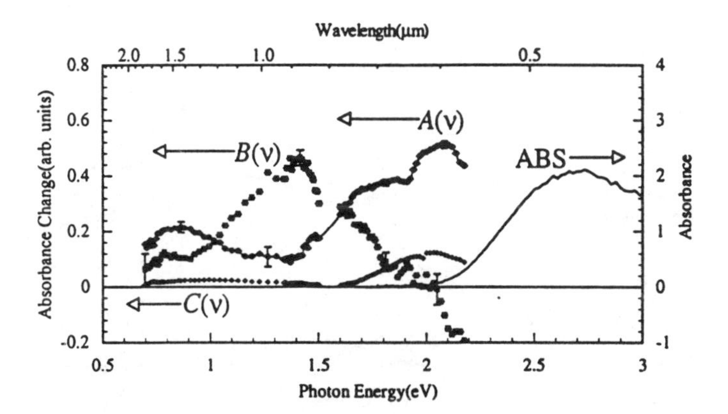

Fig. 12. Spectrum of the power-law decay component $A(\nu)$, exponential decay component $B(\nu)$, and the very long-life component $C(\nu)$, obtained from the time-resolved spectrum of the PPPA thin film sample.

5. The Features of the Spectra with Three Different Decay Kinetics

5.1 Power-Law Decay Component

5.1.1 The Power of the Decay Function

As shown in Table 1 and Fig. 13 the power n of the power-law components increases and the delay time t_d decreases with increase of gap energy E_g. The dynamics following the power-law decay is a specific behavior of species disappearing via a two-body collision after a random-walk on a single dimensional chain. In the case of ideal one-dimensional system the power n is 0.5. The time dependence of photogenerated species in *trans*-polyacetylene was found to follow the power-law decay kinetics by Shank and others [22]. The experimental result was explained in terms of the soliton and antisoliton pairwise created. The pairwise annihilation after the collision of the geminate pair was found to follow $\mathrm{erf}\left[(\sigma t)^{-\frac{1}{2}}\right]$ by both analytical and numerical calculations of one-dimensional system. The power larger than 0.5 determined from the present experiment can be explained in terms of the finite one-dimensional chain or the fractal dimensionality of the polymer chain [11]. The numerical calculation of pair annihilation in a finite-size chain shows that the decay is represented by the power law when the two species walk around along the chain within much shorter distance at first just after the pair creation, and then they come close to the ends of the finite size at longer delay time, then the decay becomes exponential [25]. However, the exponential component at longer delay times may contribute to increase the apparent value of exponent determined in a limited observation time.

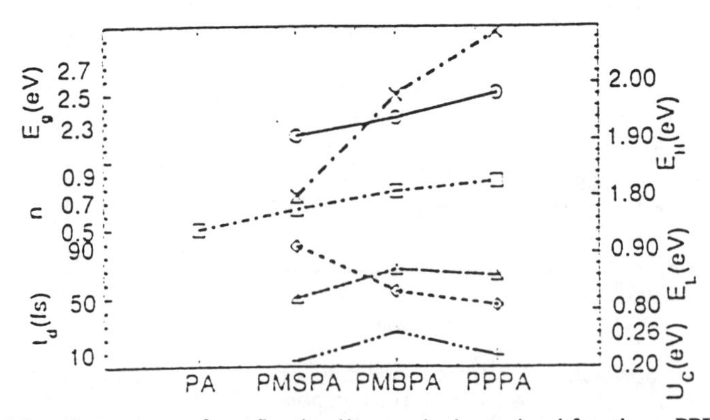

Fig. 13. Parameters of confined soliton pair determined for three PPPAs and *trans*-polyacetylene. The gap energy (E_g), circles; power of the power-law decay (n), rectangles; latent time (t_d); triangles, energy of the high-energy peak (E_K); tilted crosses, energy of the low-energy peak (E_L); diamonds, and Coulomb correlation energy (U_C); nontilted crosses.

The species exhibiting the power-law are assigned to the charged solitons (S^+ - S^-) on the basis of the following reasons. (1) In *trans*-polyacetylene the similar component of power-law decay is assigned to the charged soliton pair. (2) A new absorption band appears when the poly(phenylacetylene)s are electrochemically doped. They are attributed to the soliton-pair formation [1, 2]. · (3) It is reported that the charged soliton pairs are expected theoretically to be much more efficiently photogenerated than the neutral soliton pairs on a single chain because the photoexcitation induces the creation of an electron and a hole [26].

5.1.2 High- and Low-Energy Peaks

All of the three polymers have the dual-peak spectra with the power-law decay component in common; the lower-energy peak (E_L) appears between between 1.8 and 2.0 eV and it appears in the region of 0.8 - 0.9 eV, and the higher-energy peak (E_H) shifts to the higher energy for polymers with larger band-gap energies. The dual-peak feature is most naturally and intuitively explained in terms of the confined soliton-antisoliton pair. The observed delay-time t_d of the dual-peak spectrum can be interpreted as the formation time of the soliton pair. It ranges within 50 - 100 fs depending on the polymers and agrees quite well with the theoretical estimation [27]. These quantities E_L,E_H, t_d, and others obtained experimentally are shown together in Fig. 13. The long-life component shows a similar spectrum to that of the power-law component. It is supposed from the quadratic dependence on the pump intensity observed previously for PMSPA that the component is due to a two-photon excited species and/or any reaction product generation by the two-step one-photon excitations [28].

An electron-hole pair is initially created in a single chain by an intrachain photoexcitation, and then decays into an energetically favored oppositely-charged soliton-antisoliton pair. When a soliton-antisoliton distance is much longer than the spatial size of the soliton the two gap-states associated with the soliton-antisoliton pair are still degenerate at the gap-center. The gap-states are split off from the gap-center into two levels as experimentally observed when they come closer to each other [29] due to overlap of the wavefunctions of the soliton and antisoliton.

Both gap-states are singly occupied to form a B_u-symmetric singlet electronic configuration shown in Fig. 14. It is considered to evolve from a ^1B_u free electron-hole-pair state along the adiabatic potential surface [30]. When the soliton and antisoliton are on the same site the potential energy becomes maximum and as the distance between them increases the energy decreases asymptotically to a creation value for a well-separated soliton-antisoliton pair. After their geminate recombination a neutral soliton pair with an A_g-symmetry is formed, for which the lower gap-state is doubly-occupied and the higher gap-state is vacant (Fig. 14), resulting in the two transitions with energies of E_L and E_H as shown in Fig. 14 [31]. The neutral soliton pair is an electronic ground-state with lattice relaxation, and it rapidly decays to the ground-state configuration by emitting phonons. These two transition energies are expressed as

$$E_L = \Delta - \omega_0 + U_c, \tag{7}$$
$$E_H = \Delta + \omega_0 + U_c, \tag{8}$$

Here $\Delta = E_g/2$ is a half band-gap energy, $\pm\omega_0$ is the gap-state energy measured from the gap-center and U_c denotes an effective electronic correlation energy.

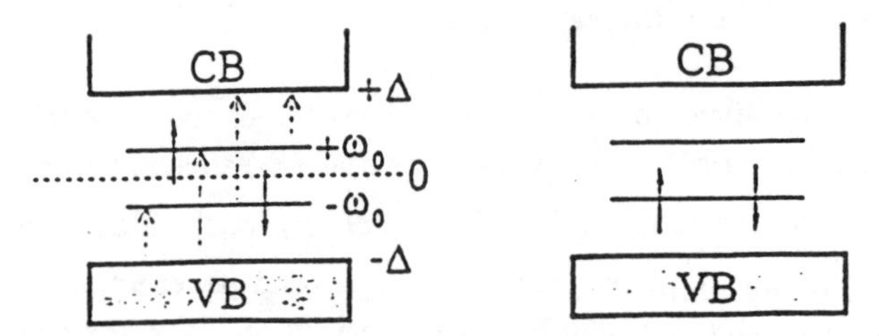

Fig. 14. Energy diagram of the confined soliton pair. Left: Energy-level diagram for a soliton-antisoliton pair. The two gap-states are singly occupied with a ^1B_u electronic configuration. The induced absorption associated with the soliton-antisoliton pair takes place at two transition energies between the gap-states at $\pm\omega_0$ and band-states at $\pm\Delta$. The lower- and higher- transition energies are indicated by dotted and dash-dotted arrows, respectively.
Right: The lower gap-state is doubly occupied and the higher one is vacant, forming an ^1A_g electronic ground state with a lattice distortion. VB : valence band, CB : conduction band.

Using the band-gap energy and two transition energies experimentally observed, the gap-states splitting ($2\omega_0$) and the effective correlation energy (U_c) are independently obtained from the two equations: $2\omega_0 = 1.11 \pm 0.02$ eV and $U_c = 0.26 \pm 0.02$ eV for PMBPA ; $2\omega_0 = 0.98 \pm 0.02$ eV and $U_c = 0.21 \pm 0.02$ eV for PMSPA; $2\omega_0 = 1.23 \pm 0.02$ eV and $U_c = 0.22 \pm 0.02$ eV for PPPA. The effective Coulomb correlation energy (U_c) was estimated [31] for a soluble polyacetylene in common solvents obtained by growing the polyenic chains onto activated sites of a flexible butadiene chain [32]. The value of U_c was found to be 0.85-1.05 eV, which is about four times larger than the poly(phenylacetylene)s studied in the present paper. This is because of increased Coulomb screening by more polarizable phenyl substituted groups than the polymer started from polybutadiene. The values of U_c determined for PPPAs are about a half of that in a polydiacetylene (PDA) [33]. This is also because of the difference in the screening effect.

5.1.3 Confinement Parameter

The gap-state splitting normalized to the band-gap energy, $p \equiv \omega_0/\Delta$, is related to the degree of confinement of a soliton-antisoliton pair, and it is often discussed in terms of a confinement parameter γ defined as [6]

$$\gamma \equiv \frac{1}{\lambda}\frac{\Delta_e}{\Delta} = \frac{p \sin^{-1} p}{\sqrt{1-p^2}} \tag{9}$$

In Eq. (9) $\lambda = 2\alpha^2/\pi T_0 K$ is a dimensionless coupling parameter with an electron-phonon coupling constant $\alpha = 4.1$ eV/A, $T_0 = 2.5$ eV is a transfer energy of π-electrons on a polyene chain with no dimerization, and $K = 21$ eV/A^2 is an elastic force constant of the main-chain, which yields $\lambda = 0.20$ [34]. Here the gap parameter Δ is the sum of intrinsic (Δ_0) and extrinsic (Δ_e) origins and the intrinsic band-gap energy is given by $E_g^{(0)} \equiv 2\Delta_0 = 2(\Delta - \Delta_e)$. From the experimentally obtained parameters p, the values of the confinement parameters γ are calculated as 0.23 ± 0.01, 0.27 ± 0.01, and 0.29 ± 0.02 for PMSPA, PMBPA, and PPPA, respectively. The values of γ of three poly(phenylacetylene)s together with those reported in literature are shown in Table 2. The parameters determined are also shown in Fig. 15.

Table 2. Confinement parameters of soliton pairs in several conjugated polymers

	γ	reference
PMSPA	0.23±0.01	[2]
PMBPA	0.27±0.01	ditto
PPPA	0.29±0.02	ditto
PT	~1	[35]
PT	0.14	[36]
PTV	0.10	ditto
PPPV	0.18	ditto

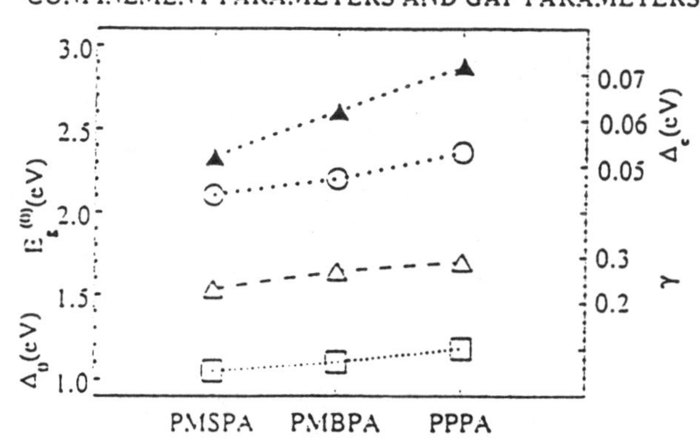

Fig. 15. Confinement parameters and gap parameters of the three PPPAs. Extrinsic gap energy (Δ_e); closed triangles, intrinsic band gap energy ($E_g^{(0)}$); circles, confinement parameters (γ); triangles, intrinsic gap parameter (Δ_0); squares.

The γ values of poly(phenylacetylene)s are smaller than γ=0.14 for PT determined by McKenzie and Wilkins [35]. It is too small to explain the phonon numbers appearing in the emission spectrum. Therefore the γ values of PT, PTV, and PPV in literature [36] are concluded to be underestimated.

5.1.4 Soliton Size and Soliton-Soliton Distance

The difference in $E_g^{(0)}$ among the three polymers results from that in the dimerization amplitude in the ground-state configuration. It therefore affects the spatial size of a soliton because the soliton is a localized excited state, of which shape is sensitive to the strength of the electron-phonon coupling. Su et al. estimated the spatial extension of chain distortion due to a soliton by minimizing the formation energy of the nonlinear excitation [34]. They estimates the half-width (ξ) of a soliton in polyacetylenes with several different band-gap energy (E_g) and hence different distortion amplitude. They assumed the soliton shape to be given as $\tanh(z/\xi_0)$ with z being a site coordinate along the chain. They obtained the values of ξ for *trans*-polyacetylenes with band-gap energies of 1.0, 1.4, and 2.0 eV to be about 9a, 7a, and 5a, respectively, with a being the carbon-carbon distance. Using the relation of inverse proportionality between the half-width at half-maximum (ξ) of the distortion due to a soliton formation and the intrinsic band-gap energy $E_g^{(0)}$, [35], and the values of ξ for the three polymers are estimated by comparing with that in *trans*-polyacetylene (PA):

$$\xi = \xi\,(\mathrm{PA}) \times \frac{\xi_g^{(0)}\,(\mathrm{PA})}{\xi_g}, \tag{10}$$

where $\xi_g^{(0)}$ (PA) = 1.4 eV and ξ (PA) = $7a$ are the intrinsic band-gap energy and the half width of a soliton in *trans*-polyacetylene. From Eq. (10) $\xi = (4.7 \pm 0.1)a$ for PMSPA, $\xi = (4.5 \pm 0.1)a$ for PMBPA and $\xi = (4.1 \pm 0.1)a$ for PPPA are obtained.

The soliton and antisoliton begin to partially overlap in space when the distance between them becomes comparable to the soliton size. In the presence of an overlap between the two solitons the bound state at the gap center was shown to split into two levels at

$$\pm\, \omega_0 = \pm\, 2\Delta \cdot e^{-d/\xi} \tag{11}$$

where d is the distance between the soliton and antisoliton in the pair [29]. Therefore from the values of ω_0 and ξ determined above, the values of the soliton distance can be determined. The results are shown in Table 3.

Table 3. The soliton size (ξ) defined by the half-width at half-maximum of soliton distortion and the soliton-antisoliton distance (d) in the unit of C-C distance (a).

sample	PPPA	PMBPA	PMSPA
ξ/a	4.1 ± 0.5	4.4 ± 0.5	4.7 ± 0.6
d/ξ	1.40 ± 0.10	1.43 ± 0.10	1.50 ± 0.10
d/a	5.7 ± 1.1	6.3 ± 1.1	7.1 ± 1.1

5.1.5 Conjugation Length

The difference between the band-gap energies with and without distortions, $E_g - E_g^{(0)}$, is related to the conjugation length λ_c defined in [37] as

$$E_g - E_g^{(0)} = 2\beta_s \left\{ 1 - \left[\frac{1}{2} + \frac{1}{2} e^{-2a/\lambda_C} \right]^{1/2} \right\} \tag{12}$$

where β_s is the π-electron transfer-energy for a single bond. The left-hand side of Eq. (12) is equal to $2\Delta_e$, and if the conjugation length is much longer than $2a$, then Eq. (12) can be approximated as:

$$\Delta_e \cong \frac{\beta_s a}{2\lambda_C} \tag{13}$$

The distance (d) is found to be nearly proportional to $1/\Delta_e$ with a best fitted slope of $\eta \approx 0.39$ eV as shown in Fig.16. Hence one can obtain the following scaling law.

$$\frac{d}{a} = \frac{\eta}{\Delta_e}. \tag{14}$$

The Eqs. (13) and (14) lead to the following expression for a ratio of the distance to the conjugation length:

$$\frac{d}{\lambda_C} = \frac{2\eta}{\beta_s} \tag{15}$$

Here $\beta_s = t_0 - \delta$ is the transfer-energy of π-electrons in a single bond. This result indicates that the soliton-antisoliton distance is proportional to the conjugation length in each polymer and that the soliton and antisoliton are confined within the conjugation length. Using $t_0 = 2.5$ eV and $\delta = E_g^{(0)}/4 = 0.52$ eV, which are the case for PMSPA, we obtain $\beta_s = 1.98$ eV and then $d/\lambda_C = 0.40$. Taking into account the spatial size of the soliton, the actual length occupied by one soliton-antisoliton pair is given by $2\xi + d$. The ratio of the overall size to the conjugation length, $(2\xi + d)/\lambda_C$ reaches as large as 0.80, 0.87, and 0.93 in PMSPA, PMBPA, and PPPA, respectively. If we consider an end-effect of the segmented chain as well, these ratios imply that the soliton-antisoliton distance is just limited by the conjugation length of the chain. The conjugation lengths are calculated as about $21a$, $17a$, and $15a$ for PMSPA, PMBPA, and PPPA, respectively, and they are much longer than the length of the unit plane ($2a$).

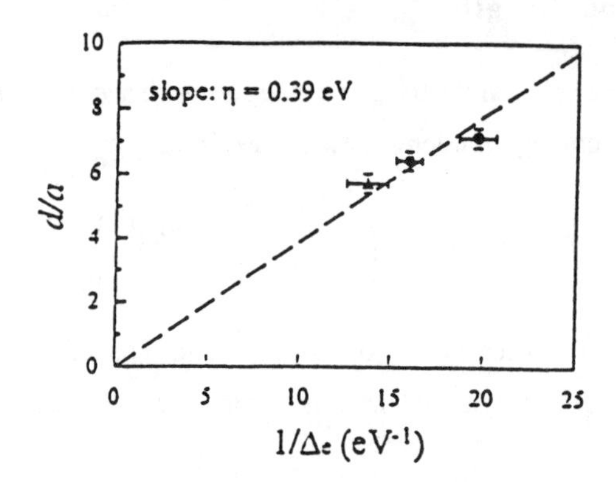

Fig. 16. Soliton-antisoliton distance in the three polymers plotted against the inverse of Δ_e. circle; PMSPA, square; PMBPA, triangle; PPPA. The broken line shows the least square fitted line including the point of origin.

Even in the limit of an infinite conjugation length, the oppositely-charged $S^{\pm}$ and $\overline{S}^{\pm}$ in *trans*-polyacetylene were theoretically predicted to form an exciton-like bound state with stabilization energy of 0.05 eV by the attractive Coulomb interaction [38]. The soliton-antisoliton distance in *trans*-polyacetylene was calculated as $d_0 \sim 12a$, and the overall size of the bound state is then estimated as $2\xi_0 + d_0 \sim 26a$. Hence the S-$\overline{S}$ distance does not increase proportionally with the conjugation length but tends to saturate above $\lambda_C \sim 26a$ as in shown in Fig. 16. In the opposite extreme case to the infinite distance is the case of the comparable conjugation length to the soliton size, where the soliton and antisoliton cannot separate due to the strong confinement within a segmented conjugation region in the main chain. The soliton-antisoliton distance is supposed to become to zero when the conjugation length becomes less than 2ξ. The photoexcited state in this situation should be regarded as a half-trapped exciton rather than a soliton pair.

5.1.6 Decay Channels

As mentioned above, the decay kinetics of a photoexcited electron-hole pair in a one-dimensional π-conjugated polymer depends upon how far the electron and hole can separate from each other. It is closely related to the ground-state degeneracy. Figure 17 illustrates schematically a potential surface in a weakly nondegenerate system. The geometrical relaxation of a photoexcited electron-hole pair, its thermalization, and the formation of a soliton-antisoliton pair are the major decay processes. At first, just after a free electron-hole pair or a free exciton is created (indicated by *1* in the figure), it couples quickly with C-C and C=C stretching vibrations along the main-chain. Recently it has been proposed that excitons are photogenerated in a highly pure *trans*-polyacetylene [39]. This self-trapping process (*a* in the figure) is supposed to take place in 10 - 20 fs after the free-exciton creation because it proceeds within a few periods of the strongly coupled vibrations to the electronic transition. The coupled modes are considered to be C-C and C=C, stretching vibrations because of the π-π^* (bonding to antibonding) excitation of the double bonds, and the oscillation periods are 20-30 fs. The STE is still vibrationally excited (hot STE, labeled 2 in the figure), and the following relaxation and thermalization processes (*b*) with emitting phonons compete with the formation of a soliton-antisoliton pair (*4* in Fig. 17).

Weakly nondegenerate case

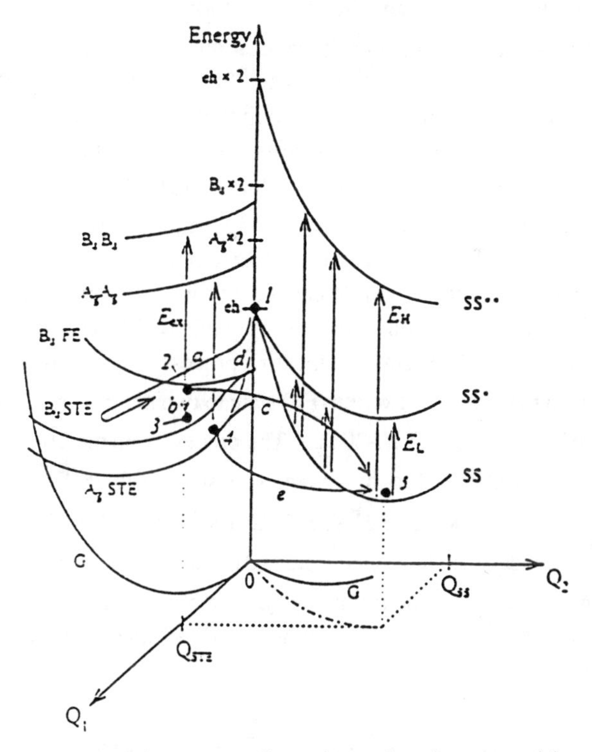

Fig. 17. Potential curves of conjugated polymers with a weakly-nondegenerate ground state.

5.1.7 Weakly Nondegenerate Ground-State System

In the following discussion of the relaxation channels in PPAs, two cases are considered, one is the case of the lowest excited state is of ^1A_g symmetry and the other is of ^1B_u symmetry.

In the case of weakly nondegenerate ground-state systems an energetically favored soliton-antisoliton pair with creation energy of $2E_g/\pi$ is preferred to be formed from the hot e-h pair or the hot STE as denoted by the decay channel (*c*) in Fig. 17. In the figure only the case of STE formation is shown. As will be discussed later, the observed femtosecond transient absorption can be explained in terms of the exciton → biexciton transition, therefore the explanation using the exciton formation is more preferable. The decay means that the system can be further stabilized by the lattice relaxation associated with the soliton-antisoliton pair formation and also with a loss of the binding energy of the hot STE. The geometrical relaxation in the process of the soliton-antisoliton pair formation, therefore, should proceed more rapidly than the thermalization and/or direct recombination of the hot STE. The proposed mechanism is thus consistent in this respect because the observed formation time of $t_d = 50 - 100$ fs is faster than the competing process (*b*) of thermalization , which takes place with a 100-150 fs time constant. Time-resolved fluorescence from the hot STE with better time resolution can give a further information on its population dynamics. As the geometrical relaxation (*c*) proceeds, the hot STE disappears and the soliton-antisoliton pair is formed in turn. It is further confirmed by our measurement that the formation time (t_d) of the soliton-antisoliton pair tends to become longer for the polymer with the longer decay-time (τ) of the hot STE. A good agreement between the two time constants is not obtained because there is another possible channel of the soliton-pair formation namely direct formation from the free exciton before self-trapping, which is not shown in Fig. 17. In case if 2^1A_g is below the 1^1B_u state, competing process is expected to exist parallely with the relaxation of the 2^1A_g free exciton (FE) state to two neutral-soliton pairs [40, 41], which is considered to take place in 10-20 fs. The process from the 2^1A_g FE state to the 2^1A_g STE state is also expected to be extremely fast (~10 fs). Therefore there is a branching from the 2^1A_g FE state to 2^1A_g STE and $S^0\ \overline{S}^0$. The relaxation process of PPAs with weakly-nondegenerate ground state and the lowest excited state with the ^1A_g symmetry is then summarized in the following three pathways (A), (B-1), and (B-2) while that in PPPAs with the lowest excited state with the ^1B_u symmetry is summarized in the two pathways (B-1) and (B-2).

$$1^1B_u\ \text{FE} \xrightarrow[<<100\ \text{fs}]{} 2^1A_g\ \text{FE} \xrightarrow[<<100\ \text{fs}]{} 2^1A_g\ \text{STE} \xrightarrow[50\text{-}100\ \text{fs}]{} (S^0\ \overline{S}^0)_2 \longrightarrow \text{G} \quad \text{(A)}$$

$$1^1B_u \text{ FE} \xrightarrow[<<100 \text{ fs}]{} \text{hot } 1^1B_u \text{ STE} \xrightarrow[(50\text{-}100) \text{ fs}]{} \text{cold } 1^1B_u \text{ STE} \xrightarrow[(100\text{-}150) \text{ fs}]{} \text{G} \quad \text{(B-1)}$$

$$\text{hot } 1^1B_u \text{ STE} \xrightarrow[(50\text{-}100) \text{ fs}]{} S^+ \overline{S}^- \text{ or } S^- \overline{S}^+ \xrightarrow[t^{-n}]{} \text{G} \quad \text{(B-2)}$$

The time constants of the processes: 1^1B_u STE $\rightarrow S^0 \overline{S}^0$ and 1^1B_u STE $\rightarrow$ G cannot be determined, since the observed time constant (100-150 fs) is determined by the contribution of the two processes. Associated with the geometrical relaxation (*c*) in Fig. 17, the lower-(E_L) and higher-(E_H) energy transitions are supposed to shift to higher and lower energies, respectively, at very early delay-times because the two associated levels in Fig. 14 move from the valence/conduction band-edges toward the gap-center. The shift of the transition energy is also indicated by a series of arrows in Fig. 17. Such a spectral evolution can be clearly seen in Fig. 18, where the lower-energy peak around 0.8 - 0.9 eV is slightly inclined toward the lower-energy side around zero delay time, although the shift of the higher-energy peak is obscured by the strong photoinduced bleaching of the ground-state absorption.

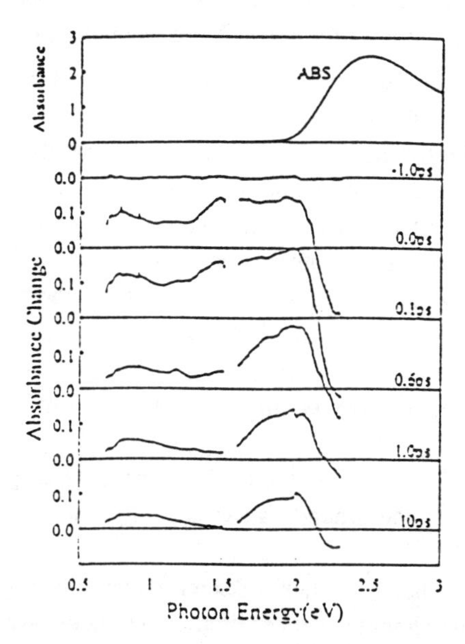

Fig. 18. Difference absorption spectra of a thin film of PMBPA from visible to NIR region plotted at several delay times. Measurements were done at room temperature, and the excitation density was ca. 2×10^{16} photons/cm^2 at 3.1 eV (λ = 0.40 μm). Stationary aborption spctrum (ABS) of the PMBPA sample film is also shown.

The above discussion was made based on the assumption that the lowest excited state in the PPPAs in 2^1A_g, but if it is 1^1B_u state then the pathway (A) does not exist. In the case of the three PPPAs studies, the latter seems to be the case as will be discussed from the values of the on-site Coulomb interaction U_0 of these polymers. Among PPPAs prepared in Prof. Masuda's group in Kyoto University these are polymers with highly fluorescent, weakly fluorescent, and nonfluorescent. Among the highly fluorescent polymers these are some which can be used as electroluminescent materials [42a, 42b, 43]. Some polymers have very low quantum efficiency which seems to be less than 10^{-15}. In this case the lowest excited state is considered to be 2^1A_g state.

The soliton-antisoliton distance is determined by various factors such as local disorders, defects in the main-chain structure, and the Coulomb attractive force. The kinetics of the soliton-antisoliton pair is given by the power-law decay exhibiting the diffusive motion of the random walk along the one dimensional chain followed by the geminate recombination. It is distinctly different from a monomolecular exponential decay of a hot STE (or a hot e-h pair) even though the spatial distance between the soliton and antisoliton is not much longer than the soliton size. Some systematic deviation of the power n from the ideal value of $n = 0.5$ in one-dimensional system [22] is found. Our experimental results show that the induced absorption by a soliton-antisoliton pairs with larger ratios $R \equiv (2\xi + d)/\lambda_C$ follows the power-law decay with a larger power n. This is considered to be due to the strong confinement effect exhibited by the large values of R. The larger ratios indicate that motion of the solitons is limited in a segmented conjugation chain and make the geminate recombination more efficient. This tendency becomes more clear in substituted poly(phenylacetylene)s with a nearly-degenerate (= weakly nondegenerate) ground state, and further to nondegenerate polythiophenes than in the degenerate *trans*-polyacetylenes. In other words, the exponential behavior gradually dominates in more strongly nondegenerate systems because the soliton formation is strongly hindered.

5.1.8 Strongly Nondegenerate Ground-State System

Extending this discussion to the other extreme case, polydiacetylene, for example, where the nondegenerate ground state causes intrinsic, strong confinement of photoexcited charges (see Fig.19). For this case the electron and hole can no longer separate from each other. The hot STE, then, relaxes and thermalizes to the bottom of the potential surface by emitting phonons (lattice vibrations) to become a thermal STE. This has been experimentally confirmed by a successive exponential behavior of the photoinduced absorption (PIA) associated with the monomolecular process of excitons [12].

The adiabatic potential [30] for a soliton-antisoliton pair decreases monotonically to approach its creation energy (see Fig.20) if the soliton-antisoliton pair is not confined at all and perfectly free. An actual soliton-antisoliton distance (Q_{ss}) in a confined system with a weakly nondegenerate ground-state is determined by a balance between an energy gain due to the lattice relaxation associated with the soliton-antisoliton pair formation and an energy loss caused by the increase in the energy of having less stable resonance configurations such as quinoid form in a polythiophene. Although there is no potential barrier between a free electron-hole pair (*1* in Fig.17) and a soliton-antisoliton pair (*5* in Fig.17), the free electron-hole pair does not relax directly, but decays through a pass (*b* in Fig.17) to a hot STE (*3* in Fig.17). The hot STE is a transient precursor of the soliton-antisoliton pair, and it is confirmed by our measurement that the exponential component peaking at 1.4 - 1.5 eV appears at early delay-times in wide varieties of conjugated polymers regardless of the ground-state degeneracy. As the lifting of the ground-state degeneracy becomes larger, the decay processes (*c* and *e* in Fig.17) are hindered, where by the decay kinetics of the PIA signal turns over from the power-law to the exponential behavior. The potential surface along the decay processes (*c* and *e* in Fig.17) is one of the most important and stimulating subjects to be further investigated in both experimental and theoretical studies to fully understand the ultrafast dynamics of the photoexcited species.

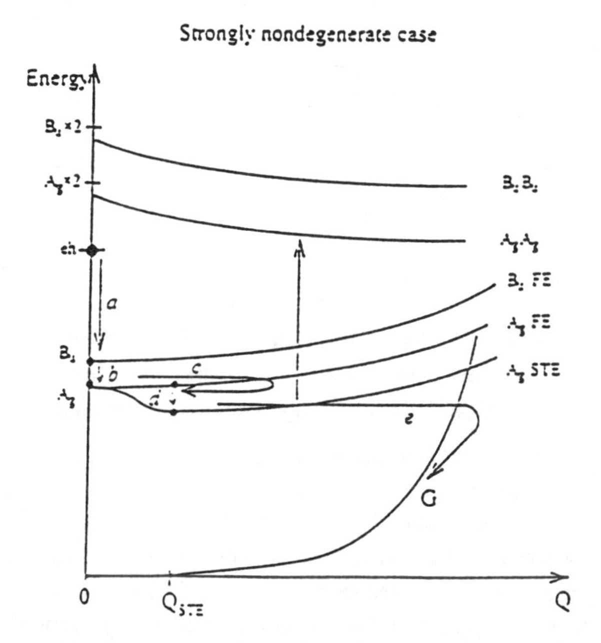

Fig. 19. Potential curves of conjugated polymers with a strongly-nondegenerate ground state.

Similar potential curves of the degenerate and strongly nondegenerate cases are shown in Figs.19 and 20, respectively. In the degenerate case as in *trans*-polyacetylene, after the creation of an electron-hole pair (labeled 1 in Fig.20) it relaxes to a charged soliton pair (labeled 2 in Fig.20) via the geometrical relaxation channel along the potential curve (*a*). Induced absorption by the formation of free *soliton pair* with a degenerate transition energy (E_H=E_L) is observed at 0.45 eV in *trans*-polyacetylene [21]. The charged soliton pair photogenerated is then converted to phonons and finally to the ground state G.

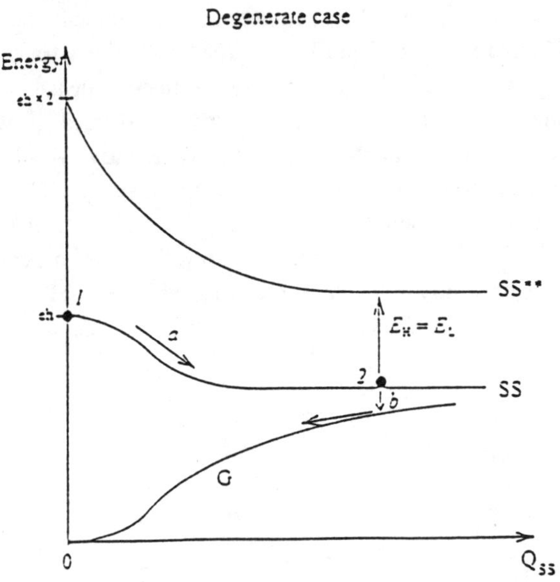

Fig. 20. Potential curves of conjugated polymers with a degenerate ground state.

In the strongly nondegenerate case (Fig.19), as in PDAs, very rapid relaxation takes place after the band-to-band excitation [10-14]. By the resonant excitation the ^1B_u exciton can also be photogenerated, which relaxes to the 2^1A_g free exciton (*b*) and then to the 2^1A_g self-trapped exciton. It decays to the ground state (G) via the tunneling process (*e*) [10-14]. By the comparison among the three systems i.e. degenerate, weakly-nondegenerate, and strongly-nondegenerate ground state systems which are classified by the degeneracy of the ground state, the confined soliton pair in a weakly-nondegenerate system can be regarded as a spatially indirect exciton [44, 45], in which the electron and hole in a pair are spatially separated after the geometrical deformation but their wave functions are still overlapping with each other even after the separation.

5.2 Single-Peak Component Exhibiting an Exponential Decay

5.2.1 Hot Self-Trapped Exciton

Figure 21 shows several parameters determined from the spectrum $B(\nu)$ having a single peak and decays exponentially. A peak is located around E_{ex} = 1.4 - 1.5 eV for all the three PPAs studied. The peak-energy has a small dependence on the polymer, while the FWHM of the spectra ranges between 0.45 and 0.72 eV. The absorption is considered to correspond to the 1^1B_u exciton because of the several consistent discussions made below with the assignment.

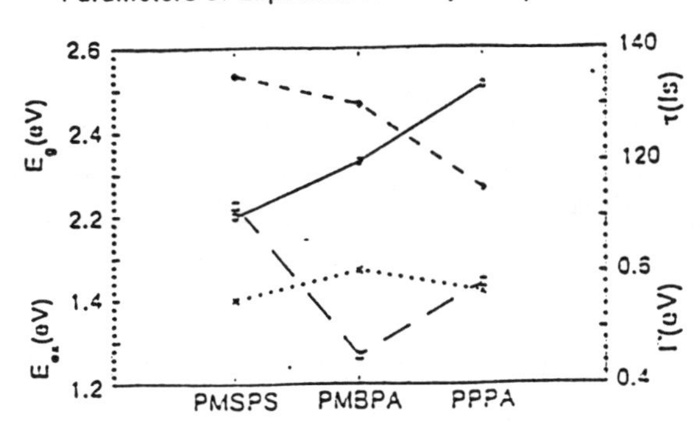

Fig. 21. Parameters of exponential-decay component of the three PPAs. Gap energy (E_g); circles, peak energy of the transition from the self-trapped exciton to the biexciton (E_{ex}); crosses, self-trapped exciton lifetime (τ); diamonds, width of the transition (Γ); squares.

The lifetime of 1^1B_u exciton is determined from the decay of $B(\nu)$ spectrum. It is shown by diamonds in Fig.21 and it decreases with increasing gap energy represented by circles (PMSPA→PMBPA→PPPA) in the figure. This can be explained as follows. The separation between the lowest excited state (2^1A_g) and the second lowest (1^1B_u) increases with the conjugation length [46]. From the energy-gap law [47] the relaxation from 1^1B_u to 2^1A_g is expected to be faster in the higher gap-energy system. Therefore the 1^1B_u exciton lifetime (τ) decreases in the order of PMSPA→PMBPA→PPPA.

Discussion on the peak-energy with reference to the strength of the Coulomb interactions is made as follows. Even though the on-site Coulomb interaction, U_0, cannot be related directly with the effective electronic correlation energy U_c which is estimated in subsection 5.1.2, it is treated as a first-order perturbation and approximate it as [48],

$$U_0 = 3\left(\frac{\xi}{a}\right) U_C. \tag{16}$$

Using the half width of a soliton in each polymer listed in Table 4, the on-site Coulomb interactions (U_0) are calculated as 2.9 ± 0.3 eV, 3.5 ± 0.3 eV, and 2.7 ± 0.3 eV for PMSPA, PMBPA, and PPPA, respectively. These are slightly larger than the transfer energy of π-electrons (t_0 ~ 2.5 eV), indicating that the π-electron system is weakly correlated ($U_0 < 4t_0$), namely, still in the Peierls regime [49]. More electron-donative side-group in PMBPA than those in PMSPA and PPPA increases an electron density in the main-chain resulting in the larger on-site Coulomb of the bulky side-group because of the largest van der Waals volume. The largest U_0 in PMBPA is also due to the smallest Coulomb screening.

Table 4. Parameters relevant to the confinement of the soliton.
Δ_e: disorder-induced increase of a half band-gap,
$E_g^{(0)}$: intrinsic band-gap energy, ξ : half-width of a soliton,
d: soliton-antisoliton distance, a: mean carbon-carbon distance.

sample	PPPA	PMBPA	PMSPA
Δ_e (meV)	73 ± 5	63 ± 3	51 ± 3
$E_g^{(0)}$ (eV)	2.36 ± 0.03	2.20 ± 0.02	2.10 ± 0.02
ξ/a	4.1 ± 0.1	4.5 ± 0.1	4.7 ± 0.1
d/a	5.7 ± 0.3	6.4 ± 0.3	7.1 ± 0.3

The on-site Coulomb interaction was estimated for soluble polyacetylene synthesized using polybutadiene. The value was 9-11 eV, which is about three times larger than those of poly(phenylacetylene). As discussed previously , this is also due to the smaller Coulomb screening in the former polymer.

As the Coulomb interaction becomes stronger, a single-photon forbidden state with an ^1A_g symmetry is also important as one of the lower-energy excited states in the decay channel. It has been theoretically investigated by Hayden and Mele [40] that the ^1A_g state becomes the lowest excited state for U_0 larger than $2t_0 \sim 5$ eV. The calculation was made for a 16-sites chain, and the chain-length is roughly the same as the conjugation lengths of the three polymers studied here. The on-site Coulomb interaction of $U_0 = 2.7 - 3.5$ eV for the three PPAs studied in the present work indicates that the one-photon allowed ^1B_u state is still the lowest excited state. If the Coulomb interaction is stronger than the above threshold of ~ 5 eV in some PPAs, relaxation from the hot ^1B_u state to the lowest ^1A_g state takes place with emitting an odd-parity phonon as discussed in Subsection 5.1.7. The ^1A_g-state may further relax to a lattice configuration similar to two pairs of neutral solitons $(S^0\ \overline{S^0})_2$ [50]. However, the creation of the neutral solitons directly from ^1B_u photoexcitation is strongly suppressed [26].

5.2.2 Transition to the Electron-Hole Threshold

The peak-energies (E_{ex}) as shown in Fig.21 seem not have with the band-gap energies but with the effective electronic correlation energies. One possible implication of this is that the peak-energy corresponds to a transition energy from the hot STE to a continuum state. This assignment is consistent with the broad PA feature which reflects the continuum band of the final state. The hot STE just after photoexcitation is vibrationally excited, and the large FWHM (Γ) of the spectrum $B(\nu)$ also results from a population distribution in the vibrational manifolds.

The transition energy (E_{ex}) is influenced by a dielectric screening of the Coulomb interaction associated with the hot STE, because an electron and a hole composing the hot STE are bound to each other by the attractive Coulomb force and the strength of the Coulomb interaction determines how much the hot STE is stabilized as compared with the electron-hole dissociation threshold (conduction band bottom). The screening is effectively included by a long-range interaction V in the theoretical treatments. Though it is difficult to uniquely determine both the on-site (U_0) and long-range (V) Coulomb interactions from the experimental results, a ratio of $U_0/V \sim 2$ was deduced by comparing the experimental and theoretical results [50]. Consequently we can estimate a relative strength of the long-range interaction (V) among the three polymers by the magnitudes of U_0 or U_c. PMSPA and PPPA have almost the same on-site Coulomb interactions (2.9 eV and 2.7 eV, respectively), and show nearly the same transition energies (1.40 eV and 1.42 eV, respectively). PMBPA with the larger on-site Coulomb interaction (3.5 eV), in contrast, has a larger transition energy (1.47 eV). Thus it is concluded that the transition energy from the hot STE tends to become larger for polymers with stronger on-site Coulomb interactions, because of the larger stabilization energy of the hot STE.

The dynamics of one-dimensional excitons and the lattice relaxation have been recently investigated using a molecular-dynamics (MD) calculation including both the electron-phonon coupling and the Coulomb interactions [26]. A two-band model was employed to simulate the self-trapping process of excitons. It was shown that a free exciton, at first, couples with the lattice vibrations and exchanges its energy back and forth with the lattice before an irreversible energy transfer to other modes or to the lattice vibration modes taking place at 50 - 100 fs after the exciton formation. The irreversible phonon-emission process, which rapidly proceeds in less than a few phonon periods was observed for polydiacetylenes [10-12]. In the present case, the exciton (hot STE) decays to the soliton-antisoliton pair before the phonon emission corresponding to the thermalization to take place, and the calculated transition energy from the hot STE before the phonon emission to the electron-hole dissociation threshold can be directly compared with the experimental results. The MD calculations were performed with the Coulomb interactions of $U_0 = 2t_e$ and $V = 1.6t_e$, where t_e is a transfer energy for electrons. The transition energy from the hot STE before the phonon emission is obtained as $E_{ex} \sim (1.30 \pm 0.05)t_e$ from the calculation. If we take $t_e = t_0/2 = 1.25$ eV, which is appropriate for polyacetylene, the transition energy of $E_{ex} \sim 1.63 \pm 0.06$ eV is obtained with the Coulomb interactions of $U_0 = 2.5$ eV and $V = 2.0$ eV. Though the on-site Coulomb interaction of $U_0 = 2.5$ eV is nearly the same as those observed for PMSPA and PPPA, the calculated transition energy is larger than the peak-energies (E_{ex}) of the two polymers (1.40 and 1.42 eV for PMSPA and PPPA, respectively). This disagreement can be attributed to the difference in V because it is taken to be larger than $U_0/2$ in the MD calculation. The large transition energy of excitons exceeding 1 eV, however, is possibly due to the one-dimensionality of excitons and further stabilization associated with the geometrical relaxation.

5.2.3 Transition to a Biexciton State

The spectrum $B(\nu)$ peaking around 1.4 - 1.5 eV can also be attributed to a transition from the hot STE to a biexciton state as shown in Fig. 17. The biexciton is one of the doubly excited states, and a pair of closely-located excitons additionally gains a binding energy between them to form a bound pair of excitons. It is a well-defined excitation in systems with moderate or stronger electronic correlations and plays an important role in the excited-state absorption because the lowest biexciton state has an even parity and hence is one-photon accessible from a ^1B_u state [51, 52]. In Fig. 17 the hot STE (2) is more stable than the free electron-hole pair (e-h) by a pure excitonic binding energy of E_b, and the exciton state energy is $E_1 = E_g - E_b$. The biexciton state (BE) gains the additional Coulomb attraction to be stabilized as compared to two independent excitons, and the biexciton state energy satisfies $E_2 < 2E_1$. But an excess Coulomb repulsion between two electrons and/or two holes in the biexciton does not make it so stabilized as in the case for one exciton. Therefore the following inequality is satisfied.

$$E_2 > 2E_1 - E_b \tag{17}$$

The transition energy from the hot STE to the biexciton, therefore, ranges

$$E_g - 2E_b < E_2 - E_1 < E_g - E_b. \tag{18}$$

If $E_b \sim 0.6$ eV is assumed and following the appropriate value for polyacetylene and poly(phenylacetylene)s, 1.3eV $< E_2 - E_1 <$ 1.9 eV is obtained for the case of $E_g = 2.5$ eV. The transition energies (E_{ex} = 1.4 - 1.5 eV) observed for the three polymers satisfy the above inequality relation, and the transition from the hot STE to the biexciton state is also a possible candidate for the observed peak around 1.4 - 1.5 eV.

6. Conclusion

The transient absorption spectra over a wide spectral range from visible to near-infrared were measured for thin films of three substituted poly(phenylacetylene)s, which have weakly nondegenerate ground state and the 1^1B_u state as the lowest excited state. The broad photoinduced absorption (PIA) observed below the stationary absorption band was clarified deep in the band-gap. Three PIA peaks were commonly found at early delay-times, and the PIA spectral shape changed within 1 ps. The middle PIA peak due to a hot self-trapped exciton (STE) rapidly decays exponentially with a time constant of τ = 115 - 135 fs depending on the polymer. The higher- and lower-energy peaks due to an oppositely-charged, overall neutral, spatially-confined soliton-antisoliton pair decay via a geminate recombination following a power-law.

The single-peak transition energy of the hot STEs (1.4 - 1.5 eV) was explained as a transition from the hot STE to the electron-hole threshold or to a biexciton. The soliton-antisoliton pair has a dual-peak spectrum in each polymer, indicating a significant overlap of the soliton- and antisoliton-wavefunctions. Using the gap-states splitting and electronic correlation energy independently evaluated from the two peak-energies, and a conjugation length (λ_C) and a soliton-antisoliton distance (*d*) were calculated. The distance was found to be nearly proportional to the conjugation length with a ratio of d/λ_C = 0.40. It was found that an overall size of the soliton-antisoliton pair is just confined within a segmented conjugation chain.

The formation time of the soliton-antisoliton pair was evaluated as 55 - 100 fs consistent with the results of the theoretical estimation [27]. As we move to more strongly nondegenerate systems the branching to the soliton-pair formation is hindered because of an inherent confinement of photoexcitations, and the relaxation and thermalization of the hot STE is followed to be observed.

Acknowledgments

The author would like to express hearty thanks to Dr. S. Takeuchi and for his essential contribution to the studies described in the present paper. He also thanks Profs. Masuda and Higashimura of Kyoto University for providing the samples of the three poly(phenylacetylene)s studied.

References

1 S. Takeuchi, T. Masuda, and T. Kobayashi, Phys. Rev. **B52**, 7166 (1995).
2 S. Takeuchi, T. Masuda, and T. Kobayashi, J. Chem. Phys. **105**, 2859 (1996).
3 A.J.Heeger, S.Kivelson, J.R.Schrieffer, and W.P.Su, Rev.Mod.Phys. **60**, 781 (1988).
4 W.P. Su, J.R. Schrieffer, and A.J. Heeger, Phys.Rev. **B22**, 2099 (1980).
5 H. Takayama, Y.R. Lin-Liu, and K. Maki, Phys.Rev.**B21**, 2388 (1980).
6 K. Fesser, A.R. Bishop, and D.K. Campbell, Phys.Rev. **B27**, 4804 (1983).
7 A.R. Bishop, Solid State Comm. **33**, 955 (1980).
8 N.A. Cade, and B. Movaghar, J.Phys.C: Solid State Phys. **16**, 539 (1983).
9 S. Abe and W.P. Su, Mol. Cryst. Liq. Cryst. **194**, 357 (1991).
10 M. Yoshizawa, M. Taiji, and T. Kobayashi, IEEE J.Quantum Electron. **QE-25**, 2532 (1989).
11 T. Kobayashi, M. Yoshizawa, U. Stamm, M. Taiji, and M. Hasegawa, J. Opt. Soc. Am. **B7**, 1558 (1990).
12 M. Yoshizawa, A. Yasuda, and T. Kobayashi, Appl. Phys. **B53**, 296 (1991).
13 M. Yoshizawa, K. Nishiyama, M. Fujihira, and T. Kobayashi, Chem. Phys. Lett. **207**, 461 (1993).
14 A. Yasuda, M. Yoshizawa, and T. Kobayashi, Chem. Phys. Lett. **209**, 281 (1993).
15 H. Tanaka, M. Inoue, and E. Hanamura, Solid State Comm. **63**, 103 (1987).
16 U. Stamm, M. Taiji, M. Yoshizawa, K. Yoshino, and T. Kobayashi, Mol. Cryst. Liq. Cryst. **182A**, 147 (1990).
17 D. McBranch, A. Heys, M. Sinclair, D. Moses, and A.J. Heeger, Phys. Rev. **B42**, 301 (1990).
18 Z. Vardeny, H.T. Grahn, A.J. Heeger, and F. Wudl, Synth. Met. **28**, C299 (1989).
19 T. Kubo, H. Takezoe, and A. Fukuda, Jpn. J. Appl. Phys. **30**, L1562 (1991).
20 L. Rothberg, T. M. Jedju, S. Etemad, and G.L. Baker, Phys. Rev. Lett. **57**, 3229 (1985).
21 L. Rothberg, T.M. Jedju, S. Etemad, and G.L. Baker, Phys. Rev. **B36**, 7529 (1987).
22 C.V. Shank, R. Yen, R.L. Fork, J. Orenstein, and G.L. Baker, Phys. Rev. Lett. **49**, 1660 (1982).
23 T. Masuda, Sec. 4 Chap. 3 in "Syntheses and reactions of polymers" in Japanese (Polymer Society) Kyoritsu, Tokyo, 1994.
24 T. Fujisaka, J. Suezaki, T. Koremoto, K. Inoue, T. Masuda, and T. Higashimura, Proceedings of Polymer Society Meeting **38**, 797 (1989).
25 S.D. Halle, M. Yoshizawa, H. Matsuda, S. Okada, H. Nakanishi, and T. Kobayashi, J. Opt. Soc. Am. **B11**, 731 (1994).
26 R. Ball, W.P. Su, and J.R. Schrieffer, J. Phys. **C3**, 429 (1983).
27 W.P. Su and J.R. Schrieffer, Proc. Natl. Acad. Sci. **77**, 5626 (1980).
28 S. Takeuchi, M. Yoshizawa, T. Masuda, T. Higashimura, and T. Kobayashi, IEEE J. Quantum Electron. **QE-28**, 2508 (1992).
29 Y.R. Lin-Liu and K. Maki, Phys. Rev. **B22**, 5754 (1980).
30 S.A. Brazovskii and N.N. Kirova, JETP Lett. **33**, 4 (1981).
31 T.W. Hagler and A.J. Heeger, Chem. Phys. Lett. **189**, 333 (1992).
32 G. Lanzani, A. Piaggi, R. Tubino, Z. V. Vardeny, and G. S. Kanner, Proceedings of Synthetic Metals (1992).

33 E. L. Pratt, K. S. Wong, W. Hayes, and D. Bloor, J. Phys. **20**. L41 (1987).
34 W.P. Su, J.R. Schrieffer, and A.J. Heeger, Phys. Rev. Lett. **42**, 1698 (1979).
35 R.H. McKenzie and J.W. Wilkins, private communications.
36 A.J. Brassett, N.F. Colaneri, D.D.C. Bradley, R.A. Lawrence, R.H. Friend, H. Murata, S. Tokito, T. Tsutsui, and S. Saito, Phys. Rev. **B41**, 10586 (1990).
37 G. Rossi, R.R. Chance, and R. Silbey, J. Chem. Phys. **90**, 7594 (1989).
38 M. Grabowski, D. Hone, and J.R. Schrieffer, Phys. Rev. **B31**, 7850 (1985).
39 T. Manaka, T. Yamada, H. Hoshi, K. Ishikawa, H. Takezoe, and A. Fukuda, preprint submitted to Jpn. J. Appl. Phys.
40 G.W.Hayden and E.J. Mele, Phys. Rev. **B34**, 5484 (1986).
41 W.P. Su, Phys. Rev. Lett. **74**, 1168 (1995).
42a R. Sun, T. Masuda, and T. Kobayashi, Jpn. J. Appl. Phys. **35**, L1673 (1996).
42b R. Sun, T. Masuda, and T. Kobayashi, Jpn. J. Appl. Phys. **35**, L1434 (1996).
43 R. Sun, J. Peng, T. Kobayashi, Y. Ma, H. Zhang, and S. Liu, Jpn. J. Appl. Phys. **35**, L1506 (1996).
44 M. Yan, L.J. Rothberg, F. Papadimitrakopoulos, M.E. Galvin, and T.M. Miller, Phys. Rev. Lett. **72**, 1104 (1994).
45 M. Yan, L.J. Rothberg, E.W. Kwock, and T.M. Miller, Phys. Rev. Lett. **75**, 1992 (1995).
46 B. S. Hudson, B. E. Kohler, K. Schulten, *Excited States*, ed. by E. C. Lim, pp.1-95 (Academic Press, New York, 1982).
47 W. Siebrand, J. Chem. Phys. **44**, 4055 (1966).
48 G. Lanzani, G. Dellepiane, A. Borghesi, and R. Tubino, Phys. Rev. **B46**, 10721 (1992).
49 W. Wu and S. Kivelson, Phys. Rev. **B33**, 8546 (1986).
50 W.P. Su, Phys. Rev. Lett. **74**, 1167 (1995).
51 S. Abe, J. Yu, and W.P. Su, Phys. Rev. **B45**, 8264 (1992).
52 K. Ishida, Solid State Comm. **90**, 89 (1994).

Characteristics and Dynamics of Superexcited States of Diatomic Molecules

Miyabi Hiyama[†,a] and Hiroki Nakamura[a,b]

[a]Department of Functional Molecular Science,
School of Mathematical and Physical Science,
The Graduate University for Advanced Studies,
Myodaiji, Okazaki 444, Japan,
and
[b]Division of Theoretical Studies, Institute for Molecular Science,
Myodaiji, Okazaki 444, Japan

Abstract. Characteristics and dynamics of superexcited states (SES) of diatomic molecules are discussed mainly from the theoretical viewpoint. A recently developed general theoretical procedure to reveal these characteristics and dynamics is briefly reviewed. The method is based on the multichannel quantum defect theory (MQDT), but effectively utilizes quantum chemical calculations of electronic states and spectroscopic experiments. This should break the bottleneck in theoretical studies and is expected to be useful in revealing SESs of diatomic molecules. The framework of MQDT is briefly explained to help in understanding the procedure. Theoretical studies of the SESs of H_2, NO, and CO made so far are also briefly summarized.

1 Introduction

The "superexcited state" (SES) is a general concept introduced by Platzman [1]. These states are the electronically highly excited states of atoms and molecules; and they play very important roles in various fields of physics and chemistry such as radiation chemistry, plasma physics, and chemical dynamics [2-3]. SES is defined as such a neutral state whose internal excitation energy is higher than the first ionization potential. Characteristics of SES are summarized as follows: (1) The optical oscillator strength distribution of a molecule is generally highest near the first ionization threshold, (2) ionization efficiency is smaller than unity in the energy region of the first ionization potential; (3) isotope effects exist in the ionization efficiency.

SES is classified mainly into two kinds; one is the two-electron, or inner-shell electron, excited state (first kind of SES), and the other is the rovibrationally excited Rydberg state (second kind of SES) [4]. Autoionization mechanisms of these states are very different. Autoionization from the first kind of SES is

---

[†]Present address : c/o Prof. Someda, Department of Pure and Applied Sciences, Graduate School of Arts and Sciences, University of Tokyo, Komaba, Meguro-ku, Tokyo 153, Japan

caused by the electron correlation interaction. Autoionization from the second kind of SES, on the other hand, is caused by energy transfer from rovibrational motion of the ion core to the outgoing Rydberg electron. It should be noted, however, that Rydberg states converging to an electronically excited molecular ion have characteristics of both the first and second kinds. They autoionize to the ionization continuum of the lower ionic state as the first kind of SES, but they can also autoionize to the ionic state to which they converge by the second rovibronic mechanism.

SESs play a key role as intermediate states in various dynamic processes such as dissociative recombination, Penning ionization, associative ionization, and chemi-ionization. Thus understanding of the characteristics and dynamics of SESs would lead to a revelation of interesting new chemical dynamics in this high-energy region. Experimental investigations of SES have been increasing not only for diatomic molecules but also for large molecules such as C_{60} [5]. There exists a useful and powerful theory, called multichannel quantum defect theory (MQDT), which can deal nicely with the infinite, although countable, number of Rydberg states. But detailed theoretical studies are very scarce even for diatomic molecules except for H_2. One main reason for this is the difficulty in estimating the electronic parameters, the quantum defects $\mu(R)$, and the electronic coupling $V(R, \varepsilon)$, crucial for the MQDT treatment [4, 6-8].

We introduce one method of solving this problem [9]. In order to obtain these parameters, not only quantum chemical calculations but also the MQDT analyses of spectroscopic experiments are carried out. This general theoretical method, useful for diatomic molecules, is summarized as follows: (1) SCF (self-consistent field) calculations of quantum defect functions; (2) restricted CI (configuration-interaction) calculations of the electronic states of the first kind of SESs; (3) MQDT analyses of spectroscopic data to improve the results of (1) and (2); (4) expansions of homo- and heteronuclear two-center Coulomb functions in terms of Gaussian functions; (5) evaluation of electronic couplings $V(R, \varepsilon)$ as functions of the electron energy and the internuclear distance; (6) nonperturbative solution of the **K**-matrix integral equation associated with the electronic coupling; (7) MQDT studies of various dynamic processes.

In this review the peculiarities of SESs are explained in Sect. 2, the framework of MQDT is outlined in Sect. 3, and the general theoretical procedure is presented in Sect. 4. Theoretical studies made so far for the molecules H_2, NO, and CO are briefly summarized in Sect. 5 followed by concluding remarks in Sect. 6.

2 Peculiarities of Superexcited States

Various kinds of dynamic processes proceed through SESs, as depicted in Fig. 1, where M^* represents an intermediate SES. This is because the two kinds of continua, ionization continuum and dissociation continuum, coexist in the energy region of SESs. Dissociative SESs, which are not well known, unfortunately, play significant roles in this world.

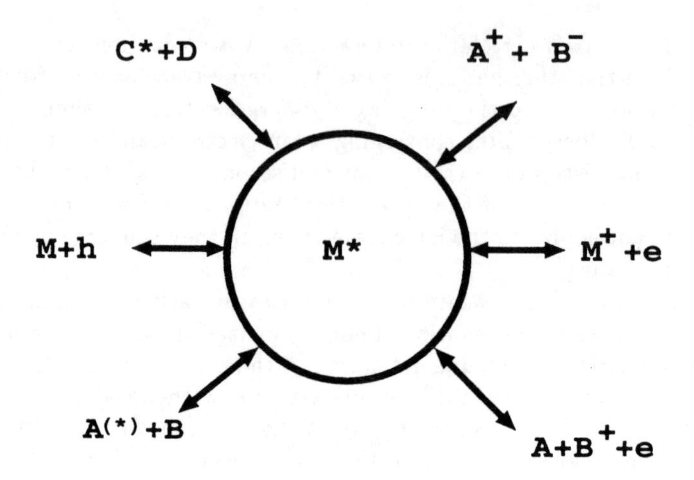

Fig. 1. A schematic unified view of the various dynamic processes, spectroscopic as well as scattering [4].

In the case of an atom, the quantum defects μ_Λ are defined by

$$E_{\rm ion} - E_{{\rm Ry},\Lambda} = \frac{1}{2(n-\mu_\Lambda)^2} \qquad \text{(in atomic units)}, \tag{1}$$

where n is the principal quantum number and Λ represents the quantum numbers other than n. $E_{\rm ion}$ and $E_{{\rm Ry},\Lambda}$ are the energies of the ion and Rydberg states, respectively. The quantum defects represent the effects of the interaction between the Rydberg electron and the molecular ion core other than the pure coulombic interaction. In the case of a diatomic molecule the quantum defect is dependent on the internuclear distance R,

$$E_{\rm ion}(R) - E_{{\rm Ry},\Lambda}(R) = \frac{1}{2[n-\mu_\Lambda(R)]^2}. \tag{2}$$

Through this dependence, vibrational states of SESs and the ionic state interact. This interaction is thus the cause of the autoionization of the second kind of SES. The autoionization of the first kind of SES is induced by the electronic coupling defined by

$$V(R,\varepsilon) = <\Psi_{\rm cont}|H_{\rm el}|\Psi_{\rm d}>, \tag{3}$$

where $\Psi_{\rm cont}$ and $\Psi_{\rm d}$ are the electronic wave functions of the ionization continuum and the bound state (first kind of SES), respectively. If a dissociative first kind of SES is involved, then this coupling $V(R,\varepsilon)$ represents the interaction between the nuclear dissociation continuum and the ionization continuum. Rydberg states converging to excited ionic states interact with the two kinds of ionization continua and also with the dissociation continuum, if a dissociative

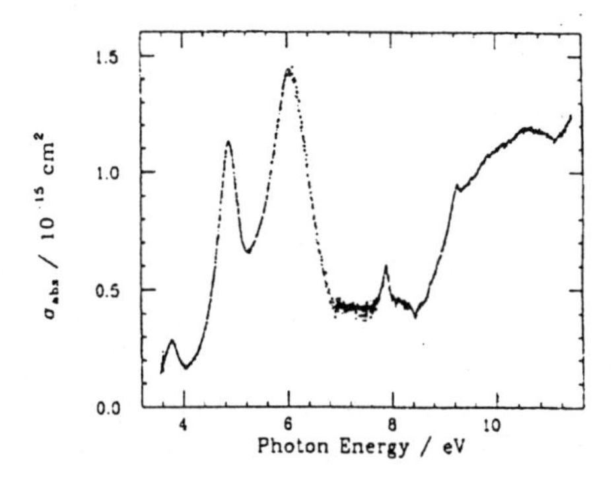

Fig. 2. Absorption spectrum of C_{60} in the gas phase measured at a temperature of 850K by Yasumatsu et al. [5].

first kind of SES coexists. This happens easily, of course, even in the case of diatomic molecules [10].

Unlike the case of diatomic molecules, large molecules have more complicated structures of nuclear continua, and the resonances may be much wider. For instance, Fig. 2 shows a photoabsorption spectrum of C_{60} observed by Yasumatsu et al. at 850K in the photon energy range, 3.5 - 11.4eV [5]. They assigned peaks at 7.87, 8.12, and 9.2eV to the Rydberg states converging to the second, third, and fourth ionic states, respectively. The competition among various kinds of decay channels represents one of the most peculiar properties of SESs. Hatano et al. measured photoionization quantum yields of organic molecules at energies higher than the lowest ionization potentials [11]. They also obtained the absolute cross sections for total photoabsorption, total ionization, and nonionization decay processes. One example is C_2H_2, and the main neutral dissociation channels are assigned as follows:

$$\begin{aligned} \mathrm{C_2H_2} + h\nu &\rightarrow \mathrm{C_2(d^3\Pi_g) + 2H} \\ &\rightarrow \mathrm{C_2(C^1\Pi_g) + 2H} \\ &\rightarrow \mathrm{CH(A^2\Delta) + CH} \\ &\rightarrow \mathrm{H(2p) + C_2H}, \end{aligned} \tag{4}$$

where the third channel is interestingly a breaking of the triple bond.

Measurements of absolute cross sections and branching ratios among various decay processes are important for comprehending the characteristics and dynamics of SESs. As is explained above, SESs are full of peculiarities compared to the ordinary valence excited states. It is strongly desired to find some systematics among the peculiarities.

3 Outline of the Multichannel Quantum Defect Theory (MQDT)

In this section the basic ideas and formulations of MQDT are briefly summarized. The theory was originally formulated to deal with the infinite number of Rydberg states and the ionization continuum in a unified way [6, 12-13]. The basic idea is that the physics of a Rydberg electron differs in the inner and outer regions of the electron coordinate space. In the outer region ($r \geq r_0$) the ion core behaves like a point charge for the Rydberg electron, and both of them move rather independently. In the inner region ($r \leq r_0$), on the other hand, the Rydberg electron couples with the ion core, and the Born-Oppenhimer (BO) approximation is a good approximation. Since the quantum numbers of a neutral molecule are good quantum numbers, the total wave function is represented by means of the BO basis as follows:

$$\Psi^{\mathrm{IN}} = \sum_{i,v,\Lambda} A_{i,v,\Lambda} |n^+\Lambda^+ v\Lambda lJM\rangle \{C_1 f_l(\nu_{v^+N^+}, r) + C_2 g_l(\nu_{v^+N^+}, r)\}, \tag{5}$$

where v is the vibrational quantum number of the molecule; J is the total angular momentum; l the angular momentum of the Rydberg electron; and v^+, N^+, Λ^+ are the vibrational quantum number, rotational quantum number, and the component along the molecular axis of the electronic angular momentum of the ion core, respectively. The function in the curly brackets represents the radial wave function of the Rydberg electron, and the ket $|\ \rangle$ represents the residual part. $f_l(\nu_{v^+N^+}, r)$ and $g_l(\nu_{v^+N^+}, r)$ are respectively the regular and irregular Coulomb functions. The parameter ν represents the energy (see (12)).

The total wave function in the outer region may be expanded in the form of a close-coupling expansion, and is given as

$$\begin{aligned} \Psi^{\mathrm{OUT}} &= \sum_{i,v^+,N^+} B|n^+\Lambda^+ v^+ N^+ lJM\rangle \{C_3 f_l(\nu_{n^+\Lambda^+ v^+ N^+}, r) \\ &+ C_4 g_l(\nu_{n^+\Lambda^+ v^+ N^+}, r)\}, \end{aligned} \tag{6}$$

where $f_l(\nu_{n^+\Lambda^+ v^+ N^+}, r)$ and $g_l(\nu_{n^+\Lambda^+ v^+ N^+}, r)$ are respectively the regular and irregular Coulomb functions in the asymptotic region. The curly brackets represent again the radial wave function of the Rydberg electron.

It is assumed that in the intermediate region the two expansions (5) and (6) hold equally well, namely, $\Psi \approx \Psi^{\mathrm{IN}} \approx \Psi^{\mathrm{OUT}}$, $\nu_{n^+\Lambda^+ v^+ N^+} \approx \nu_{n^+N^+}$. Thus the coefficients in (5) and (6) satisfy

$$\begin{aligned} BC_3 &= \sum_{v\Lambda} A_{v\Lambda} C, \\ BC_4 &= \sum_{v\Lambda} A_{v\Lambda} S, \end{aligned} \tag{7}$$

where

$$C = \langle N^+|\Lambda\rangle\langle v^+|\cos(\pi\mu_n(R))|v\rangle$$

and

$$S = \langle N^+|\Lambda\rangle\langle v^+|\sin(\pi\mu_n(R))|v\rangle. \tag{8}$$

Then the total wave function as $r \to \infty$ is expressed as

$$\Psi \sim \Psi^{cc} \to \sum_{v^+N^+} |n^+\Lambda^+v^+N^+\rangle\{[\sum_{v\Lambda} A_{v\Lambda} C_{v\Lambda}^{v^+N^+}]f_l + [\sum_{v\Lambda} A_{v\Lambda} S_{v\Lambda}^{v^+N^+}]g_l\}, \tag{9}$$

and accordingly, the reactance matrix $\mathbf{K}$ can be obtained as

$$\mathbf{K} = R_{\mathrm{oo}} - R_{\mathrm{oc}}[R_{\mathrm{cc}} + \tan \pi\nu]^{-1} R_{\mathrm{co}}, \tag{10}$$

where

$$R = SC^{-1} \tag{11}$$

and ε is the energy given by

$$\varepsilon = -(2\nu^2)^{-1}. \tag{12}$$

The suffix o (c) indicates the open (closed) channel. The scattering matrix $\mathbf{S}$ is expressed as

$$\mathbf{S} = (1 + \mathrm{i}\mathbf{K})(1 - \mathrm{i}\mathbf{K})^{-1}. \tag{13}$$

If there is no open channel, then the following secular equation holds:

$$\det|R_{\mathrm{cc}} + \tan \pi\nu| = 0. \tag{14}$$

When there is a dissociative first kind of SES, the electronic coupling in (3) plays an important role. Inclusion of the dissociative channel into the MQDT framework was made by Giusti [7], and the following expressions are obtained instead of (8):

$$\begin{aligned} C_{v^+\alpha} &\equiv \sum_v U_{v\alpha}\langle v^+|\cos(\pi\mu_\Lambda + \eta_\alpha)|v\rangle, \\ S_{v^+\alpha} &\equiv \sum_v U_{v\alpha}\langle v^+|\sin(\pi\mu_\Lambda + \eta_\alpha)|v\rangle, \\ C_{\mathrm{d}\alpha} &\equiv U_{\mathrm{d}\alpha}\cos\eta_\alpha, \end{aligned} \tag{15}$$

and

$$S_{\mathrm{d}\alpha} \equiv U_{\mathrm{d}\alpha}\sin\eta_\alpha,$$

where U is the unitary matrix that diagonalizes the $\mathbf{K}$-matrix associated with the electronic coupling $V(R, \varepsilon)$ that satisfies the Lippman–Schwinger-type integral equation, and η is the solution of the eigenvalue equation,

$$\sum_j \pi K_{ij} U_{j\alpha} = -\tan\eta_\alpha U_{i\alpha}. \tag{16}$$

4 General Theoretical Procedure

4.1 Quantum Chemical Calculations of Electronic States

It is difficult to obtain potential energy curves $E_d(R)$ and wave functions $\Psi_d(\mathbf{r}, R)$ of the first kind of SESs, since these states are not eigenstates of the electronic Hamiltonian. Since the ordinary variational principle cannot be applied directly, we employ the projection operator method. The projection operator Q is defined as the operator that projects out the resonance, i.e., the first kind of SES. Then the complement,

$$P = 1 - Q, \tag{17}$$

represents the continuum including the diffuse orbitals. The Schrödinger equation $(H - E)\Psi = 0$ is transformed as

$$(E - H_{PP})P\Psi = H_{PQ}Q\Psi \tag{18}$$

and

$$(E - H_{QQ})Q\Psi = H_{QP}P\Psi, \tag{19}$$

where $H_{PQ} = PHQ$, $H_{PP} = PHP$, and $H_{QQ} = QHQ$.

The potential energy curves of the first kind of SESs are estimated under the assumption $QHP = 0$. This neglects the energy shift due to the coupling to the continuum.

Since we want the states that diabatically cross the Rydberg manifold and enter into the ionization continuum, diffuse orbitals should not be included in the Q-space. This makes a big difference from the conventional quantum-chemical calculations that produce adiabatic potential curves avoiding crossings with Rydberg states.

In order to estimate quantum defects, the quantum-chemical calculations including diffuse orbitals are necessary. Nakashima et al. calculated the quantum defects of the NO molecule by using both SCF and CI methods [14]. They concluded that the SCF method cannot produce absolute values correctly, but can reproduce the R-dependences quite accurately. The absolute values are better corrected by using spectroscopic experiments. This is quite natural, and we can use the SCF method to evaluate the R-dependencies of quantum defects. In the case of a closed shell ion, the quantum defects, $\mu(R)$, can be estimated from the vacant orbital energies, $\epsilon(R)$, given by the SCF calculation

$$\epsilon(R) = \frac{1}{2[n - \mu(R)]^2} \quad \text{(in a.u.)}. \tag{20}$$

For an open shell ion, on the other hand, a modified SCF calculation, i.e. the hole-potential method [15-17], is needed to estimate the vacant orbital energies.

4.2 Ionization Continuum and Electronic Coupling

The two-center Coulomb functions at positive energy ($\varepsilon > 0$) are useful to represent the ionization continuum of a diatomic molecule and a long range part of the

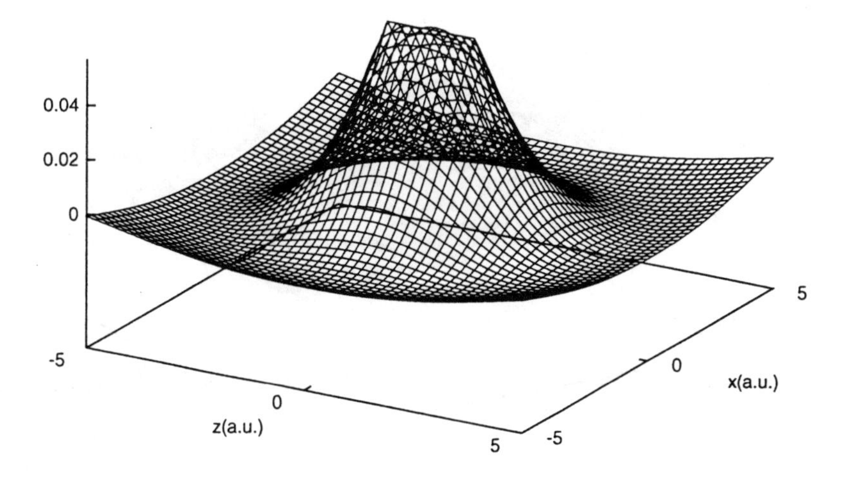

Fig. 3. Two-center sσ Coulomb function in the (x, z)-plane in the homonuclear case $(Z_1 = 0.5, Z_2 = 0.5)$. The internuclear distance R and the energy ε are $R = 2.079a_0$ and $\varepsilon = 0.001$a.u. [22].

electron scattering by a diatomic molecular ion [18-20]. We try to use the two-center Coulomb functions to evaluate the electronic coupling, which represents the bottleneck of theoretical studies of SESs.

The two-center radial and angular Coulomb functions are expressed in spheroidal coordinates,

$$\begin{aligned} \xi &= (r_1 + r_2)/R & (1 \leq \xi \leq \infty), \\ \eta &= (r_1 - r_2)/R & (-1 \leq \eta \leq 1), \\ \text{and} \quad \phi & & (0 \leq \phi < 2\pi), \end{aligned} \tag{21}$$

where r_1 and r_2 are the electron distances from the point charges Z_1 at $(x, y, z) =$ (0, 0, $-R/2$), and Z_2 at (0, 0,$R/2$). The two-center Coulomb function is written as follows,

$$\begin{aligned} \phi_{\mathbf{k}}(\mathbf{r}; R) &= \phi_{mq}(\xi, \eta, \phi; k, R) \\ &= \sqrt{\frac{2}{\pi k}} \Pi_{mq}(\xi; k, R) \Xi_{mq}(\eta; k, R) \frac{\exp(\pm im\phi)}{(2\pi)^{1/2}}, \end{aligned} \tag{22a}$$

where m is the magnetic quantum number and q represents the number of nodes

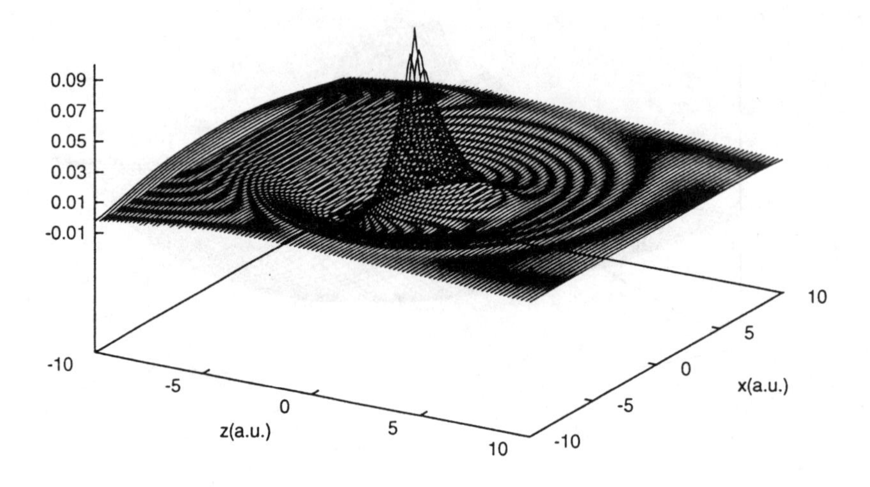

Fig. 4. The same as Fig. 3, but for the heteronuclear case, $Z_1 = 0.05017$, $Z_2 = 0.94983$, which represents a CO molecule. The internuclear distance R and the energy ε are $R = 2.079a_0$ and $\varepsilon = 0.001$a.u.[22].

of the radial function $\Xi_{mq}(\eta; k, R)$. The function $\phi_{\mathbf{k}}(\mathbf{r}; R)$ is normalized as

$$\int \mathrm{d}\mathbf{r}\phi^*_{\mathbf{k}}(\mathbf{r}; R)\phi_{\mathbf{k}}(\mathbf{r}; R) = \delta_{mm'}\delta_{qq'}\delta(\varepsilon - \varepsilon'). \tag{22b}$$

As an example, Fig. 3 shows the $s\sigma$ two-center Coulomb function for the homonuclear case, and Fig. 4 shows that for the heteronuclear case. These are numerical solutions of the corresponding Schrödinger equation [21-22]. These two-center Coulomb functions can be used to represent the wave function of the ionizing electron. In order to facilitate this usage and the evaluation of electronic coupling, we expand the numerically obtained two-center Coulomb functions in terms of Gaussian functions [21-22]. The basis functions are defined as,

$$\begin{aligned} \tilde{\chi}_{\alpha,r_c}(\vec{r}) &= x^{n_x} y^{n_y} z^{n_z} \exp\left(\alpha(\vec{r} - \vec{r}_c)^2\right) \\ &\equiv \chi_{\alpha,r_c}(\xi, \eta)\mathrm{e}^{\mathrm{i}m\phi}/\sqrt{2\pi}, \end{aligned} \tag{23}$$

where n_x, n_y, and n_z are nonnegative integers, r_c is the center of the function, and $r = \sqrt{x^2 + y^2 + z^2}$. The center r_c is taken to be on each charge center, and on the midpoint, depending on the accuracy required [21-22]. The parameters

n_x and n_y are determined easily by the magnetic quantum number m. In the case of σ-symmetry ($m = 0$), for instance, $n_x = n_y = 0$.

The electronic coupling in (3) can be estimated by using a conventional quantum-chemical CI code [23]. The discrete SES $\Psi_{\rm d}$ is evaluated from the CI calculations based on the projection operator formalism, and the continuum $\Psi_{\rm cont}$ is constructed from the two-center Coulomb function and the CI functions of the ion core.

4.3 Analysis of Spectroscopic Experiments

The relative separation between the potential curves of the first kind of SES and the ionic state are not so well reproduced by the *ab initio* calculation, since these calculations are carried out separately. In general, the ionic state is the ground state, and the first kind of SES is, of course, a highly excited state. Thus the accuracy of the *ab initio* calculations is quite different for both states.

If the experimental predissociation spectrum is available, the relative position can be estimated accurately. In [9], for instance, the relative position of $E_{\rm d}(R)$ was corrected by using the experimental data for the ratios of the predissociation and autoionization widths Γ_v for different v, where v represents the vibrational quantum number of the Rydberg state. It is, of course, desirable that the data be available for a wide range of v. The predissociation and autoionization peak widths themselves can be used to obtain a first estimate of the average electronic coupling strength $V_0 \approx \langle V(R,\varepsilon)\rangle$.

Absolute values of the quantum defects $\mu(R)$ can be determined from the experimental spectrum by using the MQDT. In this case it is better to use such a portion of the spectrum that is not affected by the dissociative state so that we can utilize the MQDT with no inclusion of the effects of the first kind of SES.

4.4 MQDT Analyses of Dynamic Processes

Various dynamic processes that proceed through the SESs can be analyzed by MQDT. The various scattering processes are evaluated directly by the scattering matrix **S** defined by (13). Examples are (1) vibrational and rotational excitation of a molecular ion by electronic impact,

$$e + AB^+(v_0^+ N_0^+) \rightarrow e + AB^+(v^+ N^+), \tag{24}$$

(2) dissociative recombination,

$$e + AB^+(v_0^+ N_0^+) \rightarrow A^*(dj_r) + B, \tag{25}$$

and (3) associative ionization

$$A^*(dj_r) + B \rightarrow e + AB^+(v_0^+ N_0^+). \tag{26}$$

In the case of photo-ionization,

$$AB(J_0 M_0) + h\nu \rightarrow AB^+(v^+ N^+) + e, \tag{27}$$

the cross section is expressed as

$$\sigma_{v+N+,J_0M_0} = 2\pi\alpha[\frac{df}{dE}]_{v+N+} \quad (\pi a_0^2), \tag{28}$$

where $\alpha = e^2(\hbar c)^{-1}$, a_0 is the Bohr radius, and the oscillator strength distribution $\mathrm{d}f/\mathrm{d}E$ is given by

$$[\frac{\mathrm{d}f}{\mathrm{d}E}]_{v+N+} = \frac{2h\nu}{2J_0+1}\sum_{M_0}|D_{v+N+,J_0M_0}|^2. \tag{29}$$

Similarly, the cross section for photodissociation,

$$AB(J_0M_0) + h\nu \rightarrow A^*(dj_r) + B, \tag{30}$$

is given by (28), where the oscillator strength $\mathrm{d}f/\mathrm{d}E$ is defined by

$$[\frac{\mathrm{d}f}{\mathrm{d}E}]_{v+N+} = \frac{2h\nu}{2J_0+1}\sum_{M_0}|D_{dj_r,J_0M_0}|^2. \tag{31}$$

The quantities D appearing in (29) and (31) represent dipole transitions by photoabsorption, and are given by [6]

$$\mathbf{D} = \mathbf{D}_\mathrm{o} - \chi_\mathrm{oc}[\chi_\mathrm{cc} - \exp(-2\pi\mathrm{i}\nu)]^{-1}\mathbf{D}_\mathrm{c}, \tag{32}$$

where χ is defined by

$$\chi = (I + \mathrm{i}R)(I - \mathrm{i}R)^{-1} \tag{33}$$

and $\mathbf{D}_\mathrm{o}$ ($\mathbf{D}_\mathrm{c}$) represents a dipole transition from the initial state to the open (closed) channel.

5 Examples

5.1 H_2

H_2 is the most extensively studied molecule. The photoionization, autoionization and photoelectron angular distribution were analyzed with the use of MQDT by Jungen and coworkers [24-28]. The potential energy curves of the various first kind of SESs are obtained by both quantum-chemical and scattering calculations. The quantum defect functions of the various symmetrics were evaluated by scattering as well as by R-matrix calculations [20, 29]. The electronic couplings were also calculated by various approximations [7, 30-35], and their R- and ε-dependencies are analytically fitted by Takagi [36]. Figure 5 shows these potential energy curves [37-38]. Full MQDT calculation of the dissociative recombination of H_2^+ and HD^+ was carried out by Takagi [39-40], and almost perfect agreement with the experiment was obtained. The theoretical calculation includes the effects not only of the vibrational states but also of the rotational states, and is also based on the nonperturbative solution of the **K**-matrix integral equation.

The experiment was carried out using an adiabatically expanded low-tempera ture electron beam in a cooler ring TARNII [40]. Figure 6 depicts the result together with the theoretical calculations for HD^+.

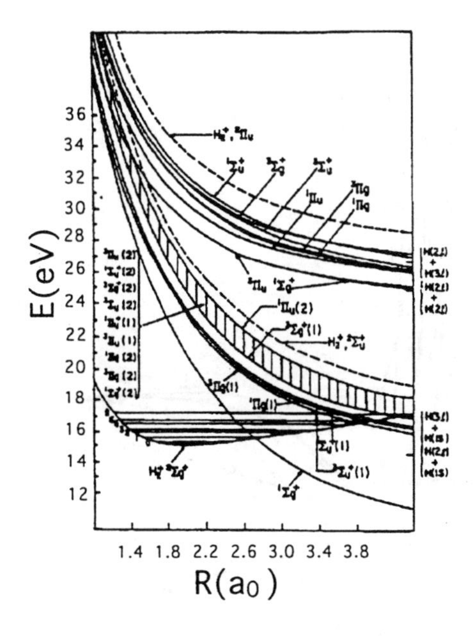

Fig. 5. Potential energy curves of H_2 amd H_2^+ [37].

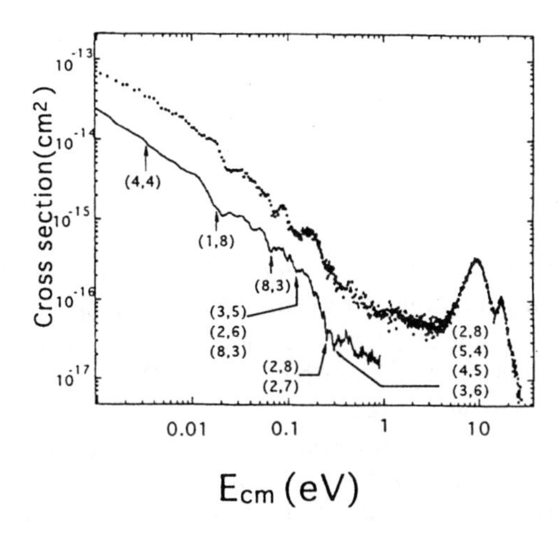

Fig. 6. Relative experimental $(\cdots)$ as well as theoretical $(-)$ cross sections of the dissociative recombination of HD^+[40]. Arrows indicate resonance positions assigned to (v, n), where v is the vibrational quantum number and n is the principal quantum number.

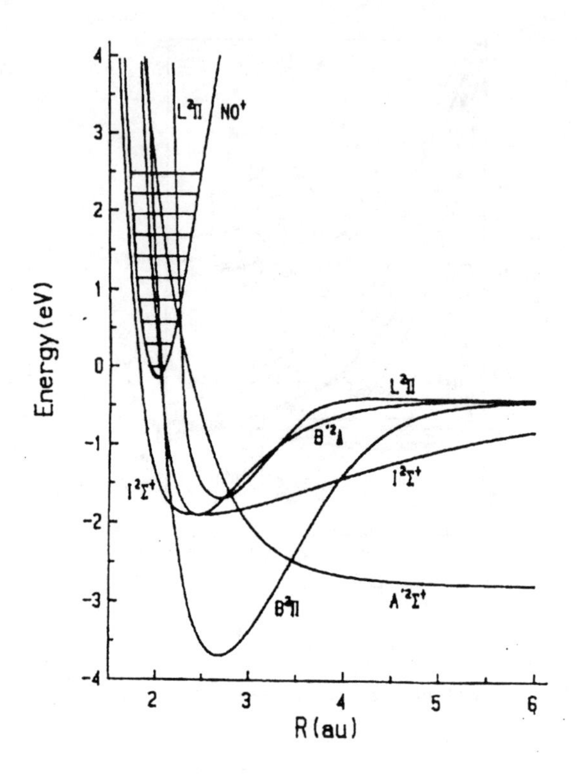

Fig. 7. Diabatic potential energy curves for the ground state of NO^+ and five dissociative first kind of superexcited states of NO [48].

5.2 NO

NO is another example that has been relatively well studied theoretically, although there is no comparison with the case of H_2. Giusti, Jungen, and Raoult have quite extensively studied photoabsorption processes of $B^2\Pi$ [41-48]. They evaluated the first derivative of $\mu(R)$ with respect to R and the average electronic coupling $\approx V_0$ from the experiments.

There are many dissociative superexcited states of various symmetries, and both vibrational and electronic autoionization are possible. Figure 7 shows the potential energy curves of NO [48]. Nakashima et al. analyzed the REMPI (resonantly enhanced multi photon ionization) of the dσ-Rydberg states and obtained the information on the potential curves of $B'^2\Delta$ and $I^2\Sigma^+$ and their average electronic couplings [14]. Since the ion core has a closed shell structure, the quantum defect functions are rather easily evaluated by an SCF method. They actually did this calculation for sσ, pσ, pπ, dσ, dπ, and dδ. With the information thus obtained they carried out the MQDT analysis of autoionization

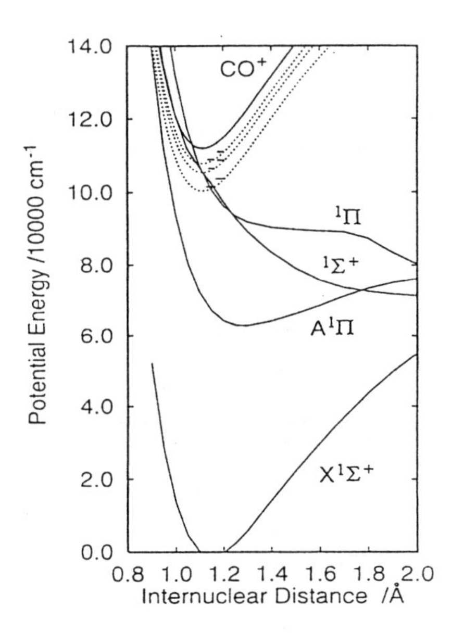

Fig. 8. Potential energy curves of CO [9]. The Rydberg states shown are those of n = 4, 5, and 6.

[14]. Sun and Nakamura studied the dissociative recombination processes via the states $B^2\Pi$, $B'^2\Delta$, and $I^2\Sigma$, $L^2\Pi$, and $A^2\Sigma$ by MQDT [48]. The electronic couplings are assumed to be constant, and thus the perturbative solutions of the **K**-matrix integral equation were used. The rotational degree of freedom was disregarded.

5.3 CO

Some of the theoretical results for CO based on the general procedure described in Sect. 4 are reported here. Figure 8 shows the diabatic potential energy curves of $^1\Sigma$ and $^1\Pi$ obtained by the restricted CI calculations [9]. The relative positions of $^1\Sigma^+$ and $^1\Pi$ near the ionic curve were determined by analyzing the ion-dip spectrum observed by Komatsu et al. and by Ebata [49-50]. These states contribute to the various dynamics near the first ionization threshold.

Figures 9 and 10 show the quantum defects of CO. Figure 9 shows their n-dependencies in comparison with the experimental data. The calculated results shown here are the values at the equilibrium internuclear distance R_e=1.1151 Å. The absolute values are not in good agreement with the experiment, but the tendency in n-dependence is in good accordance with the experiment. The n-dependence is generally quite weak. Figure 10 shows R-dependences of the

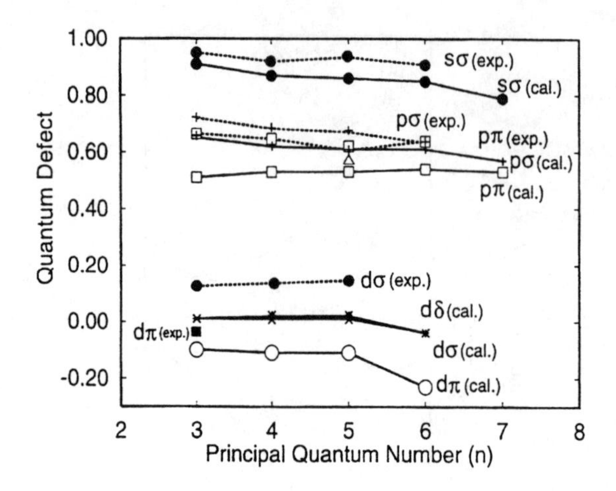

Fig. 9. Quantum defects of the CO molecule [9]. Principal quantum number (n)-dependence at the equilibrium internuclear distance. – : calculated results; ··· : experimental data. There are two experimental values for 5pπ ($\square$ and $\triangle$) that are not assigned yet.

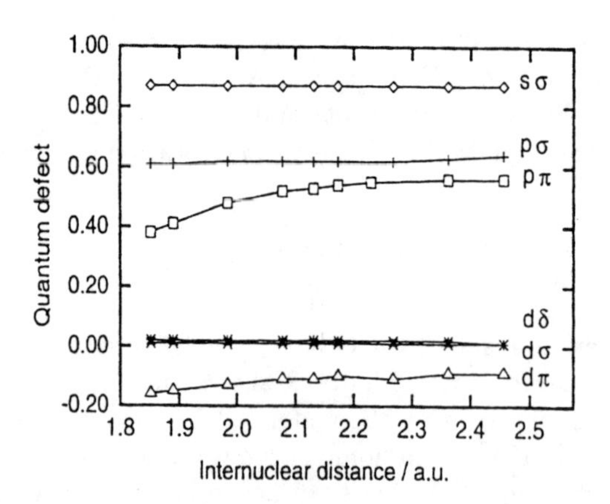

Fig. 10. Quantum defects of the CO molecule [9]. Internuclear distance (R) - dependence (calculated results) for $n = 4$.

calculated quantum defects for $n = 4$. The vibrational autoionization from the π-Rydberg manifold will be larger than that from other Rydberg manifolds, since the R-dependencies are generally weak except for pπ and dπ. It should be noted that the ion core of CO has an open shell structure, and that the hole-potential method should be employed to evaluate the quantum defects.

In order to evaluate the electronic couplings as a function of R and ε, the Gaussian expansions of the two-center Coulomb functions are carried out. The two point charges are estimated from the electron density distribution of CO^+. They are $Z_1 = 0.05017$ and $Z_2 = 0.94983$, for instance, at the equilibrium internuclear distance $R = 2.079a_0$. Figure 11 shows the Gaussian expansion of

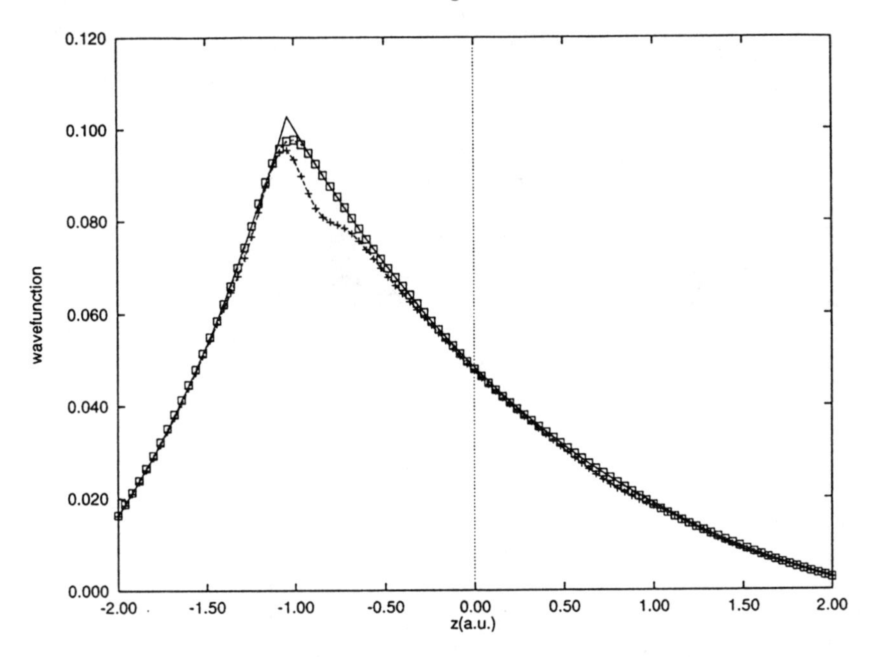

Fig. 11. The Gaussian fitting at $x = 0$ [22]. The parameters are $Z_1 = 0.05017$, $Z_2 = 0.94983$, ε=0.001a.u. and $R = 2.079a_0$. $-$: the exact Coulomb function; $-$x$-$x$-$: the fitting with use of the 90 Gaussians; $-\Box-\Box-$: the fitting with use of 96 Gaussians.

the sσ two-center Coulomb function at $\varepsilon = 0.001$a.u. . The parameter n_z in (23) was taken up to 2 ($n_z = 0, 1, 2$). The relative error $|\psi_\varepsilon - \Psi_G|/|\psi|$ is about 1×10^{-2} for the case of 96 Gaussians, where $\psi(\Psi_G)$ is the exact two-center Coulomb function (the Gaussian fit). The electronic coupling of the $^1\Sigma$ state is 0.08a.u, at $\varepsilon = 0.001$a.u. and $R = 2.079a_0$, and is smaller than the corresponding value of H_2 (≈ 0.11).

Figure 12 presents a simple example of dynamics calculations, i.e.autoionization of the sσ- and pπ-Rydberg states of CO. The upper part is the experimental multiphoton ionization spectrum measured by Ebata et al. [51-52]. The bottom part is the calculated result estimated from the second term of (32). In this calculation the direct autoionization from the dissociative SES is not included, and rotational states are completely disregarded. Thus the autoionization peaks are much narrower than the experimental ones. Since the Franck–Condon factor between the $^1\Sigma^+$ dissociative SES and the vibrational levels of sσ-Rydberg states is very small, the autoionization processes proceed predominantly by the vibrational mechanism in the case of sσ. In the case of pπ-Rydberg states the widths are larger than those of the sσ-case, since the Franck–Condon factor between the $^1\Pi$ dissociative SES and the vibrational levels of pπ-Rydberg states are much larger than that in the $^1\Sigma^+$ case, as is seen from Fig. 8. With use of the general theoretical procedure explained in Sect. 4, many more sophisticated analyses of various dynamic processes can now be carried out.

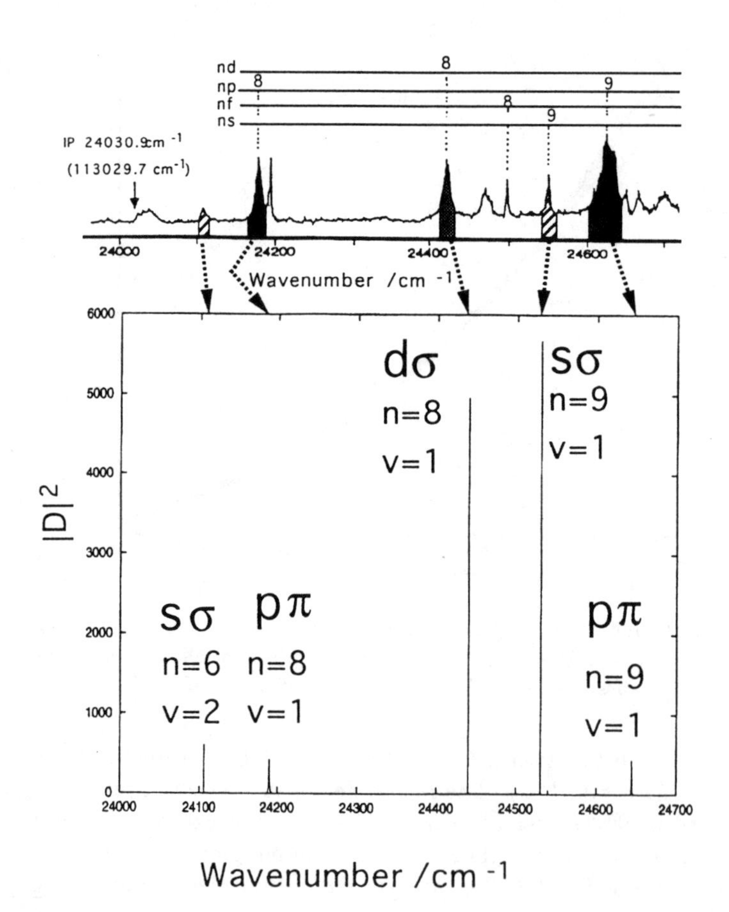

Fig. 12. Experimentally obtained photoionization spectrum (*upper panel*) and calculated autoionization peaks for the sσ-, pπ- and dσ-Rydberg states (*lower panel*) of CO.

6 Concluding Remarks

Because of the peculiarities of SESs, not only theoretical but also experimental studies are not easy and still quite scarce. Considering their intriguing properties and importance in various dynamic processes, however, both studies should be promoted more strongly. Recent remarkable progress in laser technology and synchrotron facilities may accelerate experimental studies. The general theoretical procedure presented in this review may provide a breakthrough for theoretical studies of SESs of diatomic molecules. Extensive applications are actually strongly required. Quantum-chemical studies of electronic states of SESs should also be more greatly encouraged, not only for diatomic molecules but even for larger molecules. The peculiarities of SESs can open a new world of physics and chemistry.

Acknowledgments

This work was supported in part by a Grant-in-Aid for Scientific Research on Priority Area "Theory of Chemical Reactions" and "Quantum Tunneling of Group of Atoms as Systems with Many Degrees of Freedom" from the Ministry of Education, Science, Culture, and Sports of Japan. One of the authors (M.H.) would like to thank the Japan Society for the Promotion of Science for its support by the Fellowships for Japanese Junior Scientists while she was carrying out the research at The Department of Functional Molecular Science, School of Mathematical and Physical Science, The Graduate University for Advanced Studies. The numerical computations were partly carried out by the computers at the Computer Center of IMS.

References

1 R. L. Platzman, Red. Res. **17**, 419 (1962).

2 M. Inokuti, Butsuri. **20**, 734 (1965) and **22**, 196 (1967) (in Japanese).

3 U. Fano, Science. **153**, 522 (1966).

4 H. Nakamura, Intern. Rev. Phys. Chem. **10**, 123 (1991); J. Chinese Chem. Soc. **42**, 359 (1995).

5 H. Yasumatsu, T. Kondow, H. Kitagawa, K. Tabayashi, and K. Shobatake, J. Chem. Phys. **104**, 899 (1996).

6 M. J. Seaton, Rep. Prog. Phys. **46**, 167 (1983).

7 A. Giusti, J. Phys. **B13**, 3867 (1980).

8 C. H. Green and Ch. Jungen, Adv. in Atomic and Molecular Physiscs **21**, 51 (1985).

9 M. Hiyama and H. Nakamura, Chem. Phys. Lett. **248**, 316 (1996).

10 M. Ukai, S. Machida, K. Kameta, M. Kitajima, N. Kouchi, Y. Hatano, and K. Ito Phys. Rev. Lett. **74**, 239 (1995), A. Ehresmann, S. Machida, M. Ukai, K. Kameta, M. Kitajima, N. Kouchi, Y. Hatano, K. Ito, and T. Hayashi, J. Phys. **B28**, 5283 (1995).

11 Y. Hatano, Dynamics of Excited Molecules, edited by K. Kuchitsu (Elsevier, 1994) 151; Review talk at the XIX International Conference on the Physics of Electronic and Atomic Collisions, Whistler, Canada (1995).

12 U. Fano, Phys. Rev. **A2**, 353 (1970).

13 Ch. Jungen and O. Atabek, J. Chem. Phys. **66**, 5584 (1977).

14 K. Nakashima, H. Nakamura, Y. Achiba, and K. Kimura, J. Chem. Phys. **91**, 1603 (1989).

15 K. Morokuma and S. Iwata, Chem. Phys. Lett. **16**, 192 (1972); S. Iwata and K. Morokuma, Theoret. Chim. Acta. **33**, 4972 (1974).

16 E. R. Davidson, Chem. Phys. Lett. **21**, 565 (1973).

17 S. Iwata, Chem. Phys. Lett. **83**, 134 (1981).

18 M. Iwai and H. Nakamura, Phys. Rev. **A40**, 2247 (1989).

19 S. Lee, M. Iwai, and H. Nakamura, "Molecules in Laser Fields", ed. by A. D. Bandrauk (Marul Dekker, 1994) p.217.

20 H. Takagi and H. Nakamura, Phys. Rev. **A27**, 691 (1983).

21 M. Hiyama and H. Nakamura, Comp. Phys. Comm. **103**, 197 (1997).

22 M. Hiyama and H. Nakamura, Comp. Phys. Comm. **103**, 209 (1997).

23 M. Hiyama, N. Kosugi, and H. Nakamura, J. Chem. Phys. **107**, 9370 (1997).

24 D. Dill and Ch. Jungen, J. Phys. Chem. **84**, 2116 (1980).

25 Ch. Jungen and D. Dill, J. Chem. Phys. **73**, 3338 (1989).

26 Ch. Jungen and M. Raoult, Faraday Discuss. Chem. Soc. **71**, 253 (1981).

27 M. Raoult and Ch. Jungen, J.Chem.Phys. **74**, 3388 (1981).

28 M. Raoult, Ch. Jungen, and D. Dill, J.Chim.Phys. **77**, 599 (1980).

29 I. Shimamura, C.J. Noble, P.G. Burke, Phys.Rev. **A27**, 691 (1983).

30 J. N. Bardsley, J. Phys. **B1**, 349 (1968).

31 A. Giusti-Suzor, J. N. Bardsley and C. Derkit, Phys. Rev. **A28**, 682 (1983).

32 K. Nakashima, H. Takagi, and H. Nakamura, J. Chem. Phys. **86**, 726 (1987).

33 H. Takagi, N. Kosugi, and M. LeDourneuf, J. Phys. **B24**, 711 (1991).

34 S. Hara and H. Sato, J. Phys. **B17**, 4301 (1984).

35 H. Sato and S. Hara, J. Phys. **B19**, 2611 (1986).

36 H. Takagi, *DissociativeRecombination*, Edited by Betrand R. Rowe et al. Series B; Physics **313**, 75 (1992).

37 C. Strömholm, I. F. Schneider, G. Sundström, L. Carata, H. Danared, S. Datz, O. Dulieu, A. Källberg, M. af Ugglas, X. Urbain, V. Zengin, A. Suzor-Weiner, and M. Larsson, Phys. Rev. **A52**, R4320 (1995).

38 S. L. Guberman, J. Chem. Phys. **93**, 1404 (1983).

39 H. Takagi, J. Phys. **B26**, 4815 (1993).

40 T. Tanabe, I. Katayama, H. Kamegaya, K. Chiba, Y. Arakaki, T. Watanabe, M. Yoshizawa, M. Saito, Y. Haruyama, K. Hosono, K. Hatanaka, T. Honma, K. Noda, S. Ohtani, and H. Takagi, Phys Rev. Lett. **75**, 1066 (1995).

41 A. Giusti-Suzor and Ch. Jungen, J. Chem. Phys. **80**, 986 (1984).

42 M. J. Li, 1986, Electronic and Atomic Collisions, Invited Paperts of the 15th International Conference on Physics of Electronic and Atomic Collisions (Amsterdam: North-Holland), p.621.

43 S. Fredin, D. Gauyacq, M. Horain, Ch. Jungen, and G. Lefebvre, Molec.Phys. **60**, 825 (1987).

44 M. Raoult, J. Chem. Phys. **87**, 4736 (1987).

45 A. L. Sobolewski. J. Chem. Phys. **87**, 331 (1987).

46 H. Rudolph, S. N. Dixit, V. Mckoy, and W. M. Huo, J. Chem. Phys. **88**, 1516 (1988).

47 S. Pratt, Ch. Jungen, and E. Miescher, J. Chem. Phys. **89**, 5971 (1990).

48 H. Sun and H. Nakamura, J. Chem. Phys. **93**, 6491 (1990).

49 M. Komatsu, T. Ebata, and N. Mikami, J. Chem. Phys. **99**, 9350 (1993).

50 T. Ebata, private communication.

51 N. Hosoi, T. Ebata, and M. Ito, J. Chem. Phys. **95**, 4182 (1991).

52 T. Ebata, N. Hosoi, and M. Ito, J. Chem. Phys. **97**, 9350 (1992).

Subject Index

Printing: Mercedesdruck, Berlin
Binding: Buchbinderei Lüderitz & Bauer, Berlin